This is a concise, rigorous and authoritative introduction to the information-theoretic analysis of interference management in wireless systems. It provides a principled and comprehensive reference for students and researchers with a background in communication and information theory.

Professor Osvaldo Simeone, *King's College London*

Veeravalli and El Gamal have successfully captured in this book a most productive decade of fundamental research breakthroughs in our understanding of interference management and, especially, the role of cooperation in wireless networks. The topics covered are not only some of the most exciting directions for future wireless networks, but also they showcase some of the most elegant insights that information theory has to offer.

Professor Syed Jafar, *University of California, Irvine*

With new wireless networks becoming dramatically more complex to optimize, specially from an interference point of view, the availability of solid theoretical tools lowing to study the fundamental radio access performance of future mobile systems is reat plus to engineers and researchers alike. Up to my knowledge, this book is among very first to give a solid and comprehensive analytical perspective in this area.

David Gesbert, *EURECOM*

book stands out with its comprehensive treatment of fundamental inform- -theoretic tools for interference management and with detailed descriptions of ithms that implement these tools. The book is thus very valuable for everyone who to understand how to design good practical interference management systems.

Michèle Wigger, *Telecom Paris-Tech*

Interference Management in Wireless Networks

Fundamental Bounds and the Role of Cooperation

Learn about an information-theoretic approach tc future-generation wireless networks. Focusing on coo coordinated multi-point (CoMP) technology, the boo framework for interference management that uses r design, and practical pre-coding schemes based on increased speed and reliability promised by inte into how simple, zero-forcing pre-coding scheme interference networks, and discover how sign by making cell association decisions and allo centralized (cloud) processing and knowledge between information-theoretic analyses and int easy to implement, this is an invaluable resour practicing engineers in wireless communicati

Venugopal V. Veeravalli is the Henry Magnu and Computer Engineering at the Univer Fellow of the IEEE.

Aly El Gamal is an Assistant Professor Engineering at Purdue University.

Interference Management in Wireless Networks

Fundamental Bounds and the Role of Cooperation

VENUGOPAL V. VEERAVALLI
University of Illinois, Urbana-Champaign

ALY EL GAMAL
Purdue University

CAMBRIDGE
UNIVERSITY PRESS

CAMBRIDGE
UNIVERSITY PRESS

University Printing House, Cambridge CB2 8BS, United Kingdom

One Liberty Plaza, 20th Floor, New York, NY 10006, USA

477 Williamstown Road, Port Melbourne, VIC 3207, Australia

314–321, 3rd Floor, Plot 3, Splendor Forum, Jasola District Centre,
New Delhi – 110025, India

79 Anson Road, #06-04/06, Singapore 079906

Cambridge University Press is part of the University of Cambridge.

It furthers the University's mission by disseminating knowledge in the pursuit of education, learning, and research at the highest international levels of excellence.

www.cambridge.org
Information on this title: www.cambridge.org/9781107165007
DOI: 10.1017/9781316691410

First published 2018

Printed in the United States of America by Sheridan Books, Inc.

A catalogue record for this publication is available from the British Library.

Library of Congress Cataloging-in-Publication Data
Names: Veeravalli, Venugopal V., 1963– author. | Gamal, Aly El, 1985– author.
Title: Interference management in wireless networks : fundamental bounds and the role of cooperation / Venugopal V. Veeravalli, University of Illinois, Urbana-Champaign, Aly El Gamal, Purdue University.
Description: Cambridge, United Kingdom ; New York, NY, USA : Cambridge University Press, [2017] | Includes bibliographical references and index.
Identifiers: LCCN 2017038711 | ISBN 9781107165007 (hardback)
Subjects: LCSH: Radio–Interference. | Wireless communication systems–Management.
Classification: LCC TK7867.2.V44 2017 | DDC 621.388/1–dc23
LC record available at https://lccn.loc.gov/2017038711

ISBN 978-1-107-16500-7 Hardback

Contents

Preface

Efficient interference management is the key to meeting the ever increasing demand for wireless data services in wireless networks. In this book, an information-theoretic framework for interference management is presented, along with a discussion of practically implementable schemes for interference management that approach provable bounds on performance and are applicable in large cellular networks.

A particular focus of the book is on exploiting recent technological advancements in the wireless infrastructure that enable cooperative transmission and reception. Our theoretical framework highlights the potential of this technology to deliver the rate gains promised by interference alignment, while using simple zero-forcing schemes that are easier to implement in practice. The schemes that we present rely on local cooperation, and they are shown to be optimal in locally connected interference networks, where the path loss effect allows us to neglect connections between transmitters and receivers that are far away from each other. It is also shown that cooperative communication can be used to deliver significant rate gains using simple zero-forcing with no or minimal extra load on the backhaul, by selecting cell associations and allocation of backhaul resources based on centralized processing and knowledge of the network topology. The insights obtained are also extended to dynamic interference networks that capture the effect of deep fading conditions.

This book is organized as follows:

- In Chapter 1, we provide a high-level introduction to interference management in wireless networks, including a historical perspective on wireless cellular networks, and we motivate the need for a fundamental information-theoretic understanding of this problem.
- In Chapter 2, we provide a detailed overview of the state of the art in determining the information-theoretic sum capacity of the interference channel. Determining the capacity even in the simplest setting with two users with one antenna at each of their transmitters and receivers is still an open problem, but exact results can be obtained in certain low-interference regimes that may arise in practice.
- In Chapter 3, we take an alternative approach to characterizing the rate of communication on an interference channel, based on a *degrees of freedom* (DoF) analysis, which is more analytically tractable. The DoF approach corresponds to analyzing the limiting normalized capacity as the signal-to-noise ratio goes to infinity, and the DoF measure emphasizes the loss in rate that results from the users interfering

with each other over the effect of channel noise. We also describe in Chapter 3 the important technique of interference alignment, which is justified through a DoF analysis. We follow this up with a study of iterative algorithms for approaching the interference alignment solutions for interference management in Chapter 4.

- In Chapter 5, we initiate the discussion of cooperative communication in large interference networks, and analyze the DoF of fully connected interference networks where each message can be available at more than one transmitter, as in coordinated multi-point (CoMP) transmission; this is particularly useful in locally connected networks where each of the transmitters is only connected to a set of neighboring receivers, as we show through the DoF analysis in Chapter 6. Further advantages from CoMP transmission can be gained by relaxing the cooperation constraint to one where the average number of transmitters per message cannot exceed a set value, as we show in Chapter 7. We study the dual problem of cooperative reception schemes for cellular uplink in Chapter 8, and show that similar gains in DoF with cooperation can be obtained in the uplink as in the downlink.
- In Chapter 9, we study dynamic interference networks, where we alter our interference network model to take into account deep fading conditions that can result in random link erasures. In Chapter 10, we discuss recent advances and open problems in the context of modern applications that are anticipated for upcoming generations of wireless networks.
- The book has two appendices: In Appendix A, we summarize some useful results from information theory, and in Appendix B we summarize some results in algebraic geometry.

The main audience for this book is researchers in wireless communication and information theory, with a focus on future generation cellular networks. The material in this book should be accessible to engineers working in the industry as special attention is given to bridging the gap between information-theoretic analyses and practical coding schemes.

Acknowledgments

Some of the material in Chapters 2, 5, and 6 is based on joint work with Sreekanth Annapureddy; some of the material in Chapters 4 and 5 is based on joint work with Craig Wilson; some of the material in Chapter 7 is based on joint work with Meghana Bande; some of the material in Chapter 8 is based on joint work with Manik Singhal; and some of the material in Chapter 9 is based on joint work with Yasemin Karacora and Tolunay Seyfi. We are grateful to Sreekanth Annapureddy, Craig Wilson, Meghana Bande, Manik Singhal, Yasemin Karacora, and Tolunay Seyfi for their contributions.

Our research on interference management in wireless networks over the last decade, which has led to this book, has been funded by the U.S. National Science Foundation through several grants, and also by Motorola, Intel, and Huawei, and we are grateful for this support.

Venugopal Veeravalli dedicates this book to his wife Starla Carpenter for her love and support over the years. Aly El Gamal dedicates this book to his family of seven electrical engineers, and Ali and Omar.

1 Introduction to Interference Management

In this chapter, we provide a high-level introduction to interference management in wireless networks, including a historical perspective on wireless cellular networks, and an overview of the remaining chapters in the book. We also summarize the notation used in the book.

1.1 Interference Management in Cellular Networks: A Historical Perspective

Managing interference from other users sharing the same frequency bands has been the key driver for mobile wireless communications. The first wireless phone systems served as extensions to the wired public switched telephone network [1]. These systems were "single cell" systems in the sense that mobile terminals could be connected to only one basestation during a call, with the call being lost when out of range of the basestation, akin to losing an FM radio signal while driving out of range of the station. Interference in these networks could be managed by simply orthogonalizing the users in the time–frequency plane, i.e., through the use of time-division multiple-access (TDMA) or frequency-division multiple-access (FDMA), or some combination of the two. Interference between basestations operating in the same frequency band was managed by ensuring that they are geographically far apart, again akin to the way in which radio stations operating in the same frequency band are placed.

1.1.1 Cellular Concept

A major breakthrough toward improving both the capacity and the mobility in wireless phone systems came with the introduction of the cellular concept [2]. In the cellular system design, a given geographical region is split into contiguous regions called "cells," without any gaps in coverage. The system is designed so that cells that use the same frequency band are far enough from each other to cause little interference to each other. The number of different frequency bands is called the *reuse factor* of the system. The reuse factor is a measure of spectral efficiency in the system, with a larger reuse factor corresponding to a smaller efficiency. A key innovation in the cellular concept is the introduction of *handoff* between neighboring cells operating in different frequency bands, which allows a mobile user to maintain a continuous connection while moving through the geographical region. Interference management within each of the cells is achieved by orthogonalizing the users in the time–frequency plane.

Early cellular communication systems, both analog (e.g., AMPS [2]) and digital (e.g., GSM [3]), adopted narrowband communication links within each cell. For example, in GSM the available bandwidth is divided into 200 kHz channels, with each channel serving eight users through TDMA. Since the users within a cell occupy orthogonal time–frequency resources, there is no in-cell interference. However, users in neighboring cells communicating in the same time/frequency slot cause *co-channel* interference, which is controlled through the reuse factor.

1.1.2 Code-Division Multiple-Access

An alternative way to separate the users within a cell is through code-division multiple-access (CDMA) [4], where each user's signal occupies the entire time–frequency plane, and the users are separated through the use of different code sequences. Interference from in-cell users can be controlled by using orthogonal codes, which can be implemented in the downlink since the downlink transmissions from a basestation can easily be synchronized. On the uplink, the tight time synchrony required for orthogonalization is difficult to implement, and user separation can be accomplished through simple matched filtering or through the use of more sophisticated successive interference cancellation [5]. A major advantage of CDMA cellular systems over TDMA/FDMA systems comes from the fact that it is possible through the use of pseudorandom overlay codes to randomize the interference across cells in the network so that this interference simply adds to the noise floor for a given user's communication channel. This allows for universal reuse of spectrum (i.e., a reuse factor of one), although there is a loss in spectral efficiency due to the out-of-cell interference effectively raising the noise floor within each cell. This loss in spectral efficiency is generally limited to a factor of two due to the power-law decay of transmitted power with distance [4]. Some other advantages of CDMA systems from an interference management viewpoint include: (i) no frequency planning is needed since the reuse factor is one; (ii) there is a graceful degradation of performance with the number of users in a cell; and (iii) any technique that reduces the power of interferers (e.g., soft handoff, voice activity detection, power control, etc.) increases the capacity. There are some disadvantages that offset these advantages to some extent, including the fact that in-cell interference cannot be eliminated completely and hence reduces capacity, and that tight power control is needed to manage both in-cell and out-of-cell interference, and may be difficult to implement, especially for data applications, which have a low duty cycle.

1.1.3 Orthogonal Frequency-Division Multiplexing

A question that cellular communication system designers asked in the late 1990's when wireless data applications started to grow rapidly was: Can we simultaneously have universal reuse and keep the in-cell users orthogonal? The answer came in the form of multiple access based on orthogonal frequency-division multiplexing (OFDM) [6].

The idea is to split up the bandwidth into narrowband *subchannels*, with every user having access to all the subchannels. The basic unit of resource is a *virtual channel* (i.e., a hopping sequence in time across the subchannels). A given user may be assigned one or more virtual channels for communication. The virtual channels for all users within a given cell are designed to be orthogonal in the time–frequency plane, akin to orthogonal CDMA. Due to the narrowband nature of the subchannels, the orthogonality across users can be maintained almost as easily in the uplink as it is in the downlink. Furthermore, the hopping patterns in adjacent cells are chosen so that there is minimal overlap between any pair of virtual channels across cells, thus averaging the out-of-cell interference to appear as white noise, as in CDMA, as opposed to being localized, as in FDMA/TDMA.

1.2 Additional Resources for Interference Management

In addition to time–frequency separation and geographical separation, there are a number of other resources that can be exploited to manage interference in wireless networks.

1.2.1 Multiple Antennas

The use of multiple antennas at either end of a wireless link provides resilience to fading due to the diversity in the fading seen by the different antennas. Having multiple antennas at both ends of the wireless link forms a multiple-input multiple-output (MIMO) channel, which can be exploited to create multiple parallel streams for communication [7]. This leads to multiplexing or degrees of freedom (DoF) gains. From the viewpoint of interference management, multiple antennas can be used for beamforming toward desired receivers, while minimizing the interference to other receivers. This allows for a more flexible design of interference management schemes, as we will see in Chapters 2, 4, 5, and 6.

1.2.2 Cooperation and Relaying

Cooperation among basestations equipped with multiple antennas can be used for coordinated beamforming across cells so as to maximize the signal-to-interference-plus-noise ratio (SINR) at the receivers [8]. Moreover, such cooperation can also be used for coordinated multi-point (CoMP) transmission and reception by the basestations, which can greatly enhance the DoF achievable in cellular wireless networks, a topic of particular emphasis in this book from Chapter 5 onwards.

Cooperation among mobiles in the networks can also be exploited for interference management, with one mobile relaying the information to or from another mobile [9]. This way potential interferers can become helpers in terms of relaying information to the receiver. Such relaying is particularly useful in distributed interference management in *ad hoc* and *mesh* networks [10].

1.2.3 Cognitive Radio and Dynamic Spectrum Access

Another way to manage interference in wireless networks is through active interference avoidance, as in *cognitive radio* [11]. The key idea is to use sensing to determine the times when a specific licensed band is not used at a particular place and use this band for unlicensed transmissions, without causing interference to the licensed user (referred to as the "primary user"). An important part of designing such systems is the development of a dynamic spectrum access scheme for channel selection. The cognitive radio (also called the "secondary user") needs to adopt the best strategy for selecting channels for sensing and access. The sensing and access policies should jointly ensure that the probability of interfering with the primary user's transmission meets a pre-specified constraint.

1.3 Motivation for This Book

The last few years have seen an exponential growth in data traffic over wireless networks. Wireless service providers are having to accommodate this exponential growth without any significant new useful spectrum. Spectral efficiency gains from improvements in the physical layer are quite limited, with error control coding and decoding being performed near Shannon limits. One way to accommodate the increasing demand for wireless data services is through the addition of basestations in the networks in a hierarchical manner, going from macro to micro to pico to femto basestations, but this comes with a significant cost and makes the interference management problem more difficult through the traditional means described above. Infrastructure enhancements such as cooperative transmission and reception have the potential for increasing the spectral efficiency at low cost, through efficient interference management, but new techniques for interference management are needed. The main motivation for this book is to develop a deep understanding of the fundamental limits of interference management in wireless networks with cooperative transmission and reception, and to use this understanding to develop practical schemes for interference management that approach these limits.

1.4 Overview of This Book

In Chapter 2, we introduce a mathematical model for a K-user interference channel, and use this model to develop some basic information-theoretic bounds on the rates for communication on the channel; particular cases where the sum capacity of the channel can be analyzed exactly are also discussed in this chapter. In Chapter 3, we take an alternative approach to characterizing the rate of communication on an interference channel, based on a *degrees of freedom* analysis, which will be followed in the remainder of the book. In particular, we describe the important technique of interference alignment (IA) in Chapter 3, which is justified through a DoF analysis.

In Chapter 4, we study iterative algorithms for approaching the interference alignment solutions for interference management.

In Chapter 5, we start discussing the value of cooperative communication in large interference networks by studying the DoF of fully connected interference networks when each message can be available at more than one transmitter. In Chapter 6, we extend this setting by studying locally connected networks where each of the transmitters is only connected to a set of neighboring receivers. In Chapter 7, we consider an average backhaul load constraint, where the average number of transmitters per message cannot exceed a set value. We then study cooperative reception schemes for cellular uplink in Chapter 8. In Chapter 9, we study dynamic interference networks, where we alter our interference network model to take into account the deep fading conditions that can result in random link erasures. In Chapter 10, we discuss some recent advances and open problems.

1.5 Notation

We use lower-case and upper-case letters for scalars, lower-case letters in bold font for vectors, and upper-case letters in bold font for matrices. For example, we use h, x, and K to denote scalars, $\boldsymbol{h}$ and $\boldsymbol{x}$ to denote vectors, and $\boldsymbol{H}$, $\boldsymbol{A}$ to denote matrices. Superscripts denote sequences of variables in time. For example, we use x^n and $\boldsymbol{x}^n$ to denote sequences of length n of scalars and vectors, respectively.

We use the notation $\boldsymbol{A}(d)$ for the dth column of the matrix $\boldsymbol{A}$. When we use this notation in general we will refer to a collection of matrices $\boldsymbol{A}_i$, and therefore in our notation $\boldsymbol{A}_i(d)$ is the dth column of the ith matrix $\boldsymbol{A}_i$. Also, $\boldsymbol{x}(\ell)$ is sometimes used to denote the ℓth element of the vector $\boldsymbol{x}$. The matrix $\boldsymbol{I}$ denotes the identity matrix, $\boldsymbol{A}^\dagger$ is the conjugate transpose of $\boldsymbol{A}$, and $\text{diag}(x_1,\ldots,x_N)$ is an $N \times N$ diagonal matrix with $x_1,\ldots,x_N$ on the diagonal.

We use $\boldsymbol{\Sigma}_{\boldsymbol{x}}$ and $\text{Cov}(\boldsymbol{x})$ to denote the covariance matrix of a random vector $\boldsymbol{x}$. We use $\boldsymbol{\Sigma}_{\boldsymbol{y}|\boldsymbol{x}}$ and $\text{Cov}(\boldsymbol{y}|\boldsymbol{x})$ to denote the covariance matrix of the minimum mean square estimation error in estimating the random vector $\boldsymbol{y}$ from the random vector $\boldsymbol{x}$, with similar notation for random scalars. We use $\mathcal{CN}(0,\boldsymbol{\Sigma})$ to denote the circularly symmetric complex Gaussian vector distribution with zero mean and covariance matrix $\boldsymbol{\Sigma}$, with similar notation for random scalars. We use $\mathsf{H}(.)$ to denote the entropy of a discrete random variable, $\mathsf{h}(.)$ to denote the differential entropy of a continuous random variable or vector, and $\mathsf{I}(.;.)$ to denote the mutual information. Finally, we use $[K]$ to denote the set $\{1,2,\ldots,K\}$, where the number K will be obvious from the context.

2 System Model and Sum Capacity Characterization

In this chapter, we introduce a mathematical model for the K-user interference channel, and use this model to develop some basic information-theoretic bounds on the rates for communication on the channel. The focus will be on characterizing bounds on the sum-rate (throughput) of the channel. Particular cases where the sum capacity of the channel can be analyzed exactly will be discussed.

2.1 System and Channel Model

The K-user (fully connected) Gaussian interference channel, illustrated in Figure 2.1, consists of K transmitter–receiver pairs, where every transmitter is heard by every receiver. The signal $\boldsymbol{y}_k \in \mathbb{C}^{N_\mathrm{r}}$ received by receiver k is given by

$$\boldsymbol{y}_k = \sum_{j=1}^{K} \boldsymbol{H}_{kj}\boldsymbol{x}_j + \boldsymbol{z}_k, \ \forall k \in [K], \tag{2.1}$$

where $\boldsymbol{x}_j \in \mathbb{C}^{N_\mathrm{t}\times 1}$ denotes the signal of transmitter j, $\boldsymbol{z}_k \in \mathcal{CN}\left(\boldsymbol{0}, \boldsymbol{I}_{N_\mathrm{r}}\right)$ denotes the additive white Gaussian noise at receiver k, and $\boldsymbol{H}_{kj} \in \mathbb{C}^{N_\mathrm{r}\times N_\mathrm{t}}$ denotes the channel transfer matrix from transmitter j to receiver k. Each transmitter is assumed to have N_t transmit antennas, and each receiver is assumed to have N_r receive antennas.[1] An interference channel with N_t and N_r taking arbitrary values is referred to as a multiple-input multiple-output (MIMO) interference channel. The special cases with $N_\mathrm{t} = 1$ or $N_\mathrm{r} = 1$ or $N_\mathrm{t} = N_\mathrm{r} = 1$ are referred to as the single-input multiple-output (SIMO), multiple-input single-output (MISO), and single-input single-output (SISO) interference channels, respectively. The transmitters are assumed to operate under average power constraints $\{P_k\}$; i.e., for each $k \in [K]$, the power consumed by transmitter k is not allowed to exceed P_k on average.

2.1.1 Achievable Schemes

Consider the problem of communicating K messages over the interference channel (2.1). For each $k \in [K]$, the message W_k is available at transmitter k, and is desired

[1] More generally, the number of antennas could be different at each transmitter and each receiver, as we consider in Section 2.2.

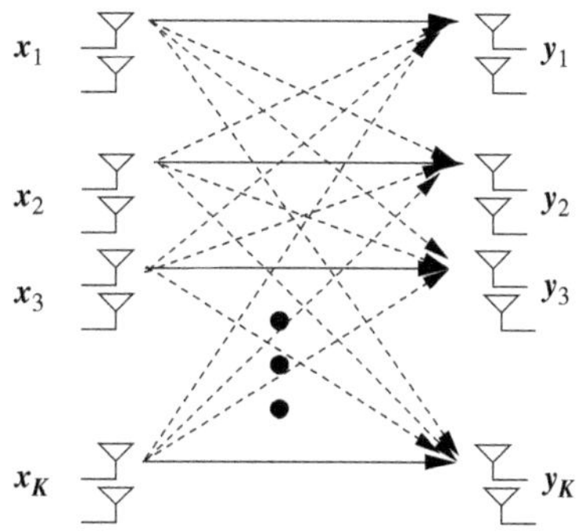

Fig. 2.1 The K-user MIMO Gaussian interference channel.

by receiver k. A communication scheme consists of an encoder–decoder pair for each message. The encoder at transmitter k maps the message W_k onto the physical signal $\boldsymbol{x}_k$ that is transmitted on the channel. The decoder at receiver k reconstructs the message W_k from the received signal $\boldsymbol{y}_k$. The communication scheme is said to be reliable if all the messages can be reconstructed at their respective receivers with high probability. For the single-user channel, Shannon [12] established that the key to reliable communication over noisy channels is coding over multiple symbols. We consider the same *block coding* framework, where the communication scheme operates over n symbols at a time. For a fixed rate tuple $(R_1, R_2, \ldots, R_K) \in \mathbb{R}_+^K$ and a block length $n \geq 1$, the message W_k takes values from the set $\mathcal{W}_k = \left\{1, 2, \ldots, \lceil 2^{nR_k} \rceil\right\}$. The block code consists of the encoders

$$\boldsymbol{x}_k^n : \mathcal{W}_k \to \mathbb{C}^{N_t \times n}, \ \forall k \in [K],$$

and the decoders

$$\hat{W}_k : \mathbb{C}^{N_r \times n} \to \mathcal{W}_k, \ \forall k \in [K].$$

Assuming that the message W_k is a uniform random variable taking values in the set $\mathcal{W}_k$, the probability of a decoding error is defined as

$$e_n = \max_{k \in [K]} \mathbb{P}\left(\hat{W}_k\left(\boldsymbol{y}_k^n\right) \neq W_k\right).$$

We say that the rate tuple $(R_1, R_2, \ldots, R_K)$ is achievable if and only if there exists a sequence of block codes satisfying the average power constraints

$$\mathbb{E}\left[\frac{1}{n}\sum_{t=1}^{n} ||\boldsymbol{x}_k(t)||^2\right] \leq P_k, \ \forall k \in [K]$$

such that the probability of error $e_n \to 0$ as $n \to \infty$. The *capacity region* $\mathcal{C}$ is defined as the closure of the set of achievable rate tuples. Except in some special cases, determining the exact capacity region of the Gaussian interference channel remains an open problem.

The *sum capacity* is defined as:

$$\mathcal{C}_{\text{sum}} = \max_{(R_1, R_2, \ldots, R_k) \in \mathcal{C}} R_1 + R_2 + \cdots + R_K. \tag{2.2}$$

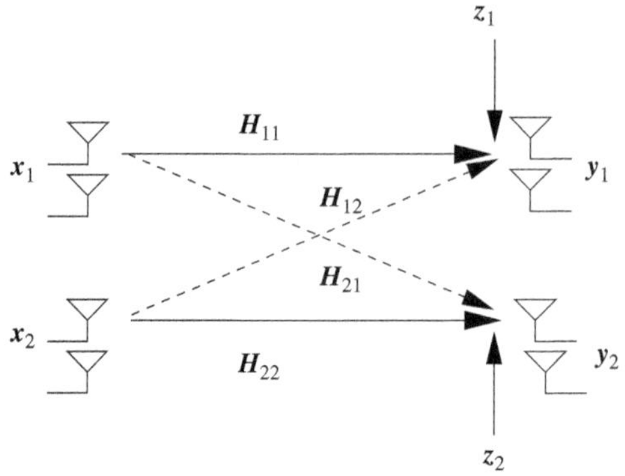

Fig. 2.2 Two-user interference channel. ©[2017] IEEE. Reprinted, with permission, from [13].

2.1.2 Channel Knowledge

We assume that the various channel coefficients are known at all the transmitters and at all the receivers at which they are required for a given achievable scheme. In practice, the channel knowledge is obtained by transmitting known signals, called pilots, at regular intervals and estimating the channel coefficients at the receivers. The estimated (local) channel coefficients are then distributed to other transmitters and receivers. Although the processes of channel estimation and distribution can incur significant overhead, it is difficult to accommodate this overhead in information-theoretic capacity analyses. The common practice, which is also followed in this book, is to perform the capacity analysis assuming channel knowledge where needed, and account for the overhead when designing practical achievable schemes.

2.2 Two-User Interference Channel

In this section, we consider the Gaussian interference channel (2.1) in the two-user case, assuming that multiple antennas are available at the transmitter and receiver:

$$\begin{aligned} \boldsymbol{y}_1 &= \boldsymbol{H}_{11}\boldsymbol{x}_1 + \boldsymbol{H}_{12}\boldsymbol{x}_2 + \boldsymbol{z}_1, \\ \boldsymbol{y}_2 &= \boldsymbol{H}_{21}\boldsymbol{x}_1 + \boldsymbol{H}_{22}\boldsymbol{x}_2 + \boldsymbol{z}_2, \end{aligned} \tag{2.3}$$

where $\boldsymbol{z}_i \in \mathcal{CN}(0, \boldsymbol{I})$, and the average power constraints at transmitters 1 and 2 are denoted by P_1 and P_2, respectively. This is depicted in Figure 2.2. Let $N_{1\text{t}}, N_{2\text{t}}$ denote the number of transmit antennas at transmitters 1 and 2, respectively, and $N_{1\text{r}}, N_{2\text{r}}$ denote the number of receive antennas at receivers 1 and 2, respectively. The dimensions of the channel matrices, the signal vectors, and the noise vectors are defined appropriately. We are interested in determining the best sum-rate achievable by using Gaussian inputs and treating interference as noise, and also the sum capacity (maximum throughput) of the two-user MIMO Gaussian interference channel.

We start by studying the problem of determining the best achievable sum-rate. The use of multiple antennas at the transmitters and the receivers provides spatial dimensions to suppress the interference and improve the achievable sum-rate. While it is easy to express the achievable sum-rate as a function of the spatial beams at the transmitter and receiver, the design of beams that maximize the achievable sum-rate is known to be a difficult problem. The main difficulty stems from the fact that the sum-rate optimization problem cannot be posed as a convex (concave) optimization problem, which makes the optimization problem difficult to solve analytically or even numerically. We introduce a technique based on convex approximation and optimization to solve this nonconvex optimization problem. Specifically, we upper-bound the achievable sum-rate with a concave function, and use this to obtain an upper bound to the original sum-rate optimization problem. We show that if the channel parameters satisfy certain conditions, then the bounds coincide, leading to an exact characterization of the best achievable sum-rate by using Gaussian inputs and treating interference as noise.

The problem of determining the best achievable sum-rate by treating interference as noise is important from a practical perspective, because coding schemes that approach the rates promised by the information-theoretic analysis can be designed in the same way as schemes for point-to-point Gaussian channels, a topic that is well understood [14]. Therefore, it is also important to understand the gap between the sum capacity and the sum-rate achievable by treating interference as noise. If the lower and upper bounds on the achievable sum-rate coincide, then the best achievable sum-rate is indeed equal to the sum capacity. Using the Karush–Kuhn–Tucker (KKT) conditions [15], we obtain necessary and sufficient conditions for the bounds to coincide, leading to an exact characterization of the sum capacity. We observe that the conditions are satisfied in a *low interference regime* where the interfering signal levels are small compared to the desired signal levels. We end the section by providing some nontrivial examples of the two-user Gaussian interference channel in the low interference regime. In particular, we consider the special cases of symmetric MISO and SIMO interference channels, and derive a simple closed-form condition on the channel parameters for the channels to be in the low interference regime. We also specialize the results to SISO interference channels.

2.2.1 Standard Form

The following assumptions can be made about the two-user MIMO Gaussian interference channel (2.3) without any loss of generality, as we establish below:

- The direct channel matrices $\boldsymbol{H}_{11}$ and $\boldsymbol{H}_{22}$ have unit (Frobenius) norm.
- The cross channel matrices $\boldsymbol{H}_{12}$ and $\boldsymbol{H}_{21}$ are diagonal with real and nonnegative entries.
- The numbers of transmit and receive antennas $(N_{1t}, N_{2t}, N_{1r}, N_{2r})$ satisfy

$$N_{1t} \leq \operatorname{rank}\left\{\begin{bmatrix} \boldsymbol{H}_{11} \\ \boldsymbol{H}_{21} \end{bmatrix}\right\}, N_{2t} \leq \operatorname{rank}\left\{\begin{bmatrix} \boldsymbol{H}_{12} \\ \boldsymbol{H}_{22} \end{bmatrix}\right\},$$

and

$$N_{1r} \leq \text{rank}\{[\boldsymbol{H}_{11}\ \boldsymbol{H}_{12}]\},\ N_{2r} \leq \text{rank}\{[\boldsymbol{H}_{21}\ \boldsymbol{H}_{22}]\}.$$

The second assumption implies that the cross channel matrices can be expressed as

$$\boldsymbol{H}_{12} = \begin{bmatrix} \tilde{\boldsymbol{H}}_{12} \\ \boldsymbol{0} \end{bmatrix},\ \boldsymbol{H}_{21} = \begin{bmatrix} \tilde{\boldsymbol{H}}_{21} \\ \boldsymbol{0} \end{bmatrix},$$

where $\tilde{\boldsymbol{H}}_{12}$ and $\tilde{\boldsymbol{H}}_{21}$ are diagonal matrices with full row rank. This is the only assumption we use in the development of the outer bound techniques presented in this chapter. The other two assumptions are used in Section 2.2.13 to simplify the presentation.

The first assumption can easily be justified by scaling the transmit power constraints P_1 and P_2 appropriately. We now justify the other two assumptions. First, consider the singular value decomposition (SVD) of $\boldsymbol{H}_{12}$ and $\boldsymbol{H}_{21}$:

$$\boldsymbol{H}_{12} = \boldsymbol{U}_1 \boldsymbol{\Lambda}_{12} \boldsymbol{V}_2^{\dagger},$$
$$\boldsymbol{H}_{21} = \boldsymbol{U}_2 \boldsymbol{\Lambda}_{21} \boldsymbol{V}_1^{\dagger},$$

where $\boldsymbol{\Lambda}_{12}, \boldsymbol{\Lambda}_{21}$ are diagonal matrices with real and nonnegative entries, and $\boldsymbol{V}_1, \boldsymbol{V}_2, \boldsymbol{U}_1, \boldsymbol{U}_2$ are unitary matrices. We obtain an equivalent Gaussian interference channel, satisfying the second assumption, by projecting the received signals along $\boldsymbol{U}_1, \boldsymbol{U}_2$, and the transmitted signals along $\boldsymbol{V}_1, \boldsymbol{V}_2$, i.e., by making the following substitutions:

$$\boldsymbol{x}_j \leftarrow \boldsymbol{V}_j^{\dagger} \boldsymbol{x}_j,$$
$$\boldsymbol{y}_i \leftarrow \boldsymbol{U}_i^{\dagger} \boldsymbol{y}_i,$$
$$\boldsymbol{z}_i \leftarrow \boldsymbol{U}_i^{\dagger} \boldsymbol{z}_i,$$
$$\boldsymbol{H}_{ij} \leftarrow \boldsymbol{U}_i^{\dagger} \boldsymbol{H}_{ij} \boldsymbol{V}_j.$$

Observe that the average transmit power constraint and the distribution of the receive noise terms remain unchanged because $\boldsymbol{U}_1, \boldsymbol{U}_2, \boldsymbol{V}_1, \boldsymbol{V}_2$ are unitary matrices.

The third assumption can be justified by appropriately choosing the unitary matrices. For example, suppose $N_{1r} > \text{rank}\{[\boldsymbol{H}_{11}\ \boldsymbol{H}_{12}]\}$. Consider the SVD of $\boldsymbol{H}_{12} = \boldsymbol{U}_1 \Lambda_{12} \boldsymbol{V}_2^{\dagger}$. Observe that the span of the first $\text{rank}\{\boldsymbol{H}_{12}\}$ columns of $\boldsymbol{U}_1$ is equal to the column space of $\boldsymbol{H}_{12}$, and we have flexibility in choosing the remaining $N_{1r} - \text{rank}\{\boldsymbol{H}_{12}\}$ columns. We may choose those columns such that the span of the first $\text{rank}\{[\boldsymbol{H}_{11}\ \boldsymbol{H}_{12}]\}$ columns of $\boldsymbol{U}_1$ is equal to the column space of $[\boldsymbol{H}_{11}\ \boldsymbol{H}_{12}]$, so that the last $N_{1r} - \text{rank}\{[\boldsymbol{H}_{11}\ \boldsymbol{H}_{12}]\}$ columns of $\boldsymbol{U}_1$ are orthogonal to the columns of $[\boldsymbol{H}_{11}\ \boldsymbol{H}_{12}]$. Therefore, the last $N_{1r} - \text{rank}\{[\boldsymbol{H}_{11}\ \boldsymbol{H}_{12}]\}$ rows of the channel matrices $\boldsymbol{H}_{11}$ and $\boldsymbol{H}_{12}$ in the new channel are equal to zero, i.e., receiver 1 sees only Gaussian noise from the last $N_{1r} - \text{rank}\{[\boldsymbol{H}_{11}\ \boldsymbol{H}_{12}]\}$ antennas. Hence, we can ignore these antennas and assume that $N_{1r} = \text{rank}\{[\boldsymbol{H}_{11}\ \boldsymbol{H}_{12}]\}$. We can repeat the same argument at receiver 2, and also at transmitters 1 and 2, to justify the other inequalities in the third assumption.

2.2.2 Achievable Sum-Rate

For the two-user MIMO Gaussian interference channel (2.3), the sum-rate achievable by using (circularly symmetric) Gaussian inputs and treating interference as noise is given by

$$\mathsf{I}\left(\boldsymbol{x}_{1\mathrm{G}};\boldsymbol{y}_{1\mathrm{G}}\right)+\mathsf{I}\left(\boldsymbol{x}_{2\mathrm{G}};\boldsymbol{y}_{2\mathrm{G}}\right), \tag{2.4}$$

where the subscript G indicates that Gaussian inputs are used. Let $\boldsymbol{Q}_1 = \boldsymbol{\Sigma}_{\boldsymbol{x}_{1\mathrm{G}}}$ and $\boldsymbol{Q}_2 = \boldsymbol{\Sigma}_{\boldsymbol{x}_{2\mathrm{G}}}$ denote the covariance matrices of the Gaussian random vectors $\boldsymbol{x}_{1\mathrm{G}}$ and $\boldsymbol{x}_{2\mathrm{G}}$, respectively. To meet the average power constraints, the covariance matrices must belong to the feasible region

$$\mathcal{Q} = \{(\boldsymbol{Q}_1,\boldsymbol{Q}_2) : \boldsymbol{Q}_i \succeq 0, \mathsf{Tr}\left(\boldsymbol{Q}_i\right) \le P, i = 1,2\}.$$

We have the option to design the covariance matrices $\boldsymbol{Q}_1$ and $\boldsymbol{Q}_2$ to maximize the achievable sum-rate, leading to the optimization problem

$$\max_{(\boldsymbol{Q}_1,\boldsymbol{Q}_2)\in\mathcal{Q}} f(\boldsymbol{Q}_1,\boldsymbol{Q}_2), \tag{2.5}$$

where $f(\boldsymbol{Q}_1,\boldsymbol{Q}_2)$ denotes the sum-rate (2.4) as a function of $\boldsymbol{Q}_1$ and $\boldsymbol{Q}_2$, i.e.,

$$\begin{aligned} f(\boldsymbol{Q}_1,\boldsymbol{Q}_2) &= \mathsf{I}\left(\boldsymbol{x}_{1\mathrm{G}};\boldsymbol{y}_{1\mathrm{G}}\right)+\mathsf{I}\left(\boldsymbol{x}_{2\mathrm{G}};\boldsymbol{y}_{2\mathrm{G}}\right) \\ &= \mathsf{h}\left(\boldsymbol{y}_{1\mathrm{G}}\right)-\mathsf{h}\left(\boldsymbol{y}_{1\mathrm{G}}|\boldsymbol{x}_{1\mathrm{G}}\right)+\mathsf{h}\left(\boldsymbol{y}_{2\mathrm{G}}\right)-\mathsf{h}\left(\boldsymbol{y}_{2\mathrm{G}}|\boldsymbol{x}_{2\mathrm{G}}\right) \\ &= \log\frac{\det\boldsymbol{\Sigma}_{\boldsymbol{y}_{1\mathrm{G}}}}{\det\boldsymbol{\Sigma}_{\boldsymbol{y}_{1\mathrm{G}}|\boldsymbol{x}_{1\mathrm{G}}}}+\log\frac{\det\boldsymbol{\Sigma}_{\boldsymbol{y}_{2\mathrm{G}}}}{\det\boldsymbol{\Sigma}_{\boldsymbol{y}_{2\mathrm{G}}|\boldsymbol{x}_{2\mathrm{G}}}}. \end{aligned} \tag{2.6}$$

The explicit dependence on $\boldsymbol{Q}_1$ and $\boldsymbol{Q}_2$ can be seen by carrying out the substitutions:

$$\begin{aligned} \boldsymbol{\Sigma}_{\boldsymbol{y}_{1\mathrm{G}}} &= \boldsymbol{I}+\boldsymbol{H}_{11}\boldsymbol{Q}_1\boldsymbol{H}_{11}^{\dagger}+\boldsymbol{H}_{12}\boldsymbol{Q}_2\boldsymbol{H}_{12}^{\dagger}, \\ \boldsymbol{\Sigma}_{\boldsymbol{y}_{2\mathrm{G}}} &= \boldsymbol{I}+\boldsymbol{H}_{22}\boldsymbol{Q}_1\boldsymbol{H}_{21}^{\dagger}+\boldsymbol{H}_{22}\boldsymbol{Q}_2\boldsymbol{H}_{22}^{\dagger}, \\ \boldsymbol{\Sigma}_{\boldsymbol{y}_{1\mathrm{G}}|\boldsymbol{x}_{1\mathrm{G}}} &= \boldsymbol{I}+\boldsymbol{H}_{12}\boldsymbol{Q}_2\boldsymbol{H}_{12}^{\dagger}, \\ \boldsymbol{\Sigma}_{\boldsymbol{y}_{2\mathrm{G}}|\boldsymbol{x}_{2\mathrm{G}}} &= \boldsymbol{I}+\boldsymbol{H}_{21}\boldsymbol{Q}_1\boldsymbol{H}_{21}^{\dagger}. \end{aligned}$$

The nonconcave nature of the objective function $f(\boldsymbol{Q}_1,\boldsymbol{Q}_2)$ makes the optimization problem (2.5) difficult to solve. To see that $f(\boldsymbol{Q}_1,\boldsymbol{Q}_2)$ is not concave in general, consider the special case of a symmetric SISO interference channel with $\boldsymbol{H}_{11} = \boldsymbol{H}_{22} = 1$ and $\boldsymbol{H}_{21} = \boldsymbol{H}_{12} = h$. Thus, we have

$$f(q_1,q_2) = \log\left(1+\frac{q_1}{1+h^2q_2}\right)+\log\left(1+\frac{q_2}{1+h^2q_1}\right).$$

Observe that

$$\frac{f(2q,0)+f(0,2q)}{2} - f(q,q) = \log(1+2q) - 2\log\left(1+\frac{q}{1+h^2q}\right) \ge 0$$

whenever q and h satisfy

$$(1+2q) \ge \left(1+\frac{q}{1+h^2q}\right)^2,$$

i.e., whenever

$$2q \geq \frac{2q}{1+h^2q} + \left(\frac{q}{1+h^2q}\right)^2,$$

which is equivalent to saying that

$$2h^2(1+h^2q) \geq 1.$$

This shows that the function $f(\boldsymbol{Q}_1,\boldsymbol{Q}_2)$ is not concave in general.

2.2.3 Locally Optimal Solution

Since a global optimal solution has to be locally optimal, we can obtain insights into the structural properties of the optimal transmit covariance matrices by analyzing the necessary KKT conditions [15]. Let $\lambda_1,\lambda_2 \geq 0$ and $\boldsymbol{M}_1,\boldsymbol{M}_2 \succeq 0$ denote the dual variables associated with the constraints $\mathrm{Tr}(\boldsymbol{Q}_1) \leq P_1, \mathrm{Tr}(\boldsymbol{Q}_2) \leq P_2$, and $\boldsymbol{Q}_1,\boldsymbol{Q}_2 \succeq 0$, respectively. The Lagrangian associated with (2.5) is given by

$$\begin{aligned}
&L(\boldsymbol{Q}_1,\boldsymbol{Q}_2,\boldsymbol{M}_1,\boldsymbol{M}_2,\lambda_1,\lambda_2)\\
&\quad = f(\boldsymbol{Q}_1,\boldsymbol{Q}_2,\Psi) + \sum_{i=1}^{2} \mathrm{Tr}(\boldsymbol{M}_i\boldsymbol{Q}_i) - \lambda_i\left(\mathrm{Tr}(\boldsymbol{Q}_i) - P_i\right).
\end{aligned}$$

The KKT conditions are given by

$$\begin{aligned}
\nabla_{\boldsymbol{Q}_1} f(\boldsymbol{Q}_1,\boldsymbol{Q}_2) &= \lambda_1 \boldsymbol{I} - \boldsymbol{M}_1,\\
\nabla_{\boldsymbol{Q}_2} f(\boldsymbol{Q}_1,\boldsymbol{Q}_2) &= \lambda_2 \boldsymbol{I} - \boldsymbol{M}_2,\\
\lambda_1(\mathrm{Tr}(\boldsymbol{Q}_1) - P_1) &= 0,\\
\lambda_2(\mathrm{Tr}(\boldsymbol{Q}_2) - P_2) &= 0,\\
\mathrm{Tr}(\boldsymbol{M}_1\boldsymbol{Q}_1) &= 0,\\
\mathrm{Tr}(\boldsymbol{M}_2\boldsymbol{Q}_2) &= 0.
\end{aligned} \tag{2.7}$$

The following fact from matrix differential calculus is useful in deriving the expressions for gradients [16, 17]: given matrices $\boldsymbol{\Sigma} = \boldsymbol{\Sigma}^\dagger$ and $\boldsymbol{H}$, it holds that

$$\nabla_{\boldsymbol{Q}} \log\det\left(\boldsymbol{\Sigma} + \boldsymbol{H}\boldsymbol{Q}\boldsymbol{H}^\dagger\right) = \boldsymbol{H}^\dagger\left(\boldsymbol{\Sigma} + \boldsymbol{H}\boldsymbol{Q}\boldsymbol{H}^\dagger\right)^{-1}\boldsymbol{H}.$$

Using the expression (2.6) for sum-rate, we obtain that

$$\begin{aligned}
\nabla_{\boldsymbol{Q}_1} f(\boldsymbol{Q}_1,\boldsymbol{Q}_2) &= \boldsymbol{H}_{11}^\dagger \boldsymbol{\Sigma}_{y_{1G}}^{-1}\boldsymbol{H}_{11} + \boldsymbol{H}_{21}^\dagger\left(\boldsymbol{\Sigma}_{y_{2G}}^{-1} - \boldsymbol{\Sigma}_{y_{2G}|x_{2G}}^{-1}\right)\boldsymbol{H}_{21},\\
\nabla_{\boldsymbol{Q}_2} f(\boldsymbol{Q}_1,\boldsymbol{Q}_2) &= \boldsymbol{H}_{22}^\dagger \boldsymbol{\Sigma}_{y_{2G}}^{-1}\boldsymbol{H}_{22} + \boldsymbol{H}_{12}^\dagger\left(\boldsymbol{\Sigma}_{y_{1G}}^{-1} - \boldsymbol{\Sigma}_{y_{1G}|x_{1G}}^{-1}\right)\boldsymbol{H}_{12}.
\end{aligned} \tag{2.8}$$

2.2.4 Convex Approximation and Optimization

We begin by introducing a convex optimization problem, the solution to which provides lower and upper bounds on the best achievable sum-rate (2.5). Suppose we upper-bound $f(\boldsymbol{Q}_1,\boldsymbol{Q}_2)$ with $\bar{f}(\boldsymbol{Q}_1,\boldsymbol{Q}_2)$, i.e., $f(\boldsymbol{Q}_1,\boldsymbol{Q}_2) \leq \bar{f}(\boldsymbol{Q}_1,\boldsymbol{Q}_2)$, such that $\bar{f}(\boldsymbol{Q}_1,\boldsymbol{Q}_2)$ is concave in $\boldsymbol{Q}_1$ and $\boldsymbol{Q}_2$. Then, we can solve the optimization problem

$$\bar{f}(\boldsymbol{Q}_1^*,\boldsymbol{Q}_2^*) = \max_{(\boldsymbol{Q}_1,\boldsymbol{Q}_2)\in\mathcal{Q}} \bar{f}(\boldsymbol{Q}_1,\boldsymbol{Q}_2)$$

to obtain lower and upper bounds to the sum-rate optimization problem (2.5):

$$f(\boldsymbol{Q}_1^*,\boldsymbol{Q}_2^*) \leq \text{Best achievable sum-rate} \leq \bar{f}(\boldsymbol{Q}_1^*,\boldsymbol{Q}_2^*).$$

The tightness of the lower and upper bounds depends on the choice of the upper bound function $\bar{f}(\boldsymbol{Q}_1,\boldsymbol{Q}_2)$. The upper bound function we use here is based on a genie giving side information to the receivers. By treating the side information as a part of the received signal, we obtain a *genie-aided* MIMO Gaussian interference channel. Let Ψ denote the genie parameters, which will be defined in Section 2.2.5. The achievable sum-rate $\bar{f}(\boldsymbol{Q}_1,\boldsymbol{Q}_2,\Psi)$ in the genie-aided channel is an obvious upper bound to the achievable sum-rate $f(\boldsymbol{Q}_1,\boldsymbol{Q}_2)$ in the original channel, i.e.,

$$f(\boldsymbol{Q}_1,\boldsymbol{Q}_2) \leq \bar{f}(\boldsymbol{Q}_1,\boldsymbol{Q}_2,\Psi).$$

We say that the genie Ψ is useful if the upper bound function $\bar{f}(\boldsymbol{Q}_1,\boldsymbol{Q}_2,\Psi)$ is concave. Let Ψ_{u} denote the usefulness set; i.e., $\bar{f}(\boldsymbol{Q}_1,\boldsymbol{Q}_2,\Psi)$ is concave in $(\boldsymbol{Q}_1,\boldsymbol{Q}_2)$ for all $\Psi \in \Psi_{\mathrm{u}}$. We obtain the best upper bound to $f(\boldsymbol{Q}_1,\boldsymbol{Q}_2)$ by optimizing over $\Psi \in \Psi_{\mathrm{u}}$; i.e.,

$$f(\boldsymbol{Q}_1,\boldsymbol{Q}_2) \leq \bar{f}(\boldsymbol{Q}_1,\boldsymbol{Q}_2) = \min_{\Psi\in\Psi_{\mathrm{u}}} \bar{f}(\boldsymbol{Q}_1,\boldsymbol{Q}_2,\Psi).$$

Therefore, the upper and lower bounds to the best sum-rate can be obtained by solving the maxmin optimization problem

$$\max_{(\boldsymbol{Q}_1,\boldsymbol{Q}_2)\in\mathcal{Q}} \min_{\Psi\in\Psi_{\mathrm{u}}} \bar{f}(\boldsymbol{Q}_1,\boldsymbol{Q}_2,\Psi).$$

We will also show that the upper bound function $\bar{f}(\boldsymbol{Q}_1,\boldsymbol{Q}_2,\Psi)$ is convex in Ψ for every $(\boldsymbol{Q}_1,\boldsymbol{Q}_2)$. Therefore, we have a convex maxmin optimization problem, which can be solved efficiently using standard convex optimization algorithms.

2.2.5 Genie-Aided Channel

Suppose the genie provides receivers 1 and 2 with side information $\boldsymbol{s}_1$ and $\boldsymbol{s}_2$, respectively. The signals $\boldsymbol{s}_1$ and $\boldsymbol{s}_2$ are defined as

$$\begin{aligned} \boldsymbol{s}_1 &= \tilde{\boldsymbol{H}}_{21}\boldsymbol{x}_1 + \boldsymbol{w}_1, \\ \boldsymbol{s}_2 &= \tilde{\boldsymbol{H}}_{12}\boldsymbol{x}_2 + \boldsymbol{w}_2, \end{aligned} \tag{2.9}$$

where $\tilde{\boldsymbol{H}}_{12},\tilde{\boldsymbol{H}}_{21}$, defined in Section 2.2.1, represent the matrices containing the nonzero rows of $\boldsymbol{H}_{12},\boldsymbol{H}_{21}$, respectively, and $\boldsymbol{w}_1,\boldsymbol{w}_2$ are random vectors denoting Gaussian noise. The genie chooses how the noise terms $\boldsymbol{w}_1$ and $\boldsymbol{w}_2$ are correlated to z_1 and z_2, respectively. We use Ψ as a shorthand notation to denote the genie parameters $\Psi = \{\boldsymbol{\Sigma}_{w_1}, \boldsymbol{\Sigma}_{w_1 z_1}, \boldsymbol{\Sigma}_{w_2}, \boldsymbol{\Sigma}_{w_2 z_2}\}$ satisfying the positive semidefinite constraints

$$\mathrm{Cov}\left(\begin{bmatrix} z_i \\ \boldsymbol{w}_i \end{bmatrix}\right) = \begin{bmatrix} \boldsymbol{I} & \boldsymbol{\Sigma}_{z_i w_i} \\ \boldsymbol{\Sigma}_{w_i z_i} & \boldsymbol{\Sigma}_{w_i} \end{bmatrix} \succeq 0, \; i = 1, 2.$$

We use the achievable sum-rate of the genie-aided interference channel as the upper bound function,

$$\bar{f}(\boldsymbol{Q}_1, \boldsymbol{Q}_2, \Psi) = \mathsf{I}\left(\boldsymbol{x}_{1\mathrm{G}}; \boldsymbol{y}_{1\mathrm{G}}, \boldsymbol{s}_{1\mathrm{G}}\right) + \mathsf{I}\left(\boldsymbol{x}_{2\mathrm{G}}; \boldsymbol{y}_{2\mathrm{G}}, \boldsymbol{s}_{2\mathrm{G}}\right). \tag{2.10}$$

Since the mutual information is nonnegative, we obtain that

$$\mathsf{I}\left(\boldsymbol{x}_{1\mathrm{G}}; \boldsymbol{y}_{1\mathrm{G}}\right) \le \mathsf{I}\left(\boldsymbol{x}_{1\mathrm{G}}; \boldsymbol{y}_{1\mathrm{G}}\right) + \mathsf{I}\left(\boldsymbol{x}_{1\mathrm{G}}; \boldsymbol{s}_{1\mathrm{G}} | \boldsymbol{y}_{1\mathrm{G}}\right) = \mathsf{I}\left(\boldsymbol{x}_{1\mathrm{G}}; \boldsymbol{y}_{1\mathrm{G}}, \boldsymbol{s}_{1\mathrm{G}}\right),$$
$$\mathsf{I}\left(\boldsymbol{x}_{2\mathrm{G}}; \boldsymbol{y}_{2\mathrm{G}}\right) \le \mathsf{I}\left(\boldsymbol{x}_{2\mathrm{G}}; \boldsymbol{y}_{2\mathrm{G}}\right) + \mathsf{I}\left(\boldsymbol{x}_{2\mathrm{G}}; \boldsymbol{s}_{2\mathrm{G}} | \boldsymbol{y}_{2\mathrm{G}}\right) = \mathsf{I}\left(\boldsymbol{x}_{2\mathrm{G}}; \boldsymbol{y}_{2\mathrm{G}}, \boldsymbol{s}_{2\mathrm{G}}\right),$$

and hence that

$$f(\boldsymbol{Q}_1, \boldsymbol{Q}_2) \le \bar{f}(\boldsymbol{Q}_1, \boldsymbol{Q}_2, \Psi) \text{ for any } \Psi.$$

To utilize the idea in Section 2.2.4, we now define the usefulness set Ψ_{u}, and show the following properties:

- The set Ψ_{u} is convex.
- The function $\bar{f}(\boldsymbol{Q}_1, \boldsymbol{Q}_2, \Psi)$ is concave in $(\boldsymbol{Q}_1, \boldsymbol{Q}_2)$ for any $\Psi \in \Psi_{\mathrm{u}}$.
- The function $\bar{f}(\boldsymbol{Q}_1, \boldsymbol{Q}_2, \Psi)$ is convex in Ψ for any $\boldsymbol{Q}_1$ and $\boldsymbol{Q}_2$.

2.2.6 Useful Genie: Concavity Property

Let Ψ_{u} be the set of genie parameters $\Psi = \left(\boldsymbol{\Sigma}_{w_1}, \boldsymbol{\Sigma}_{w_1 z_1}, \boldsymbol{\Sigma}_{w_2}, \boldsymbol{\Sigma}_{w_2 z_2}\right)$ satisfying the usefulness conditions

$$\begin{bmatrix} \boldsymbol{I} & \boldsymbol{\Sigma}_{z_1 w_1} \\ \boldsymbol{\Sigma}_{w_1 z_1} & \boldsymbol{\Sigma}_{w_1} \end{bmatrix} \succeq \begin{bmatrix} \boldsymbol{\Sigma}_{w_2} & \boldsymbol{0} \\ \boldsymbol{0} & \boldsymbol{0} \end{bmatrix},$$
$$\begin{bmatrix} \boldsymbol{I} & \boldsymbol{\Sigma}_{z_2 w_2} \\ \boldsymbol{\Sigma}_{w_2 z_2} & \boldsymbol{\Sigma}_{w_2} \end{bmatrix} \succeq \begin{bmatrix} \boldsymbol{\Sigma}_{w_1} & \boldsymbol{0} \\ \boldsymbol{0} & \boldsymbol{0} \end{bmatrix}. \tag{2.11}$$

It immediately follows that Ψ_{u} is a convex set because the convex combination of any two positive semidefinite matrices is also positive semidefinite.

LEMMA 2.1 *The function $\bar{f}(\boldsymbol{Q}_1, \boldsymbol{Q}_2, \Psi)$ is concave and nondecreasing in $(\boldsymbol{Q}_1, \boldsymbol{Q}_2)$ for any $\Psi \in \Psi_{\mathrm{u}}$.*

Proof We provide a sketch of the proof here; the details can be found in [13]. First, we expand the terms in $\bar{f}(\boldsymbol{Q}_1, \boldsymbol{Q}_2, \Psi)$:

$$\begin{aligned}
\bar{f}(\boldsymbol{Q}_1, \boldsymbol{Q}_2, \Psi) &= \mathsf{I}\left(\boldsymbol{x}_{1\mathrm{G}}; \boldsymbol{y}_{1\mathrm{G}}, \boldsymbol{s}_{1\mathrm{G}}\right) + \mathsf{I}\left(\boldsymbol{x}_{2\mathrm{G}}; \boldsymbol{y}_{2\mathrm{G}}, \boldsymbol{s}_{2\mathrm{G}}\right) \\
&= \mathsf{h}\left(\boldsymbol{y}_{1\mathrm{G}}, \boldsymbol{s}_{1\mathrm{G}}\right) - \mathsf{h}\left(\boldsymbol{y}_{1\mathrm{G}}, \boldsymbol{s}_{1\mathrm{G}} | \boldsymbol{x}_{1\mathrm{G}}\right) \\
&\quad + \mathsf{h}\left(\boldsymbol{y}_{2\mathrm{G}}, \boldsymbol{s}_{2\mathrm{G}}\right) - \mathsf{h}\left(\boldsymbol{y}_{2\mathrm{G}}, \boldsymbol{s}_{2\mathrm{G}} | \boldsymbol{x}_{2\mathrm{G}}\right) \\
&= \mathsf{h}\left(\boldsymbol{s}_{1\mathrm{G}}\right) + \mathsf{h}\left(\boldsymbol{y}_{1\mathrm{G}} | \boldsymbol{s}_{1\mathrm{G}}\right) - \mathsf{h}\left(\boldsymbol{s}_{1\mathrm{G}} | \boldsymbol{x}_{1\mathrm{G}}\right) - \mathsf{h}\left(\boldsymbol{y}_{1\mathrm{G}} | \boldsymbol{s}_{1\mathrm{G}}, \boldsymbol{x}_{1\mathrm{G}}\right) \\
&\quad + \mathsf{h}\left(\boldsymbol{s}_{2\mathrm{G}}\right) + \mathsf{h}\left(\boldsymbol{y}_{2\mathrm{G}} | \boldsymbol{s}_{2\mathrm{G}}\right) - \mathsf{h}\left(\boldsymbol{s}_{2\mathrm{G}} | \boldsymbol{x}_{2\mathrm{G}}\right) - \mathsf{h}\left(\boldsymbol{y}_{2\mathrm{G}} | \boldsymbol{s}_{2\mathrm{G}}, \boldsymbol{x}_{2\mathrm{G}}\right).
\end{aligned}$$

The terms $\mathsf{h}(\boldsymbol{s}_{1\mathrm{G}}|\boldsymbol{x}_{1\mathrm{G}})$ and $\mathsf{h}(\boldsymbol{s}_{2\mathrm{G}}|\boldsymbol{x}_{2\mathrm{G}})$ do not depend on $\boldsymbol{Q}_1$ and $\boldsymbol{Q}_2$. From Lemma A.4 in Appendix A, it immediately follows that the terms $\mathsf{h}\left(\boldsymbol{y}_{1\mathrm{G}}|\boldsymbol{s}_{1\mathrm{G}}\right)$ and $\mathsf{h}\left(\boldsymbol{y}_{2\mathrm{G}}|\boldsymbol{s}_{2\mathrm{G}}\right)$ are concave and nondecreasing in $(\boldsymbol{Q}_1, \boldsymbol{Q}_2)$. The remaining terms contribute

$$\mathsf{h}\left(\boldsymbol{s}_{1\mathrm{G}}\right) - \mathsf{h}\left(\boldsymbol{y}_{1\mathrm{G}} | \boldsymbol{s}_{1\mathrm{G}}, \boldsymbol{x}_{1\mathrm{G}}\right) + \mathsf{h}\left(\boldsymbol{s}_{2\mathrm{G}}\right) - \mathsf{h}\left(\boldsymbol{y}_{2\mathrm{G}} | \boldsymbol{s}_{2\mathrm{G}}, \boldsymbol{x}_{2\mathrm{G}}\right).$$

Using Lemma A.1 and Lemma A.6 in Appendix A, we can show that $\mathsf{h}(\boldsymbol{s}_{1\mathrm{G}}) - \mathsf{h}\left(\boldsymbol{y}_{2\mathrm{G}}|\boldsymbol{s}_{2\mathrm{G}}, \boldsymbol{x}_{2\mathrm{G}}\right)$ is a concave and nondecreasing function in $(\boldsymbol{Q}_1, \boldsymbol{Q}_2)$. Similarly, we can show that $\mathsf{h}(\boldsymbol{s}_{2\mathrm{G}}) - \mathsf{h}\left(\boldsymbol{y}_{1\mathrm{G}}|\boldsymbol{s}_{1\mathrm{G}}, \boldsymbol{x}_{1\mathrm{G}}\right)$ is also concave and nondecreasing in $(\boldsymbol{Q}_1, \boldsymbol{Q}_2)$.

2.2.7 Convexity Property

LEMMA 2.2 *For any fixed* $(\boldsymbol{Q}_1, \boldsymbol{Q}_2)$, *the function* $\bar{f}(\boldsymbol{Q}_1, \boldsymbol{Q}_2, \Psi)$ *is convex in* $\Psi = \left(\boldsymbol{\Sigma}_{w_1}, \boldsymbol{\Sigma}_{w_1 z_1}, \boldsymbol{\Sigma}_{w_2}, \boldsymbol{\Sigma}_{w_2 z_2}\right)$.

Proof We provide a sketch of the proof here; the details can be found in [13]. First, observe that

$$\bar{f}(\boldsymbol{Q}_1, \boldsymbol{Q}_2, \Psi) = \mathsf{I}\left(\boldsymbol{x}_{1\mathrm{G}}; \boldsymbol{y}_{1\mathrm{G}}, \boldsymbol{s}_{1\mathrm{G}}\right) + \mathsf{I}\left(\boldsymbol{x}_{2\mathrm{G}}; \boldsymbol{y}_{2\mathrm{G}}, \boldsymbol{s}_{2\mathrm{G}}\right).$$

We prove that $\mathsf{I}\left(\boldsymbol{x}_{1\mathrm{G}}; \boldsymbol{y}_{1\mathrm{G}}, \boldsymbol{s}_{1\mathrm{G}}\right)$ is convex in Ψ. The convexity of $\mathsf{I}\left(\boldsymbol{x}_{2\mathrm{G}}; \boldsymbol{y}_{2\mathrm{G}}, \boldsymbol{s}_{2\mathrm{G}}\right)$ follows in a similar manner. Observe that

$$\mathsf{I}\left(\boldsymbol{x}_{1\mathrm{G}}; \boldsymbol{y}_{1\mathrm{G}}, \boldsymbol{s}_{1\mathrm{G}}\right) = \mathsf{h}\left(\boldsymbol{x}_{1\mathrm{G}}\right) - \mathsf{h}\left(\boldsymbol{x}_{1\mathrm{G}} | \boldsymbol{y}_{1\mathrm{G}}, \boldsymbol{s}_{1\mathrm{G}}\right).$$

The first term $\mathsf{h}(\boldsymbol{x}_{1\mathrm{G}})$ is independent of Ψ. From Lemma A.5 in Appendix A, it follows that the second term $\mathsf{h}\left(\boldsymbol{x}_{1\mathrm{G}}|\boldsymbol{y}_{1\mathrm{G}}, \boldsymbol{s}_{1\mathrm{G}}\right)$ is concave in

$$\operatorname{Cov}\left(\begin{bmatrix} \boldsymbol{x}_{1\mathrm{G}} \\ \boldsymbol{y}_{1\mathrm{G}} \\ \boldsymbol{s}_{1\mathrm{G}} \end{bmatrix}\right) = \begin{bmatrix} \times & \times & \times \\ \times & \times & \boldsymbol{\Sigma}_{z_1 w_1} \\ \times & \times + \boldsymbol{\Sigma}_{w_1 z_1} & \times + \boldsymbol{\Sigma}_{w_1} \end{bmatrix},$$

where $\times$ denotes the terms that are independent of the genie parameters. From this, we conclude that $\mathsf{I}\left(\boldsymbol{x}_{1\mathrm{G}}; \boldsymbol{y}_{1\mathrm{G}}, \boldsymbol{s}_{1\mathrm{G}}\right)$ is convex in $\left(\boldsymbol{\Sigma}_{w_1}, \boldsymbol{\Sigma}_{w_1 z_1}, \boldsymbol{\Sigma}_{w_2}, \boldsymbol{\Sigma}_{w_2 z_2}\right)$.

Sum-Rate Upper Bound

Following the argument in Section 2.2.4, and using Lemmas 2.1 and 2.2, we obtain the following theorem.

THEOREM 2.3 *The best sum-rate achievable by treating interference as noise is bounded above and below by*

$$f(\boldsymbol{Q}_1^*, \boldsymbol{Q}_2^*) \leq \max_{(\boldsymbol{Q}_1, \boldsymbol{Q}_2) \in \mathcal{Q}} f(\boldsymbol{Q}_1, \boldsymbol{Q}_2) \leq \bar{f}(\boldsymbol{Q}_1^*, \boldsymbol{Q}_2^*, \Psi^*),$$

where $(\boldsymbol{Q}_1^*, \boldsymbol{Q}_2^*, \Psi^*)$ *is a solution to the following convex maxmin optimization problem:*

$$\max_{(\boldsymbol{Q}_1, \boldsymbol{Q}_2) \in \mathcal{Q}} \min_{\Psi \in \Psi_{\mathrm{u}}} \bar{f}(\boldsymbol{Q}_1, \boldsymbol{Q}_2, \Psi).$$

The utility of the above theorem is that we can use the standard convex optimization algorithms to efficiently solve for $(\boldsymbol{Q}_1^*, \boldsymbol{Q}_2^*, \Psi^*)$, and thus obtain computable lower and upper bounds to the best achievable sum-rate.

2.2.8 Sum Capacity Upper Bound

Thus far we have determined computable lower and upper bounds on the best achievable sum-rate when treating interference as noise. Suppose the bounds meet and we have exactly determined the best achievable sum-rate and the corresponding optimal covariance matrices. Even then, we only have an achievable sum-rate and we cannot eliminate the possibility that there may exist other simple achievable schemes that could potentially outperform the best achievable sum-rate when treating interference as noise. Interestingly, this question can be resolved, because the upper bound in Theorem 2.3 can be shown to be an upper bound on the sum capacity as well.

THEOREM 2.4 *The sum capacity* $(\mathcal{C}_{\mathrm{sum}})$ *of the two-user MIMO Gaussian interference channel satisfies*

$$f(\boldsymbol{Q}_1^*, \boldsymbol{Q}_2^*) \leq \mathcal{C}_{\mathrm{sum}} \leq \bar{f}(\boldsymbol{Q}_1^*, \boldsymbol{Q}_2^*, \Psi^*),$$

where $(\boldsymbol{Q}_1^*, \boldsymbol{Q}_2^*, \Psi^*)$ *is the solution to the convex maxmin optimization problem*

$$\max_{(\boldsymbol{Q}_1, \boldsymbol{Q}_2) \in \mathcal{Q}} \min_{\Psi \in \Psi_{\mathrm{u}}} \bar{f}(\boldsymbol{Q}_1, \boldsymbol{Q}_2, \Psi).$$

Proof We provide a sketch of the proof here; the details can be found in [13]. The lower bound is obvious since $f(\boldsymbol{Q}_1^*, \boldsymbol{Q}_2^*)$ is defined as the achievable sum-rate when the transmitters use Gaussian inputs with covariance matrices $\boldsymbol{Q}_1^*$ and $\boldsymbol{Q}_2^*$ and the receivers treat interference as noise. Note that the covariance matrices $(\boldsymbol{Q}_1^*, \boldsymbol{Q}_2^*)$ satisfy the transmit power constraints. We now prove an upper bound. Using the standard converse arguments involving Fano's inequality, we obtain that any achievable rate tuple (R_1, R_2) must satisfy

$$R_1 \leq \frac{1}{n} \mathsf{I}\left(\boldsymbol{x}_1^n; \boldsymbol{y}_1^n\right) + \epsilon_n,$$

$$R_2 \leq \frac{1}{n} \mathsf{I}\left(\boldsymbol{x}_2^n; \boldsymbol{y}_2^n\right) + \epsilon_n,$$

for some $\epsilon_n \to 0$ as $n \to \infty$. For any genie $\Psi \in \Psi_{\mathrm{u}}$, we can upper-bound the mutual information terms with the corresponding terms in the genie-aided channel to obtain

$$R_1 \leq \frac{1}{n}\mathsf{I}\left(\boldsymbol{x}_1^n;\boldsymbol{y}_1^n,\boldsymbol{s}_1^n\right) + \epsilon_n,$$

$$R_2 \leq \frac{1}{n}\mathsf{I}\left(\boldsymbol{x}_2^n;\boldsymbol{y}_2^n,\boldsymbol{s}_2^n\right) + \epsilon_n.$$

Let $\boldsymbol{Q}_1$ and $\boldsymbol{Q}_2$ denote the average covariance matrices at transmitters 1 and 2, respectively:

$$\begin{aligned} \boldsymbol{Q}_1 &= \mathbb{E}\left[\frac{1}{n}\sum_{i=1}^{n}\boldsymbol{x}_{1i}\boldsymbol{x}_{1i}^{\dagger}\right], \\ \boldsymbol{Q}_2 &= \mathbb{E}\left[\frac{1}{n}\sum_{i=1}^{n}\boldsymbol{x}_{2i}\boldsymbol{x}_{2i}^{\dagger}\right]. \end{aligned} \tag{2.12}$$

Consider the problem of maximizing $\mathsf{I}\left(\boldsymbol{x}_1^n;\boldsymbol{y}_1^n,\boldsymbol{s}_1^n\right) + \mathsf{I}\left(\boldsymbol{x}_2^n;\boldsymbol{y}_2^n,\boldsymbol{s}_1^n\right)$ over all product input distributions $p(\boldsymbol{x}_1^n)p(\boldsymbol{x}_2^n)$ satisfying the covariance constraints (2.12). Using the proof techniques of the concavity of $\bar{f}(\boldsymbol{Q}_1,\boldsymbol{Q}_2,\Psi)$ in Lemma 2.1, we can show that if the genie is useful, i.e., $\Psi \in \Psi_{\mathrm{u}}$, then independent and identically distributed (i.i.d.) Gaussian inputs are optimal; i.e.,

$$\mathsf{I}\left(\boldsymbol{x}_1^n;\boldsymbol{y}_1^n,\boldsymbol{s}_1^n\right) + \mathsf{I}\left(\boldsymbol{x}_2^n;\boldsymbol{y}_2^n,\boldsymbol{s}_1^n\right) \leq n\bar{f}(\boldsymbol{Q}_1,\boldsymbol{Q}_2,\Psi).$$

Therefore, we obtain that

$$\begin{aligned} \frac{1}{n}\mathsf{I}\left(\boldsymbol{x}_1^n;\boldsymbol{y}_1^n\right) + \frac{1}{n}\mathsf{I}\left(\boldsymbol{x}_2^n;\boldsymbol{y}_2^n\right) &\leq \min_{\Psi\in\Psi_{\mathrm{u}}} \bar{f}(\boldsymbol{Q}_1,\boldsymbol{Q}_2,\Psi) \\ &\leq \min_{\Psi\in\Psi_{\mathrm{u}}} \max_{(\boldsymbol{Q}_1,\boldsymbol{Q}_2\in\mathcal{Q}} \bar{f}(\boldsymbol{Q}_1,\boldsymbol{Q}_2,\Psi) \\ &= n\bar{f}(\boldsymbol{Q}_1^*,\boldsymbol{Q}_2^*,\Psi^*). \end{aligned}$$

Thus, we proved that any achievable rate tuple (R_1,R_2) must satisfy

$$R_1 + R_2 \leq \bar{f}(\boldsymbol{Q}_1^*,\boldsymbol{Q}_2^*,\Psi^*) + \epsilon_n.$$

The proof is completed by letting $n \to \infty$.

2.2.9 Smart Genie: Zero Gap

Thus far, we have derived computable lower and upper bounds on the sum capacity. A natural follow-up step is to check if the bounds ever meet. We start by obtaining a necessary and sufficient condition on $(\boldsymbol{Q}_1^*,\boldsymbol{Q}_2^*,\Psi^*)$ for the gap to be zero. We say that a genie Ψ is $(\boldsymbol{Q}_1,\boldsymbol{Q}_2)$-smart if

$$\bar{f}(\boldsymbol{Q}_1,\boldsymbol{Q}_2,\Psi) = f(\boldsymbol{Q}_1,\boldsymbol{Q}_2).$$

The genie just gives side information to the receivers, but it is smart enough not to leak any additional information about the respective transmitted signals that the original received signals could not provide. The total amount of additional information that a genie leaks is equal to

$$\begin{aligned}
&\bar{f}(Q_1,Q_2,\Psi) - f(Q_1,Q_2) \\
&\quad = \mathsf{I}\left(x_{1G};y_{1G},s_{1G}\right) + \mathsf{I}\left(x_{2G};y_{2G},s_{2G}\right) - \mathsf{I}\left(x_{1G};y_{1G}\right) - \mathsf{I}\left(x_{2G};y_{2G}\right) \\
&\quad = \mathsf{I}\left(x_{1G};s_{1G}|y_{1G}\right) + \mathsf{I}\left(x_{2G};s_{2G}|y_{2G}\right).
\end{aligned}$$

Since the conditional mutual information is always nonnegative, the genie is smart if and only if $\mathsf{I}\left(x_{1G};s_{1G}|y_{1G}\right) = \mathsf{I}\left(x_{2G};s_{2G}|y_{2G}\right) = 0$. The conditional mutual information $\mathsf{I}\left(x_{1G};s_{1G}|y_{1G}\right)$ is equal to zero if and only if x_{1G}–y_{1G}–s_{1G} forms a Markov chain. Since all the random variables are jointly Gaussian, x_{1G}–y_{1G}–s_{1G} forms a Markov chain if and only if the minimum mean squared error (MMSE) estimate of s_{1G} given (x_{1G},y_{1G}) is the same as the MMSE estimate of s_{1G} given y_{1G}, i.e.,

$$\mathbb{E}\left[s_{1G}|y_{1G},x_{1G}\right] = \mathbb{E}\left[s_{1G}|y_{1G}\right]. \tag{2.13}$$

Let $T_1 y_{1G}$ be the MMSE estimate of s_{1G} given y_{1G}. Using the orthogonality principle, summarized in Section A.1, we know that the MMSE estimation error $e_1 = s_{1G} - T_1 y_{1G}$ is independent of the observation y_{1G}. Since (2.13) implies that $T_1 y_{1G}$ is also the MMSE estimate of s_{1G} given y_{1G} and x_{1G}, we obtain that e_1 is also independent of x_{1G}. Therefore, we have that (2.13) is true if and only if there exists a matrix T_1 such that

$$s_{1G} = T_1 y_{1G} + e_1,$$

i.e., that

$$\tilde{H}_{21} x_{1G} + w_1 = T_1 \left(H_{11} x_{1G} + H_{12} x_{2G} + z_1\right) + e_1,$$

i.e., that

$$\left(\tilde{H}_{21} - T_1 H_{11}\right) x_{1G} + w_1 = T_1 \left(H_{12} x_{2G} + z_1\right) + e_1,$$

with e_1 being independent of y_{1G} and x_{1G}. Since x_{1G} is independent of all the other random vectors involved, the random vector $\left(\tilde{H}_{21} - T_1 H_{11}\right) x_{1G}$ must be equal to zero almost surely, which is equivalent to

$$\left(\tilde{H}_{21} - T_1 H_{11}\right) Q_1 = 0.$$

The remaining expression is equivalent to saying that $T_1 \left(H_{12} x_{2G} + z_1\right)$ is the MMSE estimate of w_1 given $H_{12} x_{2G} + z_1$, since e_1 is independent of $y_{1G} - H_{11} x_{1G} = H_{12} x_{2G} + z_1$. Therefore, we obtain the following expression for T_1:

$$T_1 = \Sigma_{w_1 z_1} \left(I + H_{12} Q_2 H_{21}^{\dagger}\right)^{-1}.$$

Thus, we can conclude that x_{1G}–y_{1G}–s_{1G} forms a Markov chain if and only if the following condition is satisfied:

$$\left(\tilde{H}_{21} - \Sigma_{w_1 z_1} \left(H_{12} Q_2 H_{12}^{\dagger} + I\right)^{-1} H_{11}\right) Q_1 = 0.$$

We can derive a similar necessary and sufficient condition for $\mathrm{I}\left(x_{2G};s_{2G}|y_{2G}\right)$ to be equal to zero, and hence we obtain the following lemma.

LEMMA 2.5 *The genie* Ψ *is* $(\boldsymbol{Q}_1,\boldsymbol{Q}_2)$*-smart, i.e.,* $\bar{f}(\boldsymbol{Q}_1,\boldsymbol{Q}_2,\Psi)=f(\boldsymbol{Q}_1,\boldsymbol{Q}_2)$*, if and only if the following conditions are satisfied:*

$$\begin{aligned}\left(\tilde{\boldsymbol{H}}_{21}-\boldsymbol{\Sigma}_{w_1z_1}\left(\boldsymbol{H}_{12}\boldsymbol{Q}_2\boldsymbol{H}_{12}^{\dagger}+\boldsymbol{I}\right)^{-1}\boldsymbol{H}_{11}\right)\boldsymbol{Q}_1&=\boldsymbol{0},\\ \left(\tilde{\boldsymbol{H}}_{12}-\boldsymbol{\Sigma}_{w_2z_2}\left(\boldsymbol{H}_{21}\boldsymbol{Q}_1\boldsymbol{H}_{21}^{\dagger}+\boldsymbol{I}\right)^{-1}\boldsymbol{H}_{22}\right)\boldsymbol{Q}_2&=\boldsymbol{0}.\end{aligned} \tag{2.14}$$

2.2.10 Low Interference Regime

We say that a two-user MIMO Gaussian interference channel is in the low interference regime if the sum capacity is achieved by using Gaussian inputs and treating interference as noise. Suppose the upper and lower bounds defined in Theorem 2.4 meet. Then the channel (2.3) is in the low interference regime. In this section, we derive necessary and sufficient conditions for the bounds in Theorem 2.4 to meet. Recall that the bounds meet if and only if

$$f(\boldsymbol{Q}_1^*,\boldsymbol{Q}_2^*)=\bar{f}(\boldsymbol{Q}_1^*,\boldsymbol{Q}_2^*,\Psi^*), \tag{2.15}$$

where $(\boldsymbol{Q}_1^*,\boldsymbol{Q}_2^*,\Psi^*)$ is an optimal solution to the following convex maxmin optimization problem:

$$\max_{(\boldsymbol{Q}_1,\boldsymbol{Q}_2)\in\mathcal{Q}}\ \min_{\Psi\in\Psi_{\mathrm{u}}}\bar{f}(\boldsymbol{Q}_1,\boldsymbol{Q}_2,\Psi). \tag{2.16}$$

We start with the following claim, proved in [13], in which we exploit the concave–convex property of $\bar{f}(\boldsymbol{Q}_1,\boldsymbol{Q}_2,\Psi)$ to simplify the above two conditions.

CLAIM 2.1 The following two statements are equivalent:

- A maxmin solution $(\boldsymbol{Q}_1^*,\boldsymbol{Q}_2^*,\Psi^*)$ to (2.16) satisfies (2.15).
- There exist $(\boldsymbol{Q}_1^*,\boldsymbol{Q}_2^*)$ and $\Psi^*\in\Psi_{\mathrm{u}}$ satisfying

$$f(\boldsymbol{Q}_1^*,\boldsymbol{Q}_2^*)=\bar{f}(\boldsymbol{Q}_1^*,\boldsymbol{Q}_2^*,\Psi^*)=\max_{(\boldsymbol{Q}_1,\boldsymbol{Q}_2)\in\mathcal{Q}}\bar{f}(\boldsymbol{Q}_1,\boldsymbol{Q}_2,\Psi^*).$$

The claim says that it is sufficient to consider only one instance of the genie (Ψ^*) to obtain the best upper bound instead of minimizing the upper bound function over all useful genies. It is easy to check that the second statement implies the first statement without invoking any special structural properties of $\bar{f}(\boldsymbol{Q}_1\boldsymbol{Q}_2,\Psi)$. Proving that the second statement is necessary requires the concave–convex property of $\bar{f}(\boldsymbol{Q}_1\boldsymbol{Q}_2,\Psi)$.

We have already derived the necessary and sufficient conditions for (2.15) to be true in Lemma 2.5. We now derive the KKT conditions which are both necessary and sufficient conditions for $(\boldsymbol{Q}_1^*,\boldsymbol{Q}_2^*)$ to be a global optimal solution to

$$\max_{(\boldsymbol{Q}_1,\boldsymbol{Q}_2)\in\mathcal{Q}}\bar{f}(\boldsymbol{Q}_1,\boldsymbol{Q}_2,\Psi^*). \tag{2.17}$$

Let $\lambda_1 \geq 0$ and $\lambda_2 \geq 0$ be the dual variables associated with the constraints $\mathrm{Tr}\left(\boldsymbol{Q}_1\right) \leq P_1$ and $\mathrm{Tr}\left(\boldsymbol{Q}_2\right) \leq P_2$. Let $\boldsymbol{M}_1 \succeq 0$ and $\boldsymbol{M}_2 \succeq 0$ be the dual variables associated with the constraints $\boldsymbol{Q}_1 \succeq 0$ and $\boldsymbol{Q}_2 \succeq 0$. The Lagrangian associated with the optimization problem (2.17) is given by

$$\bar{f}(\boldsymbol{Q}_1,\boldsymbol{Q}_2,\Psi^*)+\sum_{i=1}^{2}\mathrm{Tr}\left(\boldsymbol{M}_i\boldsymbol{Q}_i\right)-\lambda_i\left(\mathrm{Tr}\left(\boldsymbol{Q}_i\right)-P_i\right).$$

The corresponding KKT conditions are as given in (2.20). Thus, we obtain the following theorem.

THEOREM 2.6 *Suppose there exist transmit covariance matrices* $\boldsymbol{Q}_1^* \succeq \boldsymbol{0},\boldsymbol{Q}_2^* \succeq 0$, *genie parameters* $\Psi^* = \left(\boldsymbol{\Sigma}_{w_1},\boldsymbol{\Sigma}_{w_1z_1},\boldsymbol{\Sigma}_{w_2},\boldsymbol{\Sigma}_{w_2z_2}\right)$, *and dual variables* $\lambda_1 \geq 0,\lambda \geq 0,\boldsymbol{M}_1 \succeq 0,\boldsymbol{M}_2 \succeq \boldsymbol{0}$ *satisfying the following conditions:*

- *Transmit power constraints:* $\mathrm{Tr}\left(\boldsymbol{Q}_1\right) \leq P_1$ *and* $\mathrm{Tr}\left(\boldsymbol{Q}_2\right) \leq P_2$.
- *Useful genie conditions:*

$$\begin{bmatrix} \boldsymbol{I} & \boldsymbol{\Sigma}_{z_1w_1} \\ \boldsymbol{\Sigma}_{w_1z_1} & \boldsymbol{\Sigma}_{w_1} \end{bmatrix} \succeq \begin{bmatrix} \boldsymbol{\Sigma}_{w_2} & \boldsymbol{0} \\ \boldsymbol{0} & \boldsymbol{0} \end{bmatrix},$$
$$\begin{bmatrix} \boldsymbol{I} & \boldsymbol{\Sigma}_{z_2w_2} \\ \boldsymbol{\Sigma}_{w_2z_2} & \boldsymbol{\Sigma}_{w_2} \end{bmatrix} \succeq \begin{bmatrix} \boldsymbol{\Sigma}_{w_1} & \boldsymbol{0} \\ \boldsymbol{0} & \boldsymbol{0} \end{bmatrix}. \tag{2.18}$$

- *Smart genie conditions:*

$$\left(\tilde{\boldsymbol{H}}_{21}-\boldsymbol{\Sigma}_{w_1z_1}\left(\boldsymbol{H}_{12}\boldsymbol{Q}_2^*\boldsymbol{H}_{12}^\dagger+\boldsymbol{I}\right)^{-1}\boldsymbol{H}_{11}\right)\boldsymbol{Q}_1^*=\boldsymbol{0},$$
$$\left(\tilde{\boldsymbol{H}}_{12}-\boldsymbol{\Sigma}_{w_2z_2}\left(\boldsymbol{H}_{21}\boldsymbol{Q}_1^*\boldsymbol{H}_{21}^\dagger+\boldsymbol{I}\right)^{-1}\boldsymbol{H}_{22}\right)\boldsymbol{Q}_2^*=\boldsymbol{0}. \tag{2.19}$$

- *KKT conditions:*

$$\begin{aligned} \nabla_{\boldsymbol{Q}_1}\bar{f}(\boldsymbol{Q}_1^*,\boldsymbol{Q}_2^*,\Psi^*) &= \lambda_1\boldsymbol{I}-\boldsymbol{M}_1,\\ \nabla_{\boldsymbol{Q}_2}\bar{f}(\boldsymbol{Q}_1^*,\boldsymbol{Q}_2^*,\Psi^*) &= \lambda_2\boldsymbol{I}-\boldsymbol{M}_2,\\ \lambda_1(\mathrm{Tr}\left(\boldsymbol{Q}_1^*\right)-P_1) &= 0,\\ \lambda_2(\mathrm{Tr}\left(\boldsymbol{Q}_2^*\right)-P_2) &= 0,\\ \mathrm{Tr}\left(\boldsymbol{M}_1\boldsymbol{Q}_1^*\right) &= 0,\\ \mathrm{Tr}\left(\boldsymbol{M}_2\boldsymbol{Q}_2^*\right) &= 0. \end{aligned} \tag{2.20}$$

Then, the sum capacity of the two-user MIMO Gaussian interference channel (2.3) *is achieved by using Gaussian inputs and treating interference as noise, and is given by*

$$\mathcal{C}_{\text{sum}}=f(\boldsymbol{Q}_1^*,\boldsymbol{Q}_2^*).$$

Conversely, if there exist no such parameters satisfying the stated constraints, then the lower and upper bounds in Theorem 2.4 do not coincide.

The above theorem provides sufficient conditions for the two-user Gaussian interference channel (2.3). The problem now is to determine if there exists an algorithm to verify the feasibility of these conditions. Observe that the conditions in Theorem 2.6 are nothing but the necessary and sufficient conditions for the bounds in Theorem 2.4 to coincide. Therefore, we can use the standard convex optimization algorithms to solve the maxmin optimization problem in Theorem 2.4 efficiently, and thus verify the feasibility of the conditions in Theorem 2.6. In the sections to follow, we explore the possibility of verifying the feasibility of the conditions of Theorem 2.6 analytically. We provide two corollaries with simpler conditions that are sufficient but may not be necessary. In some special cases, such as symmetric MISO and SIMO interference channels, we actually simplify the conditions of Theorem 2.6 into a closed-form equation that depends only on the channel matrices and the power constraints. To achieve all these objectives, we first need to simplify the KKT conditions (2.20).

2.2.11 Simplified KKT Conditions

When the smart genie conditions (2.19) are satisfied, the gradient expressions in (2.20) can be greatly simplified. We first explain the intuition before proceeding to present the simplified expressions. Recall that $\bar{f}(\boldsymbol{Q}_1, \boldsymbol{Q}_2, \Psi)$ is an upper bound on $f(\boldsymbol{Q}_1, \boldsymbol{Q}_2)$. Let $g(\boldsymbol{Q}_1, \boldsymbol{Q}_2, \Psi)$ denote the gap

$$g(\boldsymbol{Q}_1, \boldsymbol{Q}_2, \Psi) = \bar{f}(\boldsymbol{Q}_1, \boldsymbol{Q}_2, \Psi) - f(\boldsymbol{Q}_1, \boldsymbol{Q}_2) \geq 0.$$

Suppose the genie Ψ^* is $(\boldsymbol{Q}_1^*, \boldsymbol{Q}_2^*)$-smart, i.e., $g(\boldsymbol{Q}_1^*, \boldsymbol{Q}_2^*, \Psi^*) = 0$. Then, we see that $(\boldsymbol{Q}_1^*, \boldsymbol{Q}_2^*)$ is an optimal solution to

$$\min_{\boldsymbol{Q}_i : \boldsymbol{Q}_i \succeq 0, i \in \{1,2\}} g(\boldsymbol{Q}_1, \boldsymbol{Q}_2, \Psi^*).$$

Therefore, $(\boldsymbol{Q}_1^*, \boldsymbol{Q}_2^*)$ must satisfy the corresponding necessary KKT conditions. Let $\boldsymbol{N}_1 \succeq \boldsymbol{0}$ and $\boldsymbol{N}_2 \succeq \boldsymbol{0}$ denote the dual variables corresponding to the constraints $\boldsymbol{Q}_1 \succeq \boldsymbol{0}$ and $\boldsymbol{Q}_2 \succeq \boldsymbol{0}$, respectively. The Lagrangian associated with the above minimization problem is given by

$$g(\boldsymbol{Q}_1, \boldsymbol{Q}_2, \Psi^*) - \sum_{i=1}^{2} \mathsf{Tr}\left(\boldsymbol{N}_i \boldsymbol{Q}_i\right).$$

Therefore, there must exist some dual variables $\boldsymbol{N}_1 \succeq \boldsymbol{0}$ and $\boldsymbol{N}_2 \succeq \boldsymbol{0}$ satisfying the KKT conditions

$$\begin{aligned}
\nabla_{\boldsymbol{Q}_1} g(\boldsymbol{Q}_1^*, \boldsymbol{Q}_2^*, \Psi^*) &= \boldsymbol{N}_1, \\
\nabla_{\boldsymbol{Q}_2} g(\boldsymbol{Q}_1^*, \boldsymbol{Q}_2^*, \Psi^*) &= \boldsymbol{N}_2, \\
\mathsf{Tr}\left(\boldsymbol{N}_1 \boldsymbol{Q}_1^*\right) &= 0, \\
\mathsf{Tr}\left(\boldsymbol{N}_2 \boldsymbol{Q}_2^*\right) &= 0.
\end{aligned}$$

Thus, we see that the gradients in (2.20) can be simplified as

$$\nabla_{Q_1}\bar{f}(Q_1^*,Q_2^*,\Psi^*) = \nabla_{Q_1}f(Q_1^*,Q_2^*) + N_1,$$
$$\nabla_{Q_2}\bar{f}(Q_1^*,Q_2^*,\Psi^*) = \nabla_{Q_2}f(Q_1^*,Q_2^*) + N_2$$

for some $N_1 \succeq \mathbf{0}$ and $N_2 \succeq \mathbf{0}$ satisfying $\mathrm{Tr}\left(N_1 Q_1^*\right) = \mathrm{Tr}\left(N_2 Q_2^*\right) = 0$. In the following lemma, which is proved in [13], we give the expressions for N_1 and N_2 in terms of the channel matrices, genie parameters, and the transmit covariance matrices.

LEMMA 2.7 *Suppose the smart genie conditions* (2.19) *are satisfied; i.e.,*

$$\left(\tilde{H}_{21} - T_1 H_{11}\right) Q_1^* = \mathbf{0},$$
$$\left(\tilde{H}_{12} - T_2 H_{22}\right) Q_2^* = \mathbf{0},$$

where the matrices T_1 and T_2 are defined as

$$T_1 = \Sigma_{w_1 z_1}\left(H_{12} Q_2^* H_{12}^\dagger + I\right)^{-1},$$
$$T_2 = \Sigma_{w_2 z_2}\left(H_{21} Q_1^* H_{21}^\dagger + I\right)^{-1}.$$

Then, we have

$$\nabla_{Q_1}\bar{f}(Q_1^*,Q_2^*,\Psi^*) = \nabla_{Q_1}f(Q_1^*,Q_2^*) + N_1,$$
$$\nabla_{Q_2}\bar{f}(Q_1^*,Q_2^*,\Psi^*) = \nabla_{Q_2}f(Q_1^*,Q_2^*) + N_2,$$

where the matrices N_1 and N_2 are given by

$$N_1 = \left(\tilde{H}_{21} - T_1 H_{11}\right)^\dagger \Sigma^{-1}_{w_1 | y_{1G}, x_{1G}}\left(\tilde{H}_{21} - T_1 H_{11}\right),$$
$$N_2 = \left(\tilde{H}_{21} - T_2 H_{22}\right)^\dagger \Sigma^{-1}_{w_2 | y_{2G}, x_{2G}}\left(\tilde{H}_{21} - T_2 H_{22}\right).$$

Furthermore, N_1 and N_2 satisfy $\mathrm{Tr}\left(N_1 Q_1^*\right) = \mathrm{Tr}\left(N_2 Q_2^*\right) = 0$.

REMARK *2.1* Suppose the conditions of Theorem 2.6 are satisfied; then the sum capacity is given by $f(Q_1^*,Q_2^*)$. As an obvious corollary, we also get that (Q_1^*,Q_2^*) are the optimal covariance matrices maximizing the achievable sum-rate (2.5). This means that (Q_1^*,Q_2^*) must satisfy the corresponding necessary KKT conditions (2.7). Therefore, it must be that the conditions in Theorem 2.6 imply the conditions (2.7). Lemma 2.7 makes it easier to see this connection. Observe that the KKT conditions (2.20), along with the smart genie conditions (2.19), imply that

$$\nabla_{Q_1}f(Q_1^*,Q_2^*) = \lambda_1 I - M_1 - N_1,$$
$$\nabla_{Q_2}f(Q_1^*,Q_2^*) = \lambda_2 I - M_2 - N_2,$$

where $N_1 \succeq 0$, $N_2 \succeq 0$, and $\mathrm{Tr}\left(N_1 Q_1^*\right) = \mathrm{Tr}\left(N_2 Q_2^*\right) = 0$. Therefore, by replacing M_1 by $M_1 + N_1$ and M_2 by $M_2 + N_2$, we see that the KKT conditions (2.7) are satisfied.

Full Rank Optimal Covariance Matrices

We use the insights from the previous section to simplify the conditions in Theorem 2.6 when the optimal covariance matrices $(\boldsymbol{Q}_1^*, \boldsymbol{Q}_2^*)$ have full rank. Suppose the conditions of Theorem 2.6 are satisfied and $(\boldsymbol{Q}_1^*, \boldsymbol{Q}_2^*)$ have full rank. Then, it must be that the matrices $\boldsymbol{N}_1$ and $\boldsymbol{N}_2$ defined in Lemma 2.7 are equal to zero. This is because $\boldsymbol{N}_1$ and $\boldsymbol{N}_2$ are positive semidefinite matrices satisfying $\mathsf{Tr}\left(\boldsymbol{N}_1\boldsymbol{Q}_1^*\right) = \mathsf{Tr}\left(\boldsymbol{N}_2\boldsymbol{Q}_2^*\right) = 0$. Therefore, we have that

$$\nabla_{\boldsymbol{Q}_1}\bar{f}(\boldsymbol{Q}_1^*, \boldsymbol{Q}_2^*, \Psi^*) = \nabla_{\boldsymbol{Q}_1} f(\boldsymbol{Q}_1^*, \boldsymbol{Q}_2^*),$$
$$\nabla_{\boldsymbol{Q}_2}\bar{f}(\boldsymbol{Q}_1^*, \boldsymbol{Q}_2^*, \Psi^*) = \nabla_{\boldsymbol{Q}_2} f(\boldsymbol{Q}_1^*, \boldsymbol{Q}_2^*).$$

Hence, the KKT conditions (2.20) are identical to (2.7), which are satisfied if $(\boldsymbol{Q}_1^*, \boldsymbol{Q}_2^*)$ is a local optimal solution to the optimization problem (2.5). Thus, we obtain the following corollary to Theorem 2.6.

COROLLARY 2.1 *Suppose there exists a local optimal solution* $\boldsymbol{Q}_1^* \succ \boldsymbol{0}, \boldsymbol{Q}_2^* \succ 0$ *to*

$$\max_{(\boldsymbol{Q}_1, \boldsymbol{Q}_2) \in \mathcal{Q}} f(\boldsymbol{Q}_1, \boldsymbol{Q}_2),$$

and a genie Ψ^* *that is both useful and* $(\boldsymbol{Q}_1^*, \boldsymbol{Q}_2^*)$*-smart; i.e., the conditions* (2.18) *and* (2.14) *are satisfied. Then the sum capacity of the two-user MIMO Gaussian interference channel* (2.3) *is achieved by using Gaussian inputs and treating interference as noise, and is given by*

$$\mathcal{C}_{\text{sum}} = f(\boldsymbol{Q}_1^*, \boldsymbol{Q}_2^*).$$

Concave Sum-Rate Function

As summarized in Section 2.2.4, the basic idea leading to the techniques developed in this chapter is that the achievable sum-rate function $f(\boldsymbol{Q}_1, \boldsymbol{Q}_2)$ is not necessarily concave in $(\boldsymbol{Q}_1, \boldsymbol{Q}_2)$ and so we used the genie-aided channel to develop a concave upper bound $\bar{f}(\boldsymbol{Q}_1, \boldsymbol{Q}_2, \Psi)$ to handle the optimization problem. The best concave upper bound to $f(\boldsymbol{Q}_1, \boldsymbol{Q}_2)$ is given by

$$\bar{f}(\boldsymbol{Q}_1, \boldsymbol{Q}_2) = \min_{\Psi \in \Psi_{\text{u}}} \bar{f}(\boldsymbol{Q}_1, \boldsymbol{Q}_2, \Psi).$$

Suppose $\bar{f}(\boldsymbol{Q}_1, \boldsymbol{Q}_2) = f(\boldsymbol{Q}_1, \boldsymbol{Q}_2)$ for every feasible $(\boldsymbol{Q}_1, \boldsymbol{Q}_2)$. Then, we see that $f(\boldsymbol{Q}_1, \boldsymbol{Q}_2)$ is a concave function within the region of interest, and hence we can just use the standard convex optimization algorithms to determine the global optimal solution to

$$\max_{(\boldsymbol{Q}_1, \boldsymbol{Q}_2) \in \mathcal{Q}} f(\boldsymbol{Q}_1, \boldsymbol{Q}_2).$$

Clearly, this also implies that the sum capacity is equal to $f(\boldsymbol{Q}_1^*, \boldsymbol{Q}_2^*)$. Observe that

$$\begin{aligned}
\mathcal{C}_{\text{sum}} &\le \max_{(\boldsymbol{Q}_1, \boldsymbol{Q}_2) \in \mathcal{Q}} \min_{\Psi \in \Psi_{\text{u}}} \bar{f}(\boldsymbol{Q}_1, \boldsymbol{Q}_2, \Psi) \\
&= \max_{(\boldsymbol{Q}_1, \boldsymbol{Q}_2) \in \mathcal{Q}} \bar{f}(\boldsymbol{Q}_1, \boldsymbol{Q}_2) \\
&\overset{\text{(a)}}{=} \max_{(\boldsymbol{Q}_1, \boldsymbol{Q}_2) \in \mathcal{Q}} f(\boldsymbol{Q}_1, \boldsymbol{Q}_2) \\
&= f(\boldsymbol{Q}_1^*, \boldsymbol{Q}_2^*),
\end{aligned}$$

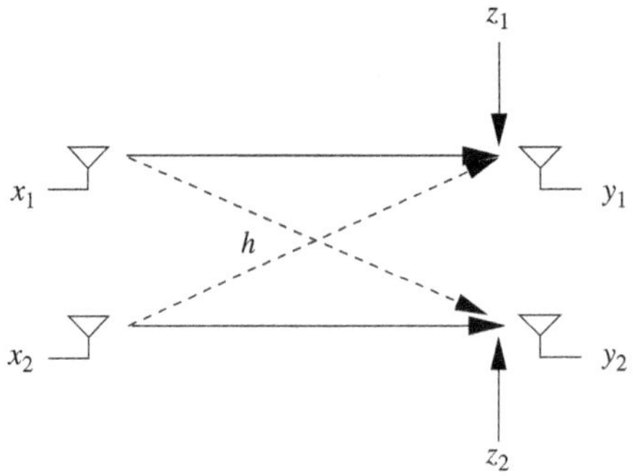

Fig. 2.3 Two-user SISO interference channel. ©[2017] IEEE. Reprinted, with permission, from [18].

where step (a) follows because we assumed that $\bar{f}(\boldsymbol{Q}_1,\boldsymbol{Q}_2) = f(\boldsymbol{Q}_1,\boldsymbol{Q}_2)$ for every feasible $(\boldsymbol{Q}_1,\boldsymbol{Q}_2)$. Since the condition $\bar{f}(\boldsymbol{Q}_1,\boldsymbol{Q}_2) = f(\boldsymbol{Q}_1,\boldsymbol{Q}_2)$ is equivalent to the existence of a genie Ψ that is both useful and $(\boldsymbol{Q}_1,\boldsymbol{Q}_2)$-smart, we obtain the following corollary to Theorem 2.4.

COROLLARY 2.2 *Suppose that for every feasible transmit covariance matrices $(\boldsymbol{Q}_1,\boldsymbol{Q}_2)$ there exists a genie Ψ that is both useful and $(\boldsymbol{Q}_1,\boldsymbol{Q}_2)$-smart; i.e., the conditions* (2.18) *and* (2.14) *are satisfied. Then the achievable sum-rate function $f(\boldsymbol{Q}_1,\boldsymbol{Q}_2)$ is concave in $(\boldsymbol{Q}_1,\boldsymbol{Q}_2)$ in the feasible region $\mathcal{Q}$, and the sum capacity is achievable by using Gaussian inputs and treating interference as noise, and is given by $f(\boldsymbol{Q}_1^*,\boldsymbol{Q}_2^*)$, the optimal solution to the convex optimization problem*

$$\max_{(\boldsymbol{Q}_1,\boldsymbol{Q}_2)\in\mathcal{Q}} f(\boldsymbol{Q}_1,\boldsymbol{Q}_2).$$

2.2.12 SISO Interference Channel

We now specialize the results for the MIMO interference channel to the case of the SISO interference channel illustrated in Figure 2.3:

$$\begin{aligned} y_1 &= x_1 + h_{12}x_2 + z_1, \\ y_2 &= h_{21}x_1 + x_2 + z_2, \end{aligned} \tag{2.21}$$

with the transmit power constraints P_1 and P_2. We can simplify the low interference regime conditions of Theorem 2.6 to obtain a simple closed-form condition. In the SISO case, the KKT conditions of Theorem 2.6 are automatically satisfied. Recall that the role of the KKT conditions is to make sure that (q_1^*,q_2^*) is the global optimal solution to

$$\max_{(q_1,q_2)\in\mathcal{Q}} \bar{f}(q_1,q_2,\Psi^*).$$

However, we have already proved in Lemma 2.1 that $\bar{f}(q_1,q_2,\Psi^*)$ is a nondecreasing function in (q_1,q_2). Therefore, $(q_1^*,q_2^*) = (P_1,P_2)$ must be a global optimal solution to the above optimization problem, and hence there must exist dual variables satisfying the KKT conditions (2.20). Therefore, it only remains to verify the existence of a genie $\Psi^* = \left(\Sigma_{w_1}, \Sigma_{w_1 z_1}, \Sigma_{w_2}, \Sigma_{w_2 z_2}\right)$ that is useful and (P_1,P_2)-smart, i.e., satisfying the conditions (2.18) and (2.19). The following result was first derived directly (without going through Theorem 2.6) in [18–20].

THEOREM 2.8 *The sum capacity of the two-user SISO Gaussian interference channel* (2.21) *is achieved by using Gaussian inputs and treating interference as noise, and is given by*

$$\mathcal{C}_{\text{sum}} = \log\left(1+\frac{P_1}{1+|h_{12}|^2 P_2}\right)+\log\left(1+\frac{P_2}{1+|h_{21}|^2 P_1}\right)$$

if the channel parameters satisfy the low interference regime *condition*

$$|h_{21}|\left(1+|h_{12}|^2 P_2\right)+|h_{12}|\left(1+|h_{21}|^2 P_1\right) \le 1. \tag{2.22}$$

Proof The smart genie conditions (2.19) are given by

$$\left(h_{21} - \Sigma_{w_1 z_1}\left(h_{12} P_2 h_{12}^{\dagger}+1\right)^{-1}\right) P_1 = 0,$$

$$\left(h_{12} - \Sigma_{w_2 z_2}\left(h_{21} P_1 h_{21}^{\dagger}+1\right)^{-1}\right) P_2 = 0,$$

which are equivalent to

$$\Sigma_{w_1 z_1} = h_{21}\left(1+|h_{12}|^2 P_2\right),$$

$$\Sigma_{w_2 z_2} = h_{12}\left(1+|h_{21}|^2 P_1\right).$$

The useful genie conditions (2.18) are given by

$$\begin{bmatrix} 1 & \Sigma_{z_1 w_1} \\ \Sigma_{w_1 z_1} & \Sigma_{w_1} \end{bmatrix} \succeq \begin{bmatrix} \Sigma_{w_2} & 0 \\ 0 & 0 \end{bmatrix},$$

$$\begin{bmatrix} 1 & \Sigma_{z_2 w_2} \\ \Sigma_{w_2 z_2} & \Sigma_{w_2} \end{bmatrix} \succeq \begin{bmatrix} \Sigma_{w_1} & 0 \\ 0 & 0 \end{bmatrix},$$

which are equivalent to

$$0 \le \Sigma_{w_1}, \Sigma_{w_2} \le 1,$$

$$\Sigma_{w_1}(1-\Sigma_{w_2}) \le |\Sigma_{w_1 z_1}|^2,$$

$$\Sigma_{w_2}(1-\Sigma_{w_1}) \le |\Sigma_{w_2 z_2}|^2.$$

Substituting $\Sigma_{w_1} = \cos^2\phi_1$ and $\Sigma_{w_2} = \sin^2\phi_2$, where $\phi_1,\phi_2 \in [0,\pi/2]$, the above equations can be simplified to

$$\cos\phi_1 \cos\phi_2 \le |\Sigma_{w_1 z_1}|,$$

$$\sin\phi_1 \sin\phi_2 \le |\Sigma_{w_2 z_2}|.$$

It is easy to check that a solution $\phi_1,\phi_2 \in [0,\pi/2]$ exists if and only if

$$|\Sigma_{w_1 z_1}| + |\Sigma_{w_2 z_2}| \le 1,$$

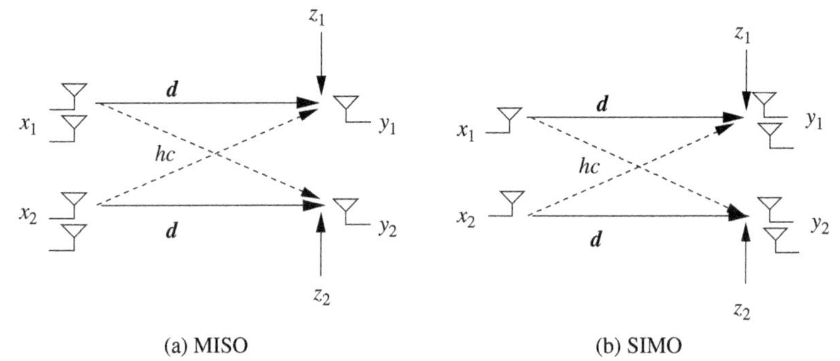

Fig. 2.4 Symmetric MISO and SIMO interference channels.

i.e., if and only if

$$|h_{21}|\left(1+|h_{12}|^2P_2\right)+|h_{12}|\left(1+|h_{21}|^2P_1\right)\leq 1.$$

2.2.13 Symmetric MISO and SIMO Interference Channels

We now specialize the results for the MIMO interference channel to the symmetric MISO and SIMO interference channels shown in Figure 2.4, and simplify the conditions in Theorem 2.6 to derive a simple closed-form equation for the low interference regime.

Symmetric MISO interference channel:

$$y_1=\boldsymbol{d}^\dagger\boldsymbol{x}_1+h\boldsymbol{c}^\dagger\boldsymbol{x}_2+z_1,$$
$$y_2=\boldsymbol{d}^\dagger\boldsymbol{x}_2+h\boldsymbol{c}^\dagger\boldsymbol{x}_1+z_2.$$

Symmetric SIMO interference channel:

$$\boldsymbol{y}_1=\boldsymbol{d}x_1+h\boldsymbol{c}x_2+\boldsymbol{z}_1,$$
$$\boldsymbol{y}_2=\boldsymbol{d}x_2+h\boldsymbol{c}x_1+\boldsymbol{z}_2.$$

In both the above cases, we assume that the transmitters satisfy an average transmit power constraint of P. We assume that $h\geq 0$ is a real number, and the vectors $\boldsymbol{d}$ and $\boldsymbol{c}$ have unit norm, and are defined as

$$\boldsymbol{d}=\begin{bmatrix}\cos\theta\\ \sin\theta\end{bmatrix},\quad \boldsymbol{c}=\begin{bmatrix}1\\ 0\end{bmatrix} \tag{2.23}$$

for some $\theta\in[0,\pi/2]$. See Section 2.2.1 for a justification for these assumptions. In particular, observe that the third assumption in Section 2.2.1 states that we can restrict the study of the MISO (respectively, SIMO) Gaussian interference channels to the case with only two transmit (respectively, receive) antennas.

Observe that both the MISO and SIMO channels with $\theta=0$ are equivalent to the classical two-user SISO Gaussian interference channel. In Section 2.2.12, we showed that the channel is in the low interference regime, and hence treating interference as noise achieves a sum capacity equal to

$$2\log\left(1+\frac{P}{1+h^2P}\right),$$

if the parameters h and P satisfy $h(1+h^2P) \leq 0.5$. On the other extreme, with $\theta = \pi/2$, we obtain the scenario where the users do not interfere with each other, and hence the sum capacity is given by $2\log(1+P)$ for any h. We vary θ from 0 to $\pi/2$, and analyze the behavior of the sum capacity and the low interference regime as a function of θ.

Achievable Sum-Rate

First, consider the SIMO interference channel. The achievable sum-rate (2.6) obtained by using Gaussian inputs and treating interference as noise is given by

$$f(q_1,q_2) = \log\frac{|\boldsymbol{I}+q_1\boldsymbol{d}\boldsymbol{d}^\dagger + h^2q_2\boldsymbol{c}\boldsymbol{c}^\dagger|}{|\boldsymbol{I}+h^2q_2\boldsymbol{c}\boldsymbol{c}^\dagger|} + \log\frac{|\boldsymbol{I}+q_2\boldsymbol{d}\boldsymbol{d}^\dagger + h^2q_1\boldsymbol{c}\boldsymbol{c}^\dagger|}{|\boldsymbol{I}+h^2q_1\boldsymbol{c}\boldsymbol{c}^\dagger|},$$

where q_1 and q_2 denote the transmit powers. Recall that q_1 and q_2 must satisfy the average power constraints of $q_1, q_2 \leq P$. The achievable sum-rate by using the maximum power, i.e., by setting $q_1 = q_2 = P$, is given by

$$\begin{aligned}
f(P,P) &= 2\log\frac{|\boldsymbol{I}+P\boldsymbol{d}\boldsymbol{d}^\dagger + h^2P\boldsymbol{c}\boldsymbol{c}^\dagger|}{|\boldsymbol{I}+h^2P\boldsymbol{c}\boldsymbol{c}^\dagger|} \\
&= 2\log\left|\boldsymbol{I}+\boldsymbol{J}^{-1}P\boldsymbol{d}\boldsymbol{d}^\dagger\right| \\
&= 2\log\left(1+P\boldsymbol{d}^\dagger\boldsymbol{J}^{-1}\boldsymbol{d}\right) && \text{(2.24a)} \\
&= 2\log\left(1+\frac{P\cos^2\theta}{1+h^2P}+P\sin^2\theta\right), && \text{(2.24b)}
\end{aligned}$$

where the matrix $\boldsymbol{J}$ denotes

$$\boldsymbol{J} = \boldsymbol{I}+h^2P\boldsymbol{c}\boldsymbol{c}^\dagger = \begin{bmatrix} 1+h^2P & 0 \\ 0 & 1 \end{bmatrix}.$$

The above sum-rate can be shown to be achievable with the receivers projecting the received vector along a beamforming direction, denoted by a unit norm vector $\boldsymbol{b}$, i.e.,

$$\begin{aligned}
\tilde{y}_1 &= \boldsymbol{b}^\dagger\boldsymbol{y}_1 = \boldsymbol{b}^\dagger\boldsymbol{d}x_1 + h\boldsymbol{b}^\dagger\boldsymbol{c}x_2 + \boldsymbol{b}^\dagger z_1, \\
\tilde{y}_2 &= \boldsymbol{b}^\dagger\boldsymbol{y}_2 = \boldsymbol{b}^\dagger\boldsymbol{d}x_2 + h\boldsymbol{b}^\dagger\boldsymbol{c}x_1 + \boldsymbol{b}^\dagger z_2.
\end{aligned}$$

We choose the beamforming direction $\boldsymbol{b}$ as

$$\boldsymbol{b} = \frac{\boldsymbol{J}^{-1}\boldsymbol{d}}{||\boldsymbol{J}^{-1}\boldsymbol{d}||} \tag{2.25}$$

in order to achieve the best SINR:

$$\text{SINR} = \frac{P|\boldsymbol{b}^\dagger\boldsymbol{d}|^2}{1+h^2P|\boldsymbol{b}^\dagger\boldsymbol{c}|^2} = \frac{P|\boldsymbol{b}^\dagger\boldsymbol{d}|^2}{\boldsymbol{b}^\dagger\boldsymbol{J}\boldsymbol{b}} = P\boldsymbol{d}^\dagger\boldsymbol{J}^{-1}\boldsymbol{d}.$$

This interpretation of receive beamforming helps in understanding the best achievable sum-rate of the dual MISO interference channel. Observe that the achievable sum-rate of the MISO interference channel is given by

$$f(\boldsymbol{Q}_1,\boldsymbol{Q}_2) = \log\left(1+\frac{P\boldsymbol{d}^\dagger\boldsymbol{Q}_1\boldsymbol{d}}{1+h^2P\boldsymbol{c}^\dagger\boldsymbol{Q}_2\boldsymbol{c}}\right)+\log\left(1+\frac{P\boldsymbol{d}^\dagger\boldsymbol{Q}_2\boldsymbol{d}}{1+h^2P\boldsymbol{c}^\dagger\boldsymbol{Q}_1\boldsymbol{c}}\right).$$

Using the insight from the SIMO interference channel, we let the transmitters transmit along the beamforming direction $\boldsymbol{b}$; i.e., we set

$$\boldsymbol{Q}_1^* = \boldsymbol{Q}_2^* = \boldsymbol{Q}^* = P\boldsymbol{b}\boldsymbol{b}^\dagger.$$

The corresponding achievable sum-rate is given by

$$\begin{aligned} f(\boldsymbol{Q}^*,\boldsymbol{Q}^*) &= 2\log\left(1+\frac{P\boldsymbol{d}^\dagger\boldsymbol{Q}^*\boldsymbol{d}}{1+h^2P\boldsymbol{c}^\dagger\boldsymbol{Q}^*\boldsymbol{c}}\right) \\ &= 2\log\left(1+\frac{P\cos^2\theta}{1+h^2P}+P\sin^2\theta\right). \end{aligned}$$

Low Interference Regime

THEOREM 2.9 *The sum capacity of the symmetric MISO and SIMO Gaussian interference channels described in Section 2.2.13 is achieved by using Gaussian inputs and treating interference as noise at the receivers, and is given by*

$$\mathcal{C}_{\text{sum}} = 2\log\left(1+\frac{P\cos^2\theta}{1+h^2P}+P\sin^2\theta\right)$$

if the channel parameters satisfy the threshold condition $h \le h_0(\theta,P)$, where $h_0(\theta,P)$ is defined as the unique positive solution to the implicit equation

$$h^2-\sin^2\theta = \left(\frac{\cos\theta}{1+h^2P}-h\right)_+^2, \tag{2.26}$$

where we use the notation x_+^2 to denote $(\max(0,x))^2$.

The above theorem is obtained by specializing the conditions in Theorem 2.6 for the special case of symmetric MISO and SIMO channels. Before we go into the proof details, we first prove some properties of the threshold function $h_0(\theta,P)$. The threshold $h_0(\theta,P)$ is plotted as a function of θ for different values of P in Figure 2.5. It can be observed that the threshold curve is always above the $\sin\theta$ curve and approaches the $\sin\theta$ curve as P becomes larger.

We summarize the observations from Figure 2.5 in the following claim.

CLAIM 2.2 The threshold $h_0(\theta,P)$ satisfies

- $h_0(\theta,P) > \sin\theta$ for all $P < P_0(\theta)$,
- $h_0(\theta,P) = \sin\theta$ for all $P \ge P_0(\theta)$,

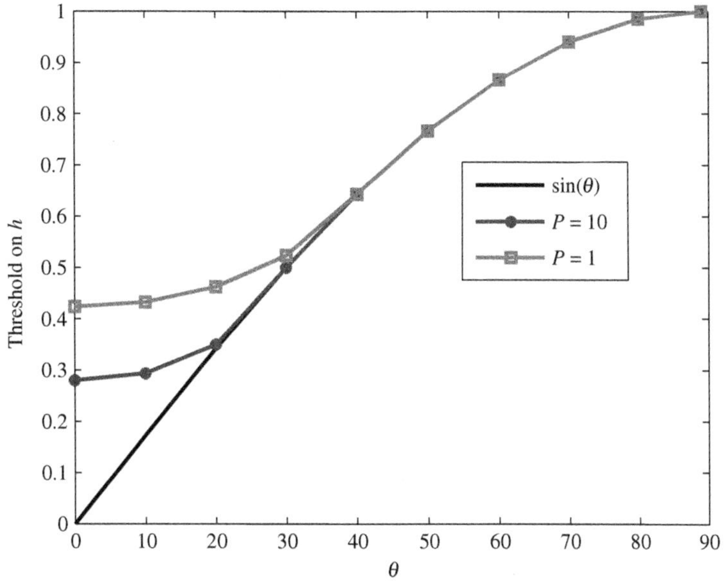

Fig. 2.5 Threshold on h characterizing the low interference regime of the symmetric MISO and SIMO interference channels. ©[2017] IEEE. Reprinted, with permission, from [13].

where $P_0(\theta)$ is defined as

$$P_0(\theta) = \begin{cases} \dfrac{\cos\theta - \sin\theta}{\sin^3\theta} & \text{when } 0 \leq \theta < \pi/4, \\ 0 & \text{when } \pi/4 \leq \theta < \pi/2. \end{cases}$$

Proof Observe that the left-hand side of (2.26) is strictly increasing in h, whereas the right-hand side is decreasing in h. This verifies that (2.26) has a unique positive solution. Note that the left-hand side is strictly negative when $h < \sin\theta$, whereas the right-hand side is always nonnegative. This immediately implies that $h_0(\theta,P) \geq \sin\theta$. It can be easily checked that the right-hand side is equal to zero at $h = \sin\theta$ when $P \geq P_0(\theta)$. Hence, we obtain the second statement. Similarly, it can be easily checked that the right-hand side is greater than zero at $h = \sin\theta$ when $P < P_0(\theta)$. Hence, we obtain the first statement.

SIMO Interference Channel

Recall from Section 2.2.1 that $\tilde{\boldsymbol{H}}_{12}$ and $\tilde{\boldsymbol{H}}_{21}$ denote the nonzero rows of the respective matrices $\boldsymbol{H}_{12}$ and $\boldsymbol{H}_{21}$. Therefore, for the special case of the SIMO interference channel, we have $\tilde{\boldsymbol{H}}_{12} = \tilde{\boldsymbol{H}}_{21} = h$, and the genie signals (2.9) are given by

$$s_1 = hx_1 + w_1,$$

$$s_2 = hx_2 + w_2.$$

We now simplify the conditions in Theorem 2.6 and show that they are equivalent to the threshold condition in Theorem 2.9. As explained in Section 2.2.12 for the

SISO interference channel, we do not have to explicitly check for the KKT conditions (2.20) because they are automatically satisfied at $(q_1^*, q_2^*) = (P, P)$. This is true because $\bar{f}(q_1, q_2, \Psi^*)$ is nondecreasing in (q_1, q_2). Therefore, it only remains to verify the existence of a genie $\Psi^* = \left(\Sigma_{w_1}, \mathbf{\Sigma}_{w_1 z_1}, \Sigma_{w_2}, \mathbf{\Sigma}_{w_2 z_2}\right)$ that is useful and (P, P)-smart, i.e., satisfying the conditions (2.18) and (2.19). Since we are working with a symmetric interference channel, we restrict ourselves to a symmetric genie; i.e., we assume $\mathbf{\Sigma}_{w_1 z_1} = \mathbf{\Sigma}_{w_2 z_2} = \mathbf{\Sigma}_{wz} = [a_1 \; a_2]$ and $\Sigma_{w_1} = \Sigma_{w_2} = \Sigma_w$.

The usefulness condition (2.18) when specialized to the SIMO channel is given by

$$\begin{bmatrix} 1 & 0 & a_1^\dagger \\ 0 & 1 & a_2^\dagger \\ a_1 & a_2 & \Sigma_w \end{bmatrix} \succeq \begin{bmatrix} \Sigma_w & 0 & 0 \\ 0 & 0 & 0 \\ 0 & 0 & 0 \end{bmatrix},$$

which is equivalent to

$$\begin{bmatrix} 1 - \Sigma_w & 0 & a_1^\dagger \\ 0 & 1 & a_2^\dagger \\ a_1 & a_2 \Sigma_w & \end{bmatrix} \succeq 0.$$

Using the fact that a Hermitian matrix is positive semidefinite if and only if all the principal minors are nonnegative, the above condition is equivalent to the conditions

$$\begin{aligned} 0 \leq \Sigma_w \leq 1, \\ \Sigma_w - |a_2|^2 \geq 0, \\ (1 - \Sigma_w)(\Sigma_w - |a_2|^2) - |a_1|^2 \geq 0. \end{aligned} \tag{2.27}$$

The smartness condition (2.19) when specialized to the SIMO channel is given by

$$\left(h - \mathbf{\Sigma}_{wz} \left(h^2 P \mathbf{c}\mathbf{c}^\dagger + \mathbf{I}\right)^{-1} \mathbf{d}\right) P = \mathbf{0},$$

which is equivalent to

$$h - \mathbf{\Sigma}_{wz} \begin{bmatrix} 1 + h^2 P & 0 \\ 0 & 1 \end{bmatrix}^{-1} \mathbf{d} = \mathbf{0},$$

which is further equivalent to

$$h - \frac{a_1 \cos\theta}{1 + h^2 P} - a_2 \sin\theta = 0. \tag{2.28}$$

Therefore, we obtain that the SIMO interference channel is in the low interference regime if there exist parameters a_1, a_2, Σ_w satisfying the conditions (2.27) and (2.28). The following claim, which is proved in [13], completes the proof of the SIMO part of Theorem 2.9.

CLAIM 2.3 There exist parameters a_1, a_2, Σ_w satisfying the conditions (2.27) and (2.28) if and and only if the following *low interference regime* condition is true:

$$h^2 - \sin^2\theta \leq \left(\frac{\cos\theta}{1 + h^2 P} - h\right)_+^2. \tag{2.29}$$

MISO Interference Channel

Observe that Theorem 2.9 is obtained by specializing Theorem 2.6 for the special case of the symmetric MISO interference channel. The genie signals (2.9) are given by

$$s_1 = h\boldsymbol{c}^\dagger \boldsymbol{x}_1 + w_1,$$
$$s_2 = h\boldsymbol{c}^\dagger \boldsymbol{x}_2 + w_2.$$

Theorem 2.9 follows if we show that there exist genie parameters and dual variables satisfying the conditions of Theorem 2.6 at

$$\boldsymbol{Q}_1^* = \boldsymbol{Q}_2^* = \boldsymbol{Q}^* = P\boldsymbol{b}\boldsymbol{b}^\dagger.$$

Since the channel is symmetric across the users, we restrict the genie parameters $\Psi^* = \left(\Sigma_{w_1}, \Sigma_{w_1 z_1}, \Sigma_{w_2}, \Sigma_{w_2 z_2}\right)$ and the dual variables $\lambda_1, \lambda_2, \boldsymbol{M}_1, \boldsymbol{M}_2$ to be symmetric, i.e.,

$$\Sigma_{w_1 z_1} = \Sigma_{w_2 z_2} = \Sigma_{wz},$$
$$\Sigma_{w_1} = \Sigma_{w_2} = \Sigma_w,$$
$$\lambda_1 = \lambda_2 = \lambda,$$
$$\boldsymbol{M}_1 = \boldsymbol{M}_2 = \boldsymbol{M}.$$

Therefore, we need to prove the existence of the parameters $\Sigma_w, \Sigma_{wz}, \lambda \geq 0, \boldsymbol{M} \succeq 0$ satisfying the following conditions:

- Useful genie condition:

$$\begin{bmatrix} 1 & \Sigma_{zw} \\ \Sigma_{wz} & \Sigma_w \end{bmatrix} \succeq \begin{bmatrix} \Sigma_w & 0 \\ 0 & 0 \end{bmatrix}.$$

- Smart genie condition:

$$\left(h\boldsymbol{c}^\dagger - \frac{\Sigma_{wz}}{1 + h^2 P |\boldsymbol{b}^\dagger \boldsymbol{c}|^2} \boldsymbol{d}^\dagger \right) \boldsymbol{b} = 0.$$

- KKT conditions:

$$\nabla_{\boldsymbol{Q}_1} \bar{f}(\boldsymbol{Q}^*, \boldsymbol{Q}^*, \Psi^*) = \lambda \boldsymbol{I} - \boldsymbol{M},$$
$$\operatorname{Tr}\left(\boldsymbol{M}\boldsymbol{Q}^*\right) = 0.$$

Note that the useful genie condition is equivalent to

$$\Sigma_w (1 - \Sigma_w) \geq |\Sigma_{wz}|^2,$$

and that the smart genie condition is equivalent to

$$h\boldsymbol{c}^\dagger \boldsymbol{b} = \frac{\Sigma_{wz}}{\boldsymbol{b}^\dagger \boldsymbol{J} \boldsymbol{b}} \boldsymbol{d}^\dagger \boldsymbol{b},$$

which is equivalent to

$$\Sigma_{wz} = h \frac{\boldsymbol{c}^\dagger \boldsymbol{b}}{\boldsymbol{d}^\dagger \boldsymbol{b}} \boldsymbol{b}^\dagger \boldsymbol{J} \boldsymbol{b}.$$

Recall that the matrix $\boldsymbol{J}$ is defined as

$$\boldsymbol{J} = \boldsymbol{I} + h^2 P \boldsymbol{c}\boldsymbol{c}^\dagger = \begin{bmatrix} 1+h^2P & 0 \\ 0 & 1 \end{bmatrix},$$

and the beamforming vector $\boldsymbol{b}$ is defined as the unit norm vector in the direction of

$$\boldsymbol{b} = \frac{\boldsymbol{J}^{-1}\boldsymbol{d}}{||\boldsymbol{J}^{-1}\boldsymbol{d}||} = \frac{1}{||\boldsymbol{J}^{-1}\boldsymbol{d}||} \begin{bmatrix} \frac{\cos^2\theta}{1+h^2P} \\ \sin\theta \end{bmatrix}.$$

Thus, we see that the smart genie condition is equivalent to

$$\begin{aligned} \Sigma_{wz} &= h \frac{\boldsymbol{c}^\dagger \boldsymbol{b}}{\boldsymbol{d}^\dagger \boldsymbol{b}} \boldsymbol{b}^\dagger \boldsymbol{J} \boldsymbol{b} \\ &= h \frac{\boldsymbol{c}^\dagger \boldsymbol{J}^{-1}\boldsymbol{d}}{\boldsymbol{d}^\dagger \boldsymbol{J}^{-1}\boldsymbol{d}} \frac{\boldsymbol{d}^\dagger \boldsymbol{J}^{-1}\boldsymbol{d}}{||\boldsymbol{J}^{-1}\boldsymbol{d}||^2} \\ &= h \frac{\boldsymbol{c}^\dagger \boldsymbol{J}^{-1}\boldsymbol{d}}{||\boldsymbol{J}^{-1}\boldsymbol{d}||^2}, \end{aligned}$$

i.e.,

$$\Sigma_{wz} = \frac{h\dfrac{\cos\theta}{1+h^2P}}{\dfrac{\cos^2\theta}{(1+h^2P)^2} + \sin^2\theta}. \tag{2.30}$$

We now simplify the KKT conditions. Observe that $\mathsf{Tr}\big(\boldsymbol{M}\boldsymbol{Q}^*\big) = 0$ is equivalent to saying that $\boldsymbol{M}\boldsymbol{b} = 0$. Therefore, we see that $\boldsymbol{b}$ is an eigenvector of the gradient matrix $\nabla_{\boldsymbol{Q}_1}\bar{f}(\boldsymbol{Q}^*, \boldsymbol{Q}^*, \Psi^*)$, with $\lambda \geq 0$ as the eigenvalue value. The condition $\boldsymbol{M} \succeq 0$ implies that the other eigenvalue is smaller than λ. Thus, we see that the KKT conditions are equivalent to saying that $\boldsymbol{b}$ is the dominant eigenvector of $\nabla_{\boldsymbol{Q}_1}\bar{f}(\boldsymbol{Q}^*, \boldsymbol{Q}^*, \Psi^*)$ with eigenvalue $\lambda \geq 0$.

We can use Lemma 2.7 to simplify the derivation of the gradient matrix. Observe that

$$\nabla_{\boldsymbol{Q}_1}\bar{f}(\boldsymbol{Q}^*, \boldsymbol{Q}^*, \Psi^*) = \nabla_{\boldsymbol{Q}_1} f(\boldsymbol{Q}^*, \boldsymbol{Q}^*) + \boldsymbol{N},$$

where the matrix $\boldsymbol{N}$ is given by

$$\boldsymbol{N} = \frac{(h\boldsymbol{c} - t\boldsymbol{d})^\dagger (h\boldsymbol{c} - t\boldsymbol{d})}{\Sigma_{w_1|y_{1G}, \boldsymbol{x}_{1G}}},$$

and the parameter t is given by

$$t = \frac{\Sigma_{wz}}{h^2 P |\boldsymbol{b}^\dagger \boldsymbol{c}|^2 + 1} = \frac{\Sigma_{wz}}{\boldsymbol{b}^\dagger \boldsymbol{J} \boldsymbol{b}} = h \frac{\boldsymbol{c}^\dagger \boldsymbol{b}}{\boldsymbol{d}^\dagger \boldsymbol{b}}. \tag{2.31}$$

Observe that the matrix $\boldsymbol{N}$ satisfies $\boldsymbol{Nb} = 0$. Therefore, we need to show that $\boldsymbol{b}$ is an eigenvector of $\nabla_{Q_1} f(\boldsymbol{Q}^*, \boldsymbol{Q}^*)$. Note that

$$\begin{aligned}
\nabla_{Q_1} f(\boldsymbol{Q}^*, \boldsymbol{Q}^*) &= \frac{\boldsymbol{dd}^\dagger}{\Sigma_{y_{1G}}} + \frac{h^2 \boldsymbol{cc}^\dagger}{\Sigma_{y_{2G}}} - \frac{h^2 \boldsymbol{cc}^\dagger}{\Sigma_{y_{2G}|\boldsymbol{x}_{2G}}} \\
&= \frac{1}{\Sigma_{y_{1G}}} \left(\boldsymbol{dd}^\dagger - h^2 \left(\frac{\Sigma_{y_{2G}}}{\Sigma_{y_{2G}|\boldsymbol{x}_{2G}}} - 1 \right) \boldsymbol{cc}^\dagger \right) \\
&= \frac{1}{\Sigma_{y_{1G}}} \left(\boldsymbol{dd}^\dagger - h^2 \, \mathrm{SINR} \, \boldsymbol{cc}^\dagger \right).
\end{aligned} \tag{2.32}$$

Recall that the expression for SINR is given by

$$\mathrm{SINR} = \frac{P \cos^2 \theta}{1 + h^2 P} + P \sin^2 \theta.$$

Using the expansion

$$\begin{aligned}
\boldsymbol{dd}^\dagger - h^2 \, \mathrm{SINR} \, \boldsymbol{cc}^\dagger &= \begin{bmatrix} \cos^2\theta - h^2 \mathrm{SINR} & \cos\theta \sin\theta \\ \cos\theta \sin\theta & \sin^2\theta \end{bmatrix} \\
&= \begin{bmatrix} \frac{\cos^2\theta}{1+h^2P} - h^2 P \sin^2\theta & \cos\theta \sin\theta \\ \cos\theta \sin\theta & \sin^2\theta \end{bmatrix},
\end{aligned}$$

we can easily check that $\boldsymbol{b}$ is an eigenvector of the matrix $\nabla_{Q_1} f(\boldsymbol{Q}^*, \boldsymbol{Q}^*)$, with the corresponding eigenvalue given by

$$\lambda = \frac{1}{\Sigma_{y_{1G}}} \left(\frac{\cos^2\theta}{1 + h^2 P} + \sin^2\theta \right). \tag{2.33}$$

Since $\boldsymbol{Nb} = 0$, we also obtain that $\boldsymbol{b}$ is an eigenvector of the matrix $\nabla_{Q_1} \bar{f}(\boldsymbol{Q}^*, \boldsymbol{Q}^*, \Psi^*)$ with the same eigenvalue λ. Since the sum of eigenvalues is equal to the trace of the matrix, the other eigenvalue is equal to

$$\mathsf{Tr}\left(\nabla_{Q_1} \bar{f}(\boldsymbol{Q}^*, \boldsymbol{Q}^*, \Psi^*) \right) - \lambda.$$

Therefore, we have that $\boldsymbol{b}$ is the dominant eigenvector if and only if

$$\begin{aligned}
2\lambda &\geq \mathsf{Tr}\left(\nabla_{Q_1} \bar{f}(\boldsymbol{Q}^*, \boldsymbol{Q}^*, \Psi^*) \right) \\
&= \mathsf{Tr}\left(\nabla_{Q_1} f(\boldsymbol{Q}^*, \boldsymbol{Q}^*) \right) + \mathsf{Tr}(\boldsymbol{N}).
\end{aligned}$$

Note that

$$\begin{aligned}
\mathsf{Tr}(\boldsymbol{N}) &= \frac{||h\boldsymbol{c} - t\boldsymbol{d}||^2}{\Sigma_{w_1|y_{1G}, \boldsymbol{x}_{1G}}} \\
&= \frac{||h\boldsymbol{c} - t\boldsymbol{d}||^2}{\Sigma_w - \dfrac{\Sigma_{wz}^2}{1 + h^2 |\boldsymbol{b}^\dagger \boldsymbol{c}|^2}}.
\end{aligned}$$

Therefore, $\boldsymbol{b}$ is the dominant eigenvector if and only if

$$\Sigma_w \geq \frac{\Sigma_{wz}^2}{1+h^2|\boldsymbol{b}^\dagger \boldsymbol{c}|^2} + \frac{||h\boldsymbol{c} - t\boldsymbol{d}||^2}{2\lambda - \mathrm{Tr}\left(\nabla_{Q_1} f(\boldsymbol{Q}^*, \boldsymbol{Q}^*)\right)}. \tag{2.34}$$

Observe that all the variables other than Σ_w are known and can be expressed as a function of h, θ, and P. By substituting the corresponding expressions, we can simplify the right-hand side of (2.34), as in the following claim, which is proved in [13].

CLAIM 2.4 The condition (2.34) is equivalent to

$$\Sigma_w \geq \frac{\dfrac{h^2(1+P\sin^2\theta)}{1+h^2P}}{\dfrac{\cos^2\theta}{(1+h^2P)^2} + \sin^2\theta}. \tag{2.35}$$

It now remains to verify the existence of the parameter Σ_w satisfying the usefulness condition,

$$\Sigma_w(1-\Sigma_w) \geq \Sigma_{wz}^2,$$
$$\Sigma_w^2 - \Sigma_w + \Sigma_{wz}^2 \leq 0,$$

and (2.35), where Σ_{wz} is given by the smartness condition (2.30). Observe that the above quadratic inequality has a solution if and only if $\Sigma_{wz} \leq 0.5$, and that the largest possible value for Σ_w satisfying the quadratic inequality is given by $0.5(1+\sqrt{1-4\Sigma_{wz}^2})$. Since the condition (2.35) requires Σ_w to be larger than a threshold, without any loss of generality we can set $\Sigma_w = 0.5(1+\sqrt{1-4\Sigma_{wz}^2})$. Therefore, it remains to verify whether h, θ, and P satisfy the following two conditions:

$$\Sigma_{wz} = \frac{h\dfrac{\cos\theta}{1+h^2P}}{\dfrac{\cos^2\theta}{(1+h^2P)^2} + \sin^2\theta} \leq 0.5,$$

$$\Sigma_w = \frac{-1+\sqrt{1-4\Sigma_{wz}^2}}{2} \geq \frac{\dfrac{h^2(1+P\sin^2\theta)}{1+h^2P}}{\dfrac{\cos^2\theta}{(1+h^2P)^2} + \sin^2\theta}.$$

We now show that the above two conditions are satisfied when $h \leq h_0(\theta, P)$ by dividing the proof into two cases.

Case 1: The channel parameters satisfy $h \leq \sin\theta$. Recall that Σ_{wz} is given by

$$\Sigma_{wz} = \frac{h\dfrac{\cos\theta}{1+h^2P}}{\dfrac{\cos^2\theta}{(1+h^2P)^2} + \sin^2\theta}$$

$$\leq \frac{\sin\theta \dfrac{\cos\theta}{1+h^2P}}{\dfrac{\cos^2\theta}{(1+h^2P)^2}+\sin^2\theta}$$
$$= \frac{ab}{a^2+b^2},$$

where we used a and b to denote $\cos\theta/(1+h^2P)$ and $\sin\theta$, respectively. Using the fact that $2ab \leq a^2+b^2$, we obtain that $\Sigma_{wz} \leq 0.5$. Using the fact that $(a^2+b^2)^2 - 4a^2b^2 = (a^2-b^2)^2$, we see that the corresponding Σ_w satisfies

$$\Sigma_w = \frac{1+\sqrt{1-4\Sigma_{wz}^2}}{2}$$
$$\geq \frac{\max(a^2,b^2)}{a^2+b^2}$$
$$\geq \frac{b^2}{a^2+b^2}.$$

Now, observe that

$$b^2 = \sin^2\theta$$
$$\geq \sin^2\theta - \frac{\sin^2\theta - h^2}{1+h^2P}$$
$$= \frac{h^2(1+P\sin^2\theta)}{1+h^2P}.$$

Thus, we see that the KKT condition (2.35) is satisfied.
Case 2: The channel parameters satisfy

$$\sin\theta < h \leq \frac{\cos\theta}{1+h^2P}. \tag{2.36}$$

Observe that the condition

$$\Sigma_w = \frac{2h\dfrac{\cos\theta}{1+h^2P}}{\dfrac{\cos^2\theta}{(1+h^2P)^2}+\sin^2\theta} \leq 1$$

is equivalent to

$$h^2 - \sin^2\theta \leq \left(\frac{\cos\theta}{1+h^2P} - h\right)^2,$$

and is hence satisfied because $h \leq h_0(\theta,P)$. Observe that the condition (2.35) is satisfied because

$$\Sigma_w = 0.5(1+\sqrt{1-4\Sigma_{wz}^2}) \geq 0.5 \geq \Sigma_{wz}$$

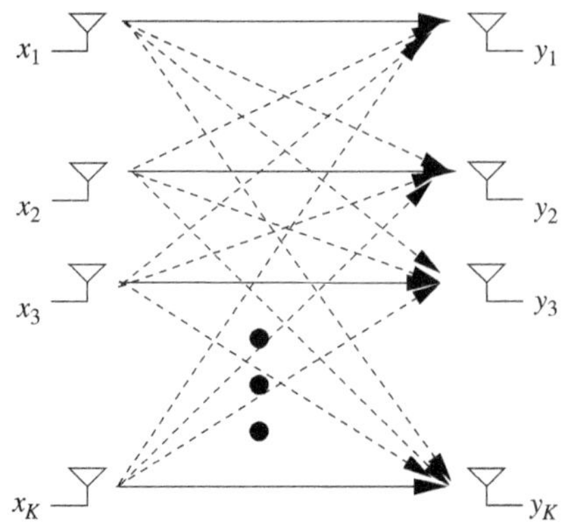

Fig. 2.6 K-user interference channel.

and

$$h\cos\theta \geq h^2(1+h^2P) \geq h^2(1+P\sin^2\theta).$$

2.2.14 Two-User Gaussian Interference Channel – Summary

We studied the nonconvex optimization problem of determining the best achievable sum-rate, using Gaussian inputs and treating interference as noise, in the two-user MIMO Gaussian interference channel. We used the idea of the genie-aided channel to relax the nonconvex optimization problem, and proposed a related convex maxmin optimization problem. The corresponding saddle point solution provides lower and upper bounds for the best achievable sum-rate. We then showed that the resulting upper bound is indeed an upper bound to the sum capacity as well. We also derived necessary and sufficient conditions for the bounds to coincide, leading to an exact characterization of the the best achievable sum-rate when treating interference as noise, and the sum capacity. We then simplified the conditions in the special cases of symmetric MISO and SIMO Gaussian interference channels, and showed that the conditions are equivalent to a threshold condition on the cross-channel gain. Interestingly, the threshold is identical for the symmetric MISO and the dual SIMO interference channels.

2.3 *K*-User Interference Channels

In the previous section, we established the sum capacity of the two-user MIMO Gaussian interference channel in a low interference regime. The intuition is that if the interference is low enough, the receiver will not be able to exploit the structure in the interference, and hence treating interference as noise achieves the sum capacity. In this section, we extend the low interference regime results from the two-user case to the K-user case. We focus on the SISO case, where each transmitter and receiver is equipped with a single antenna. The K-user SISO Gaussian interference channel illustrated in

Figure 2.6 is given by

$$y_i = \sum_{j=1}^{K} h_{ij}x_j + z_i, i \in [K], \tag{2.37}$$

with P_j denoting the average transmit power constraint on the transmitters j. Without any loss of generality, by appropriately scaling the power constraints, we assume that the direct channel gains are equal to unity; i.e.,

$$h_{ii} = 1, \forall i \in [K].$$

Since we assumed a single transmit antenna, it is easy to determine the lower bound on the sum capacity obtained by using Gaussian inputs (with maximum power) and treating interference as noise:

$$\mathcal{C}_{\text{sum}} \geq \sum_{i=1}^{K} \mathsf{I}(x_{i\mathrm{G}}; y_{i\mathrm{G}}) = \sum_{i=1}^{K} \log\left(1 + \frac{P_i}{1 + \sum_{j \neq k} |h_{ij}|^2 P_j}\right),$$

where the subscript G indicates that the inputs are Gaussian distributed (with maximum power). We say that the *K*-user Gaussian interference channel (2.37) is in the low interference regime if the above lower bound is equal to the sum capacity. The objective of this section is to derive conditions such that the *K*-user interference channel belongs to the low interference regime.

We first consider two special cases of the *K*-user interference channel, introduced in [21, 22]: the *many-to-one interference channel*, where only one user experiences interference, and the *one-to-many interference channel*, where the interference is generated by only one user. For these two special cases, we derive conditions under which the channels belong to the low interference regime. The genie-aided channel concept, used in the previous section, is not required in the upper bound proofs of these two special cases. For the general *K*-user interference channel, however, the upper bounds are based on the genie-aided channel concept. The basic idea behind the proof is the same as in the previous section, and can be summarized as follows:

$$\begin{aligned} \mathcal{C}_{\text{sum}} &\leq \text{sum capacity of the genie-aided channel} \\ &\overset{\text{(a)}}{=} \sum_{i=1}^{K} \mathsf{I}(x_{i\mathrm{G}}; y_{i\mathrm{G}}, s_{i\mathrm{G}}) \\ &\overset{\text{(b)}}{=} \sum_{i=1}^{K} \mathsf{I}(x_{i\mathrm{G}}; y_{i\mathrm{G}}), \end{aligned} \tag{2.38}$$

where s_i denotes the side information given to the receiver i. The subscript G indicates that Gaussian inputs (with maximum power) are used. We use the same terminology as in the previous section. We say that a genie is useful if step (a) is satisfied; i.e., treating interference as noise with Gaussian inputs achieves the sum capacity of the genie-aided channel. We say that a genie is smart if step (b) is satisfied; i.e., the genie does not improve the achievable sum-rate when Gaussian inputs are used. Therefore,

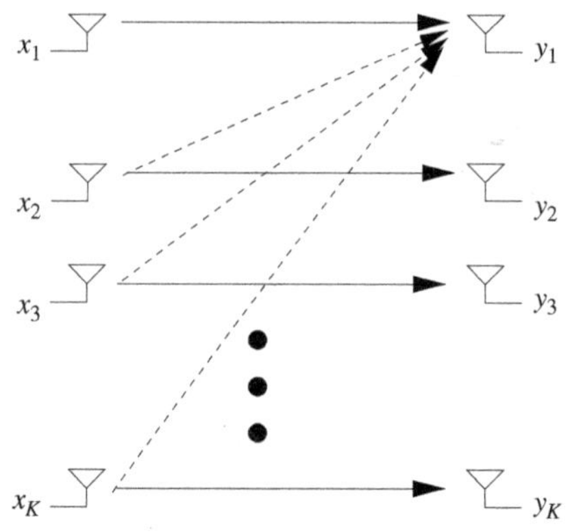

Fig. 2.7 Many-to-one interference channel.

the objective is to determine conditions under which there exists a genie that is both useful and smart.

2.3.1 Many-to-One Interference Channel

Consider the many-to-one Gaussian interference channel shown in Figure 2.7, where only one user experiences the interference:

$$y_1 = x_1 + \sum_{j=2}^{K} h_{1j} x_j + z_1,$$

$$y_i = x_i + z_i,\ i = 2,3,\ldots,K.$$

THEOREM 2.10 *The sum capacity of the many-to-one interference channel is achieved by using Gaussian inputs and treating interference as noise, and is given by*

$$C_{\text{sum}} = \log\left(1 + \frac{P_1}{\sum_{i=2}^{K} |h_{1i}|^2 P_i}\right) + \sum_{i=2}^{K} \log(1+P_i)$$

if the channel parameters satisfy the low interference regime condition

$$\sum_{i=2}^{K} |h_{1i}|^2 \le 1.$$

Proof The achievability is based on the transmitters using Gaussian inputs and the receivers treating interference as noise. We now prove the converse. Using Fano's inequality, we have

$$n(C_{\text{sum}} - K\epsilon_n) \le \sum_{i=1}^{K} \mathrm{I}\left(x_i^n; y_i^n\right).$$

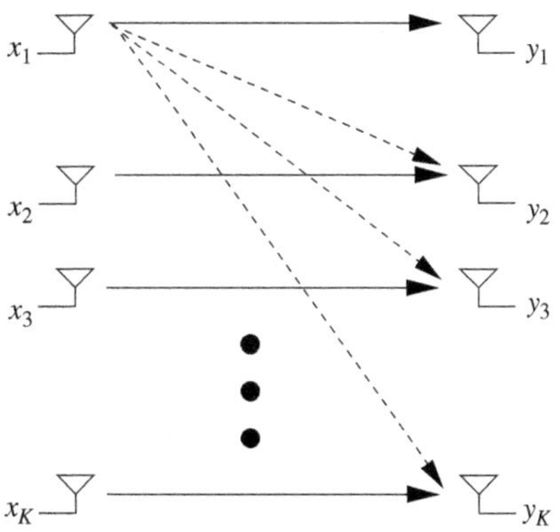

Fig. 2.8 One-to-many interference channel.

Therefore, it is sufficient to prove that the right-hand side of the above equation is maximized by i.i.d. Gaussian inputs (with maximum power). Observe that

$$\begin{aligned}\sum_{i=1}^{K} \mathsf{I}\left(x_i^n; y_i^n\right) &= \mathsf{I}\left(x_1^n; y_1^n\right) + \sum_{i=2}^{K} \mathsf{I}\left(x_i^n; y_i^n\right) \\ &= \mathsf{h}\left(y_1^n\right) - \mathsf{h}\left(y_1^n | x_1^n\right) + \sum_{i=2}^{K} \mathsf{h}\left(y_i^n\right) - \sum_{i=2}^{K} \mathsf{h}\left(z_i^n\right).\end{aligned}$$

The terms $\mathsf{h}\left(z_i^n\right)$ are independent of the input distributions. From Lemma A.8 in Appendix A, it follows that the term $\mathsf{h}\left(y_1^n\right)$ is maximized by i.i.d. Gaussian inputs with maximum power. The remaining terms contribute

$$\sum_{i=2}^{K} \mathsf{h}\left(y_i^n\right) - \mathsf{h}\left(y_1^n | x_1^n\right) = \sum_{i=2}^{K} \mathsf{h}\left(x_i^n + z_i^n\right) - \mathsf{h}\left(\sum_{i=2}^{K} h_{1i} x_i^n + z_1^n\right).$$

From Lemma A.11, it follows that the above expression is maximized by i.i.d. Gaussian inputs with maximum power when the condition $\sum_{i=2}^{K} |h_{1i}|^2 \leq 1$ is satisfied.

2.3.2 One-to-Many Interference Channel

Consider the one-to-many Gaussian interference channel shown in Figure 2.8, where only one user causes the interference:

$$\begin{aligned} y_1 &= x_1 + z_1, \\ y_i &= x_i + h_{i1} x_1 + z_i, \ i = 2, 3, \ldots, K. \end{aligned} \tag{2.39}$$

THEOREM 2.11 *The sum capacity of the one-to-many interference channel* (2.39) *is achieved by using Gaussian inputs and treating interference as noise, and is given by*

$$C_{\text{sum}} = \log\left(1 + P_1\right) + \sum_{i=2}^{K} \log\left(1 + \frac{P_i}{1 + |h_{i1}|^2 P_1}\right)$$

if the channel parameters satisfy the low interference regime condition

$$\sum_{i=2}^{K} \frac{h_{i1}^2 P_1 + h_{i1}^2}{h_{i1}^2 P_1 + 1} \leq 1. \tag{2.40}$$

Proof The achievability proof is based on the transmitters using Gaussian inputs and the receivers treating interference as noise. We now prove the converse. Using Fano's inequality, we have

$$n(C_{\text{sum}} - K\epsilon_n) \leq \sum_{i=1}^{K} \mathsf{I}\left(x_i^n; y_i^n\right).$$

Therefore, it is sufficient to prove that the right-hand side of the above equation is maximized by i.i.d. Gaussian inputs (with maximum power). Observe that

$$\begin{aligned}\sum_{i=1}^{K} \mathsf{I}\left(x_i^n; y_i^n\right) &= \mathsf{I}\left(x_1^n; y_1^n\right) + \sum_{i=2}^{M} \mathsf{I}\left(x_i^n; y_i^n\right) \\ &= \mathsf{h}\left(y_1^n\right) - \mathsf{h}\left(z_1^n\right) + \sum_{i=2}^{K} \mathsf{h}\left(y_i^n\right) - \sum_{i=2}^{K} \mathsf{h}\left(y_i^n | x_i^n\right).\end{aligned}$$

The term $\mathsf{h}\left(z_1^n\right)$ is independent of the input distributions. From Lemma A.8, it follows that the terms $\mathsf{h}\left(y_i^n\right)$ are maximized by i.i.d. Gaussian inputs with maximum power. The remaining terms contribute

$$\begin{aligned}\mathsf{h}\left(y_1^n\right) - \sum_{i=2}^{K} \mathsf{h}\left(y_i^n | x_i^n\right) &= \mathsf{h}\left(x_1^n + z_1^n\right) - \sum_{i=2}^{K} \mathsf{h}\left(h_{i1} x_1^n + z_i^n\right) \\ &= \sum_{i=2}^{K} \left(\lambda_i \mathsf{h}\left(x_1^n + z_1^n\right) - \mathsf{h}\left(h_{i1} x_1^n + z_i^n\right)\right),\end{aligned}$$

where the λ_is are nonnegative real numbers satisfying $\sum_{i=2}^{K} \lambda_i = 1$. If the condition (2.40) is satisfied, then we can choose λ_is satisfying

$$\lambda_i \geq \frac{h_{i1}^2 P_1 + h_{i1}^2}{h_{i1}^2 P_1 + 1}.$$

Observe that the condition (2.40) immediately implies that $|h_{i1}| \leq 1$ for each $i \in [K]$. Therefore, from Lemma A.11, it follows that the expression

$$\lambda_i \mathsf{h}\left(x_1^n + z_1^n\right) - \mathsf{h}\left(h_{i1} x_1^n + z_i^n\right)$$

is maximized by Gaussian inputs with maximum power. This completes the proof.

2.3.3 Scalar Genie

As mentioned in the introduction, the two-user genie provides each receiver with a noisy and interference-free observation of the desired signal. Generalizing this observation to the K-user case, we obtain the genie

$$s_k = x_k + w_k, \forall k \in [K],$$

where the genie controls how the Gaussian noise random variable w_i is correlated to z_i. We call this genie a *scalar genie* because it provides each receiver with a scalar signal. Shang et al. [23] derived the conditions under which there exists a scalar genie that is both useful and smart.

THEOREM 2.12 *The sum capacity of the K-user SISO Gaussian interference channel* (2.37) *is achieved by using Gaussian inputs and treating interference as noise, and is given by*

$$\mathcal{C}_{\text{sum}} = \sum_{i=1}^{K} \log\left(1 + \frac{P_i}{1 + \text{INR}_i}\right)$$

if the channel parameters satisfy the conditions

$$\begin{aligned} &\sum_{j \neq i} |h_{ij}|^2 \frac{(1+\text{INR}_j)^2}{\rho_j^2} + \rho_i^2 \leq 1, \forall i \in [K], \\ &\left(\sum_{i \neq j} \frac{|h_{ij}|^2}{\text{INR}_i + 1 - \rho_i^2}\right)\left(P_j + \frac{(1+\text{INR}_j)^2}{\rho_j^2}\right) \leq 1, \forall j \in [K] \end{aligned} \tag{2.41}$$

for some $\{\rho_i \in [0\ 1]\}_{i=1}^{K}$. *Here we use* INR_i *to denote the total interference-to-noise ratio at receiver i:*

$$\text{INR}_i = \sum_{j \neq i} |h_{ij}|^2 P_j.$$

We can verify that Theorem 2.12, when specialized to the many-to-one and the one-to-many interference channels, simplifies to Theorem 2.10 and Theorem 2.11, respectively. For the K-user symmetric interference channel obtained by setting $h_{ij} = h, \forall i \neq j$, and $P_j = P, \forall j$, Theorem 2.12 simplifies to the following corollary.

COROLLARY 2.3 *The sum capacity of the K-user symmetric Gaussian interference channel is achieved by using Gaussian inputs and treating interference as noise, and is given by*

$$\mathcal{C}_{\text{sum}} = K \log\left(1 + \frac{P}{1 + |\hat{h}|^2 P}\right)$$

if the channel parameters satisfy the low interference regime condition

$$|\hat{h}|(1 + |\hat{h}|^2 P) \leq 0.5,$$

where $\hat{h} = h\sqrt{K-1}$.

2.3.4 Vector Genie

We now explore an alternative way of generalizing the two-user genie to the K-user Gaussian interference channel. For simplicity, we restrict the presentation to the SISO case, but the genie construction can be extended to the MIMO case in a straightforward fashion. First, we provide the intuition behind the choice of our genie. Mathematically speaking, the reason for employing a genie is to combat interference. As explained in Appendix A, the positive differential entropy and conditional differential entropy terms are always maximized by i.i.d. Gaussian inputs (with maximum power), and are concave in the covariance matrices. On the other hand, the negative terms, which arise whenever there is interference, are minimized by i.i.d. Gaussian inputs, and are convex in the covariance matrices. Therefore, it is not clear in general if the sum of the positive and negative terms is maximized by i.i.d. Gaussian inputs or not. However, using the worst-case noise lemma (Lemma A.9), the negative terms can be shown to be maximized by i.i.d. Gaussian inputs, and concave in the covariance matrices, if they are coupled with appropriate positive terms. However, the worst-case noise lemma requires the positive terms to have the same signal structure in both the positive and negative terms. For this reason, the two-user genie in Section 2.2 was chosen as

$$\begin{aligned} s_1 &= \boldsymbol{H}_{21}\boldsymbol{x}_1 + \boldsymbol{w}_1, \\ s_2 &= \boldsymbol{H}_{12}\boldsymbol{x}_2 + \boldsymbol{w}_2, \end{aligned}$$

so that the genie signal s_1 provides the positive term to combat the interference seen at receiver 2, and the genie signal s_2 provides the positive term to combat the interference seen at receiver 1. Generalizing this idea, we can provide a signal similar to the interference seen at receiver $i-1$ as side information to receiver i:

$$s_i = \sum_{j \neq (i-1)} h_{i-1,j} x_j + w_{i-1} \qquad (\sim y_{i-1} | x_{i-1}).$$

For the three-user case, the corresponding genie signals are given by

$$\begin{aligned} s_1 &= h_{31}x_1 + h_{32}x_2 + w_1 \qquad (\sim y_3 | x_3), \\ s_2 &= h_{12}x_2 + h_{13}x_3 + w_2 \qquad (\sim y_1 | x_1), \\ s_3 &= h_{23}x_3 + h_{21}x_1 + w_3 \qquad (\sim y_2 | x_2). \end{aligned}$$

Unlike in the two-user case, the above construction does not suffice in the K-user case because the genie signals are not interference free. The genie signal s_{i+1} helps in combating the interference seen at receiver i, but in the process it also creates a new (negative) interference term at receiver $i+1$. We can fix this problem by repeating the above process $K-1$ times, which results in the following *vector genie*. For each receiver i, the genie signal $\boldsymbol{s}_i$ is a vector of length $K-1$:

$$s_{i\ell} = \sum_{j \notin \{i-1, i-2, \ldots, i-\ell\}} h_{i-\ell,j} x_j + w_{i\ell}, \ 1 \leq \ell \leq K-1.$$

Observe that $s_{i\ell}$ has the same structure as $y_{i-\ell}|x_{i-\ell},x_{i-\ell+1},\ldots,x_{i-1}$. For the three-user Gaussian interference channel, the vector genie is given by

$$\begin{aligned}
s_{11} &= h_{31}x_1 + h_{32}x_2 + w_{11} && (\sim y_3|x_3),\\
s_{12} &= h_{21}x_1 + w_{12} && (\sim s_{31}|x_3 \sim y_2|x_2x_3),\\
s_{21} &= h_{12}x_2 + h_{13}x_3 + w_{21} && (\sim y_1|x_1),\\
s_{22} &= h_{32}x_2 + w_{22} && (\sim s_{11}|x_1 \sim y_3|x_3x_1),\\
s_{31} &= h_{23}x_3 + h_{21}x_1 + w_{31} && (\sim y_2|x_2),\\
s_{32} &= h_{13}x_3 + w_{32} && (\sim s_{21}|x_2 \sim y_1|x_1x_2).
\end{aligned}$$

The genie controls how the Gaussian noise random vector $\mathbf{w}_i$ is correlated to z_i. As in Section 2.2, we use Ψ to denote the genie parameters collectively:

$$\Psi = \left(\boldsymbol{\Sigma}_{\mathbf{w}_1}, \boldsymbol{\Sigma}_{\mathbf{w}_1 z_1}, \boldsymbol{\Sigma}_{\mathbf{w}_2}, \boldsymbol{\Sigma}_{\mathbf{w}_2 z_2}, \ldots, \boldsymbol{\Sigma}_{\mathbf{w}_K}, \boldsymbol{\Sigma}_{\mathbf{w}_K z_K}\right).$$

2.3.5 Useful Genie

As in Section 2.2, we say that a genie is useful if the sum capacity of the genie-aided channel is achieved by using Gaussian inputs and treating interference as noise. The following lemma provides conditions on the genie parameters such that the genie is useful.

LEMMA 2.13 *If the genie parameters Ψ satisfy the usefulness conditions*

$$\begin{bmatrix} 1 & \boldsymbol{\Sigma}_{z_i \mathbf{w}_i} \\ \boldsymbol{\Sigma}_{\mathbf{w}_i z_i} & \boldsymbol{\Sigma}_{\mathbf{w}_i} \end{bmatrix} \succeq \begin{bmatrix} \boldsymbol{\Sigma}_{\mathbf{w}_{i-1}} & \mathbf{0}_{K-1\times 1} \\ \mathbf{0}_{1\times K-1} & 0 \end{bmatrix}, \forall i \in [K], \tag{2.42}$$

then the genie is useful; i.e., the sum capacity of the genie-aided interference channel is achieved by using Gaussian inputs with maximum power and treating interference as noise; i.e.,

$$\mathcal{C}_{\text{sum}} \le \mathcal{C}_{\text{sum}}^{\text{ga-ic}} = \sum_{i=1}^{K} \mathsf{I}\left(x_{iG}; y_{iG}, \mathbf{s}_{iG}\right),$$

where $x_{iG} \sim \mathcal{CN}(0,P_i)$, and y_{iG} and $\mathbf{s}_{iG}$ are the corresponding received signal and genie signal, respectively.

Proof The proof is similar to the proof of Theorem 2.4 in Section 2.2. See [18] for details.

2.3.6 Smart Genie

As in Section 2.2, we say that a genie is smart if the achievable sum-rate in the genie-aided channel is equal to the achievable sum-rate in the original interference channel. In this section, we derive conditions on the genie parameters such that the genie is smart. Before we present the smartness conditions, we summarize the discussion in Section 2.2.9 in the following lemma.

LEMMA 2.14 *Suppose $\boldsymbol{x}_{\mathrm{G}} \sim \mathcal{CN}(0, \boldsymbol{\Sigma}_{\boldsymbol{x}_{\mathrm{G}}})$ and $\boldsymbol{y}_{\mathrm{G}}$ and $\boldsymbol{s}_{\mathrm{G}}$ are noisy observations of $\boldsymbol{x}_{\mathrm{G}}$:*

$$\boldsymbol{y}_{\mathrm{G}} = \boldsymbol{H}_1 \boldsymbol{x}_{\mathrm{G}} + \boldsymbol{n}_1,$$
$$\boldsymbol{s}_{\mathrm{G}} = \boldsymbol{H}_2 \boldsymbol{x}_{\mathrm{G}} + \boldsymbol{n}_2,$$

where $\boldsymbol{n}_1$ and $\boldsymbol{n}_2$ are jointly circularly symmetric, and jointly Gaussian complex random vectors. Then,

$$\mathsf{I}\left(\boldsymbol{x}_{\mathrm{G}}; \boldsymbol{y}_{\mathrm{G}}, \boldsymbol{s}_{\mathrm{G}}\right) = \mathsf{I}\left(\boldsymbol{x}_{\mathrm{G}}; \boldsymbol{y}_{\mathrm{G}}\right)$$

if and only if the following condition is satisfied:

$$\left(\boldsymbol{H}_2 - \boldsymbol{\Sigma}_{\boldsymbol{n}_2\boldsymbol{n}_1} \boldsymbol{\Sigma}_{\boldsymbol{n}_1}^{-1} \boldsymbol{H}_1\right) \boldsymbol{\Sigma}_{\boldsymbol{x}_{\mathrm{G}}} = 0.$$

Proof The lemma follows immediately by replacing $\boldsymbol{x}_{1\mathrm{G}}, \boldsymbol{y}_{1\mathrm{G}}, \boldsymbol{s}_{1\mathrm{G}}, \boldsymbol{H}_{12}\boldsymbol{x}_{2\mathrm{G}} + z_1, \boldsymbol{w}_1$ in the discussion of Section 2.2.9 by $\boldsymbol{x}_{\mathrm{G}}, \boldsymbol{y}_{\mathrm{G}}, \boldsymbol{s}_{\mathrm{G}}, \boldsymbol{n}_1, \boldsymbol{n}_2$ respectively.

We now use the above lemma to derive the smartness conditions.

LEMMA 2.15 *The genie is smart, i.e.,*

$$\sum_{i=1}^{K} \mathsf{I}\left(x_{i\mathrm{G}}; y_{i\mathrm{G}}, \boldsymbol{s}_{i\mathrm{G}}\right) = \sum_{i=1}^{K} \mathsf{I}\left(x_{i\mathrm{G}}; y_{i\mathrm{G}}\right),$$

if the genie parameters Ψ satisfy the smartness conditions

$$\Sigma_{w_{i\ell} z_i} = h_{i-\ell,i}\left(1 + \mathrm{INR}_i\right) - \sum_{j \notin \{i, i-1, \ldots, i-\ell\}} h_{i-\ell,j} h_{ij}^{\dagger} P_j \tag{2.43}$$

for all $1 \le i \le K$ and $1 \le \ell \le K-1$. Recall that we use INR_i to denote the total interference-to-noise ratio at receiver i:

$$\mathrm{INR}_i = \sum_{j \neq i} |h_{ij}|^2 P_j.$$

Proof First, observe that the genie is smart if and only if

$$\mathsf{I}\left(x_{i\mathrm{G}}; y_{i\mathrm{G}}, \boldsymbol{s}_{i\mathrm{G}}\right) = \mathsf{I}\left(x_{i\mathrm{G}}; y_{i\mathrm{G}}\right), \forall i \in [K].$$

Recall that

$$y_{i\mathrm{G}} = x_{i\mathrm{G}} + \underbrace{\sum_{j \neq i} h_{ij} x_{j\mathrm{G}} + z_i}_{},$$

and that

$$s_{i\ell\mathrm{G}} = h_{i-\ell,i} x_{i\mathrm{G}} + \underbrace{\sum_{j \notin \{i, i-1, \ldots, i-\ell\}} h_{i-\ell,j} x_{j\mathrm{G}} + w_{i\ell}}_{}.$$

Using Lemma 2.14, we see that $\mathsf{I}(x_{iG}; y_{iG}, s_{iG}) = \mathsf{I}(x_{iG}; y_{iG})$ if and only if

$$h_{i-\ell,i} = \frac{\Sigma_{w_{i\ell}z_i} + \sum_{j \notin \{i,i-1,\ldots,i-\ell\}} h_{i-\ell,j} h_{ij}^\dagger P_j}{1 + \mathrm{INR}_i}$$

is satisfied for each $1 \le \ell \le K-1$.

2.3.7 Low Interference Regime

Combining Lemmas 2.13 and 2.15, we obtain the following theorem.

THEOREM 2.16 *The sum capacity of the K-user SISO Gaussian interference channel is achieved by using Gaussian inputs and treating interference as noise, and is given by*

$$C_{\text{sum}} = \sum_{i=1}^{K} \log\left(1 + \frac{P_i}{1 + \sum_{j \neq i} |h_{ij}|^2 P_j}\right)$$

if there exist genie parameters

$$\Psi = \left(\boldsymbol{\Sigma}_{w_1}, \boldsymbol{\Sigma}_{w_1 z_1}, \boldsymbol{\Sigma}_{w_2}, \boldsymbol{\Sigma}_{w_2 z_2}, \ldots, \boldsymbol{\Sigma}_{w_K}, \boldsymbol{\Sigma}_{w_K z_K}\right)$$

satisfying the usefulness conditions (2.42) *and the smartness conditions* (2.43).

2.3.8 Symmetric Interference Channel

In this section, we consider the K-user symmetric interference channel

$$y_i = x_i + h \sum_{j \neq i} x_j + z_i,$$

with symmetric power constraint, i.e., $P_j = P$, $\forall j \in [K]$, and simplify the conditions in Theorem 2.16. We restrict our attention to only symmetric genie parameters; i.e., we assume

$$\boldsymbol{\Sigma}_{w_i} = \boldsymbol{\Sigma}_{w},$$
$$\boldsymbol{\Sigma}_{w_i z_i} = \boldsymbol{\Sigma}_{wz}.$$

We now proceed to simplify the conditions in Theorem 2.16. The smartness conditions (2.43) determine the parameter $\boldsymbol{\Sigma}_{wz} = \boldsymbol{a}$, where

$$a_\ell = h(1 + (K-1)|h|^2 P) - (K - \ell - 1)|h|^2 P, \quad 1 \le \ell \le K-1. \tag{2.44}$$

Therefore, it remains to verify the existence of $\boldsymbol{\Sigma}_w \succeq 0$ satisfying the usefulness conditions

$$\begin{bmatrix} 1 & \boldsymbol{a}^\dagger \\ \boldsymbol{a} & \boldsymbol{\Sigma}_w \end{bmatrix} \succeq \begin{bmatrix} \boldsymbol{\Sigma}_w & \boldsymbol{0} \\ \boldsymbol{0} & 0 \end{bmatrix}.$$

Thus, we obtain the following theorem.

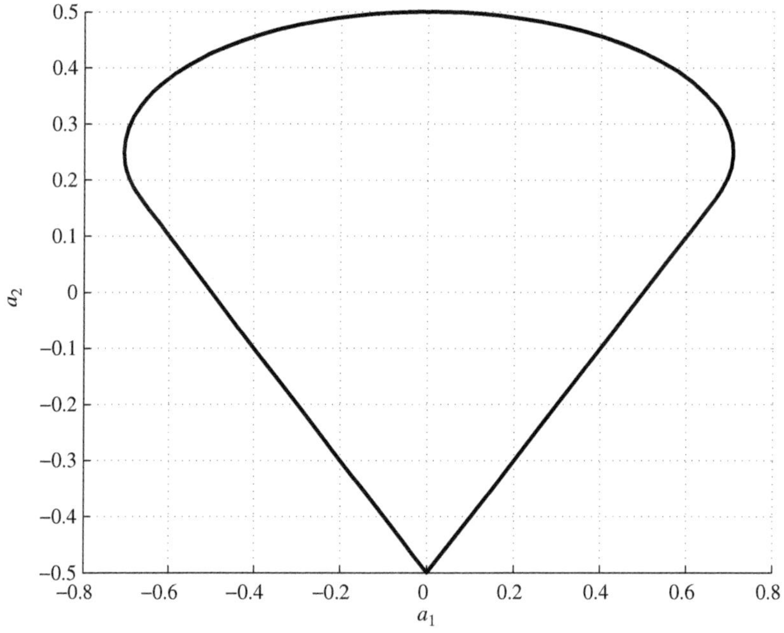

Fig. 2.9 The boundary of the feasible set $\mathcal{A}$, defined as the set of parameters (a_1,a_2) for which there exists $\mathbf{\Sigma} \succeq 0$ satisfying condition (2.45) in Theorem 2.17.

THEOREM 2.17 *The sum capacity of the three-user symmetric Gaussian interference channel is achieved by using Gaussian inputs and treating interference as noise, and is given by*

$$\mathcal{C}_{\text{sum}} = K \log\left(1 + \frac{P}{1+(K-1)|h|^2 P}\right)$$

if there exists a $(K-1)\times(K-1)$ *positive semidefinite matrix* $\mathbf{\Sigma}$ *satisfying the condition*

$$\begin{bmatrix} 1 & \boldsymbol{a}^\dagger \\ \boldsymbol{a} & \mathbf{\Sigma} \end{bmatrix} \succeq \begin{bmatrix} \mathbf{\Sigma} & \mathbf{0} \\ \mathbf{0} & 0 \end{bmatrix}, \tag{2.45}$$

where the vector $\boldsymbol{a}$ *is as defined in* (2.44).

We now assume that $K = 3$ and h is a real number, and determine the range of h such that the three-user symmetric Gaussian interference channel is in the low interference regime. Let $\mathcal{A}$ denote the set of (a_1,a_2) such that there exists $\mathbf{\Sigma} \succeq 0$ satisfying (2.45). Clearly, the set $\mathcal{A}$ must be convex. We would like to determine the implicit equations in a_1 and a_2 characterizing the boundary of the the set $\mathcal{A}$ so that the conditions in Theorem 2.17 can be simplified further. It is not clear if it is possible to do so. Since we assumed that h is a real number, we have that a_1 and a_2 are also real numbers. In Figure 2.9, we plot the boundary of the feasible set $\mathcal{A}$.

Using this region, we numerically determine the range of real h such that the three-user symmetric Gaussian interference channel is in the low interference regime.

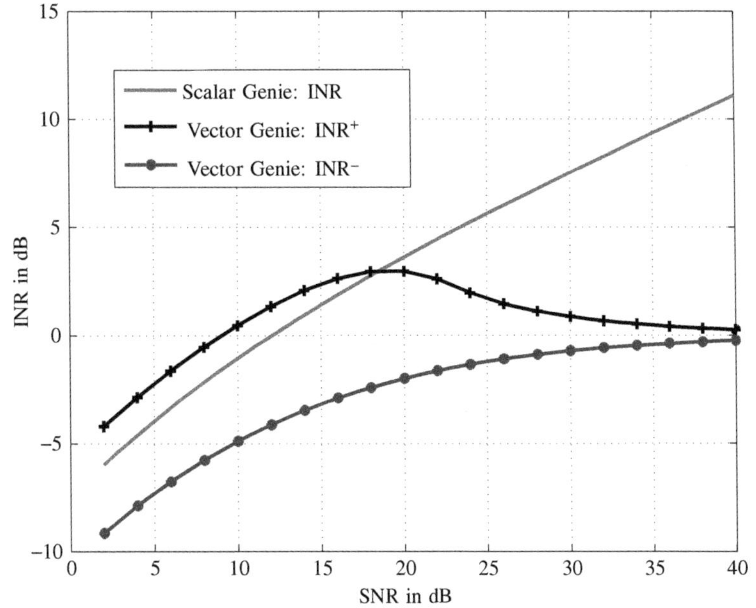

Fig. 2.10 The INR threshold as a function of the SNR, below which treating interference as noise achieves the sum capacity of the three-user symmetric Gaussian interference channel.

We observe that for every fixed $P \geq 0$, the channel is in the low interference regime if h satisfies the threshold criterion

$$-h_0^-(P) \leq h \leq h_0^+(P).$$

Interestingly, the positive and negative thresholds $h_0^+(P)$ and $h_0^-(P)$ are not equal. The positive threshold $h_0^+(P)$ is in general greater than the negative threshold $h_0^-(P)$. This is in contrast to the threshold criterion obtained by the scalar genie. Recall from Section 2.3.3 that the scalar genie provides a threshold criterion of the form

$$|h| \leq h_0(P),$$

where $h_0(P)$ is the (unique) positive solution to the equation

$$2\sqrt{2}h(1+2h^2P) = 1.$$

The results are summarized in Figure 2.10, where we plot the three curves corresponding to $\mathrm{INR}^+ = 2\left(h_0^+\right)^2 P$, $\mathrm{INR}^- = 2\left(h_0^-\right)^2 P$, and $\mathrm{INR} = 2h_0^2 P$ as a function of $\mathrm{SNR} = P$. It can be observed that neither the vector genie nor the scalar genie is strictly better compared to the other.

2.3.9 *K*-User Interference Channel – Summary

We extended the sufficient conditions for the low interference regime, presented in the previous section for the two-user case, for the K-user case. We proposed the vector genie construction in order to obtain tight upper bounds on the sum capacity of the K-user SISO Gaussian interference channel. The advantage of the vector genie construction is that it is intuitive, and can be easily generalized to the MIMO case. Following the steps in the previous section, such a generalization can be used to obtain a convex maxmin optimization problem that facilitates the numerical computation of the optimal covariance matrices in the low interference regime. The disadvantage of the vector genie construction is that, in the symmetric case, the low interference regime condition depends on the phase of the cross-channel coefficient, which is counterintuitive. Also, the low interference regime condition obtained by the vector genie is not uniformly better than that obtained by the scalar genie. A hybrid genie construction combining the good features of the scalar and vector genies may lead to further insights on the optimality of treating interference as noise in the K-user interference channel. An interesting question that remains to be answered is: *How does the optimal interference threshold scale as a function of the number of users in the symmetric K-user Gaussian interference channel?*

2.4 Notes and Further Reading

The study of the two-user interference channel was initiated by Shannon in 1961 [24]. Carleial made an interesting and counterintuitive observation that interference does not reduce the capacity of the two-user SISO Gaussian interference channel in the *very strong interference* regime [25]. If the interference level is very strong compared to the signal level, the receivers can first decode the interfering message to subtract its contribution from the received signal, thus achieving the same rate as if there was no interference. Subsequently, the capacity region was determined in the *strong interference* regime [26, 27], where interference reduces the capacity, but, like in the very strong interference regime, the optimal strategy requires the receivers to again decode the interfering message. Characterizing the capacity region of the two-user SISO Gaussian interference channel in the general setting still remains an open problem. The best known achievable region is based on the Han–Kobayashi (HK) scheme [26, 28]. The HK region was shown to be the capacity region for a class of discrete-memoryless deterministic interference channels in 1982 [29]. Little was known about the optimality of the HK region in the Gaussian case until more recently. The concept of a genie giving side information to the receivers was used in [30, 31] to derive outer bounds on the capacity region of the two-user SISO Gaussian interference channel. The outer bound region in [31] is within one bit of the HK region, thus leading to an approximate characterization of the capacity region of the two-user SISO Gaussian interference channel. The genie-based outer bound technique was further extended in [18–20] to prove that treating interference as noise achieves the sum capacity in a low interference regime (referred to as the noisy interference regime in [19]).

The two-user MIMO Gaussian interference channel was studied in [17, 32–37] from the point of view of determining the best achievable rate region obtained by using Gaussian inputs and treating interference as noise. Several iterative algorithms including the water-filling algorithm [32, 33], the gradient projection algorithm [17], the interference pricing algorithm [34], and the Max-SINR algorithm [38, 39] were proposed to find good lower bounds for the best achievable sum-rate. For the MISO Gaussian interference channel, it was proved in [35–37] that rank-one covariance matrices are optimal. However, the problem of determining the best achievable sum-rate remains open and is known to be a difficult problem even in the SISO case [40, 41] due to the nonconvex nature of the sum-rate.

The two-user MIMO Gaussian interference channel was studied in [42–44] from the point of view of determining the capacity region. In [43], the authors showed that the sum DoF of the MIMO Gaussian interference channel is equal to

$$\min\left(N_{1t}+N_{2t}, N_{1r}+N_{2r}, \max(N_{1t}, N_{2r}), \max(N_{2t}, N_{1r})\right),$$

and that the optimal sum DoF is achieved by treating interference as noise. In [44], the authors extended the approximate capacity characterization of the two-user SISO Gaussian interference channel in [31] to the MIMO case.

Unlike the two-user interference channel where one user causes interference at each receiver, the K-user interference channel has multiple interfering signals at each receiver. This turns out to be a key bottleneck in extending the approximate capacity characterization result in [31, 44] for the general K-user case. In [21, 22], the special cases of many-to-one and one-to-many interference channels are studied. In the many-to-one channel, only one user experiences interference, and in the one-to-many channel, only one user causes interference. The approximate capacity region of these two special cases is characterized in [21, 22]. For the fully connected interference channel, only the DoF is known. The approximate capacity characterization for the two-user case is not yet extended for the general K-user case. The sum-rate algorithms such as the water-filling algorithm [32, 33] and the gradient projection algorithm [17] are designed for the K-user interference channel.

3 Degrees of Freedom and Interference Alignment

In Chapter 2, we studied bounds on the sum-rate (throughput) of K-user interference channels in both single- and multiple-antenna settings, and characterized the sum capacity in some special cases. In particular, we observed that it was difficult to obtain the exact sum capacity, except in a *low interference* regime, where the interference level is below a certain threshold at each receiver. In this chapter, we take an alternative approach to characterizing the rate of communication on an interference channel, based on a *degrees of freedom* analysis.

3.1 Degrees of Freedom

Recall from Section 2.1.1 that the *capacity region* $\mathcal{C}$ of a K-user interference channel is defined as the closure of the set of achievable rate tuples $(R_1, R_2, \ldots, R_K)$. Roughly speaking, the degrees of freedom (DoF) region is equal to the capacity region scaled by $\log \text{SNR}$ at high SNR (signal-to-noise ratio). More specifically, we say that the DoF tuple $(d_1, d_2, \ldots, d_K)$ is achievable if there exists an achievable rate tuple $(R_1(P_1), R_2(P_2), \ldots, R_K(P_K))$ such that

$$d_k = \limsup_{P_k \to \infty} \frac{R_k(P_k)}{\log P_k}, \forall k \in [K], \tag{3.1}$$

where P_k is the power constraint for user k. The DoF region $\mathcal{D}$ is defined as the closure of the set of achievable DoF tuples.

The sum DoF is defined as

$$\eta = \max_{(d_1, d_2, \ldots, d_k) \in \mathcal{D}} d_1 + d_2 + \cdots + d_K. \tag{3.2}$$

Unless otherwise noted, when we refer to the DoF of an interference channel in this book, we will be referring to the sum DoF. Note that the DoF captures the number of interference-free sessions that can be supported simultaneously in an interference channel.

The DoF criterion provides an approach to studying the fundamental limits of communication on an interference channel, with some immediate advantages. First, by considering the limit as the SNR goes to infinity, the DoF measure emphasizes the loss in rate that results from the users interfering with each other over the effect of channel noise. Secondly, the DoF analysis does not require an explicit calculation of the channel capacity, and is therefore analytically tractable.

As an example, consider a basic strategy for communicating on the interference channel using time-division multiple-access (TDMA). In this approach, each unit of time is divided into K equal slots and each user is allowed to communicate during one slot. As a result, each user achieves $\frac{1}{K}$ degrees of freedom; this can be seen more formally using the definition of DoF given in (3.1) as follows:

$$d_k^{\text{TDMA}} = \lim_{P\to\infty} \frac{R_k(P)}{\log P} = \lim_{P\to\infty} \frac{\frac{1}{K}\log(1+KP)}{\log(P)} = \frac{1}{K},$$

where we have assumed an average power constraint over each unit of time of P for each user. Therefore, a sum DoF equal to one can be achieved by TDMA, which implies the lower bound

$$\eta \geq 1. \tag{3.3}$$

Now the natural question to ask is whether η can be larger than one for a K-user interference channel.

The first part of the answer to this question comes from an upper bound on η given in [45]. This work demonstrated that for a two-user interference channel, the maximum possible sum DoF is one. This result, combined with the lower bound in (3.3), yields that for the two-user interference channel $\eta = 1$. With this result, by considering each pair of users in a K-user interference channel with $K > 2$, it can be seen that η can be no greater than $\frac{K}{2}$. This follows since, for any pair of users $k \neq \ell$,

$$d_k + d_\ell \leq 1$$

by [45]. Then, summing over all such equations gives

$$\sum_{k=1}^{K}\sum_{\substack{\ell=1\\ \ell\neq k}}^{K} (d_k + d_\ell) \leq K(K-1). \tag{3.4}$$

Each transmitter/receiver appears in exactly $2(K-1)$ pairs, and therefore (3.4) is equivalent to

$$2(K-1)\eta \leq K(K-1).$$

Thus, it follows that

$$\eta \leq \frac{K}{2}. \tag{3.5}$$

It was conjectured in [45] that the upper bound is not tight and that in fact η equals one. It seems likely that the $\frac{K}{2}$ bound would be loose, since it only considers pairs of users at a time, and including more than two users at a time could lead to tighter bounds. Surprisingly, though, it was proved in [46] that the upper bound is in fact tight, and that

$$\eta = \frac{K}{2}.$$

The result is based on an innovative technique called *interference alignment*, which we describe in detail in the following section.

3.2 Vector Space Interference Alignment for a Fully Connected Interference Channel

To understand the concept of interference alignment [46], consider first the special case of a three-user SISO interference channel. Interference alignment is facilitated through the use of *symbol extensions*, i.e., through communicating over n different channel uses and letting $n \to \infty$ as in Shannon's random block coding argument. A key assumption that is needed here is the following:

ASSUMPTION 3.1 The channel coefficients change from one channel use to the next, and the collection of all the channel coefficients over time comes from a continuous well-defined (non-degenerate) joint distribution.

Interference alignment is a linear coding strategy in signal space, i.e., it is achieved using linear precoding at the transmitters and linear processing at the receivers. Denote the linear precoding matrices at the transmitters by $\boldsymbol{V}_k$ and the linear processing matrices at the receivers by $\boldsymbol{U}_k$, $k \in [K]$. With n symbol extensions the end-to-end system model can be written as:

$$\begin{bmatrix} \boldsymbol{U}_1 & \mathbf{0} & \mathbf{0} \\ \mathbf{0} & \boldsymbol{U}_2 & \mathbf{0} \\ \mathbf{0} & \mathbf{0} & \boldsymbol{U}_3 \end{bmatrix}^{\dagger} \begin{bmatrix} \boldsymbol{H}_{11} & \boldsymbol{H}_{12} & \boldsymbol{H}_{13} \\ \boldsymbol{H}_{21} & \boldsymbol{H}_{22} & \boldsymbol{H}_{23} \\ \boldsymbol{H}_{31} & \boldsymbol{H}_{32} & \boldsymbol{H}_{33} \end{bmatrix} \begin{bmatrix} \boldsymbol{V}_1 & \mathbf{0} & \mathbf{0} \\ \mathbf{0} & \boldsymbol{V}_2 & \mathbf{0} \\ \mathbf{0} & \mathbf{0} & \boldsymbol{V}_3 \end{bmatrix}, \tag{3.6}$$

where $\boldsymbol{U}_k$ is a $d_k \times n$ matrix, $\boldsymbol{V}_k$ is an $n \times d_k$ matrix, and $\boldsymbol{H}_{kj}$ is an $n \times n$ diagonal matrix of the form

$$\boldsymbol{H}_{kj} = \begin{bmatrix} h_{kj}(1) & & 0 \\ & \ddots & \\ 0 & & h_{kj}(n) \end{bmatrix}.$$

Then the idea behind interference alignment is to find $\boldsymbol{V}_k$ and $\boldsymbol{U}_k$, $k \in [K]$, such that the end-to-end model is diagonal. If the end-to-end matrix is diagonal, then at each receiver the desired signals lie in a subspace separate from the interfering signals. That is to say, the interfering signals are *simultaneously* aligned at each receiver in an interference subspace. So if it is possible to diagonalize (3.6), then it is possible to achieve a sum DoF of

$$\frac{1}{n} \sum_{k=1}^{3} d_k.$$

The following simple example shows that is possible to achieve a sum DoF of $\frac{4}{3}$, which is larger than the DoF achieved by TDMA, for the three-user IC with three symbol extensions ($n = 3$).

1. Transmitter 2 selects any transmit vector $\boldsymbol{v}_2$, and is received as $\boldsymbol{H}_{k2}\boldsymbol{v}_2$ at receiver k, $k \in \{1,2,3\}$. This is represented in Figure 3.1(a).
2. Transmitter 3 selects a transmit vector $\boldsymbol{v}_3$ such that $\boldsymbol{H}_{12}\boldsymbol{v}_2$ is in the subspace spanned by $\boldsymbol{H}_{13}\boldsymbol{v}_3$ at receiver 1, e.g., we can set

$$\boldsymbol{v}_3 = \boldsymbol{H}_{13}^{-1}\boldsymbol{H}_{12}\boldsymbol{v}_2.$$

   When this condition is satisfied, $\boldsymbol{v}_2$ is said to be aligned with $\boldsymbol{v}_3$ at receiver 1. This is always possible due to Assumption 3.1. Again by Assumption 3.1, it follows that $\boldsymbol{H}_{22}\boldsymbol{v}_2$ is not in the subspace spanned by $\boldsymbol{H}_{23}\boldsymbol{v}_3$ at receiver 2. Similarly, $\boldsymbol{H}_{33}\boldsymbol{v}_3$ is not in the subspace spanned by $\boldsymbol{H}_{32}\boldsymbol{v}_2$ at receiver 3. Up to this point we have used up one dimension at receiver 1, and two dimensions at receivers 2 and 3. This is represented in Figure 3.1(b).
3. Now, transmitter 1 selects a vector $\boldsymbol{v}_{11}$ that is aligned with $\boldsymbol{v}_3$ at receiver 2, e.g., by setting

$$\boldsymbol{v}_{11} = \boldsymbol{H}_{21}^{-1}\boldsymbol{H}_{23}\boldsymbol{v}_3.$$

   Note that $\boldsymbol{v}_{11}$ appears in an arbitrary direction at receivers 1 and 3, as shown in Figure 3.1(c). Up to this point we have used two dimensions at receivers 1 and 2, and three dimensions at receiver 3.
4. To complete the procedure, transmitter 1 chooses a second transmit vector $\boldsymbol{v}_{12}$, such that $\boldsymbol{v}_{12}$ is aligned with $\boldsymbol{v}_2$ at receiver 3, e.g., by setting

$$\boldsymbol{v}_{12} = \boldsymbol{H}_{31}^{-1}\boldsymbol{H}_{32}\boldsymbol{v}_2.$$

   Note that $\boldsymbol{v}_{11}$ appears in an arbitrary direction at receivers 1 and 2, as shown in Figure 3.1(c). At this point all three dimensions are used at all receivers.
5. The interference occupies two dimensions at receivers 2 and 3, with the signal occupying one dimension. Therefore, zero-forcing vectors $\boldsymbol{u}_2$ and $\boldsymbol{u}_3$ can be chosen at receivers 2 and 3, respectively, to obtain one interference-free dimension at each of these receivers. Similarly, zero-forcing vectors $\boldsymbol{u}_{11}$ and $\boldsymbol{u}_{12}$ can be chosen at receiver 1 to obtain two interference-free dimensions at receiver 1.

A total of four interference-free dimensions are obtained across the three users in three channel uses, which results in a sum DoF of $\frac{4}{3}$. It is shown in [46] that the above procedure can be extended to the case of an arbitrary number of symbol extensions n, with the sum DoF approaching $\frac{3}{2}$ as $n \to \infty$. For a K-user interference channel, a more systematic way to establish that a sum DoF equal to $\frac{K}{2}$ can be achieved is given in the following section.

3.3 Asymptotic Interference Alignment

In order to establish that a sum DoF equal to $\frac{K}{2}$ can be achieved for a K-user interference channel with a symbol extension length of n, we need to show that at each receiver (on average) the interference is aligned to occupy only $\frac{n}{2}$ dimensions, with the remaining

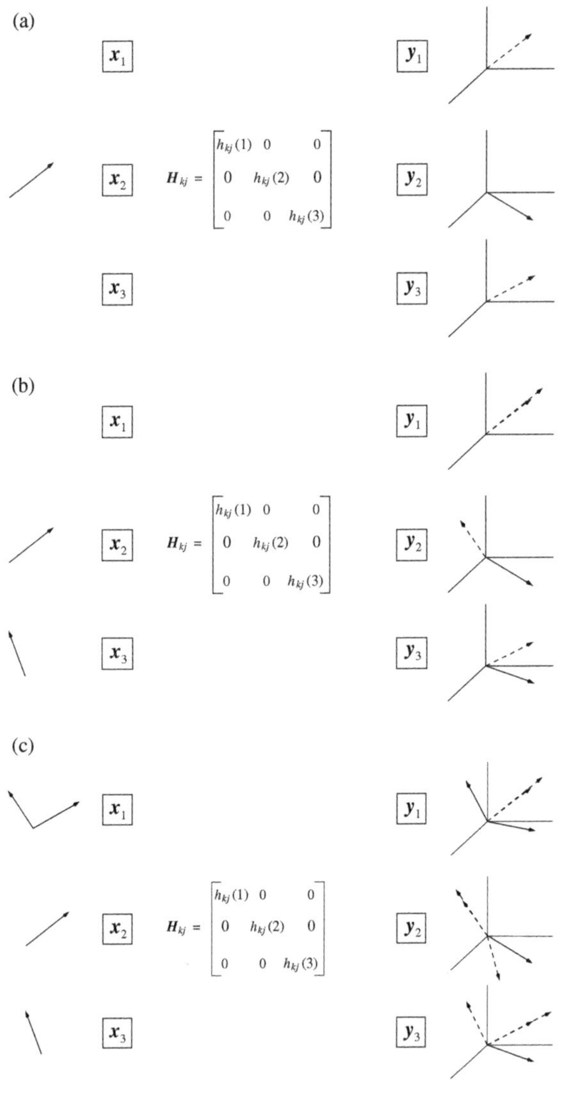

Fig. 3.1 Achieving a sum DoF of $\frac{4}{3}$ in a three-user interference channel.

$\frac{n}{2}$ dimensions being available for the intended signal. This may not be possible for any fixed n, for an arbitrary choice of the channel coefficients. We now show that a sum DoF approaching $\frac{K}{2}$ can be achieved as $n \to \infty$ using the technique introduced in [46].

First note that all the channel matrices $\{\boldsymbol{H}_{kj}\}$ are diagonal matrices and therefore products of subsets of these matrices satisfy the commutativity property. For convenience, we index all the cross channel matrices $\{\boldsymbol{H}_{kj} : k \neq j\}$ by a single index. Denote the re-indexed matrices by $\{G_\ell\}$, and let $L = K(K-1)$ be the number of cross channel matrices. Then, the set of cross channel matrices is given by:

$$\mathcal{G} = \{\boldsymbol{G}_\ell : \ell = 1,2,\ldots,L\} = \{\boldsymbol{H}_{kj} : k,j \in \{1,2,\ldots,K\}, j \neq k\}. \tag{3.7}$$

We use the same signal set at all the transmitters, i.e., the columns of the linear precoding matrices $\boldsymbol{V}_k$ are the same for all k. In order to define the signal space corresponding to a certain symbol extension length, we use an intermediate integer index variable m, and define the symbol extension length in terms of m as:

$$n(m) = \binom{m+L}{m} + \binom{m+1+L}{m+1}. \tag{3.8}$$

This choice of $n(m)$ will become clear later in the analysis of the DoF. The signal set used at the transmitters at stage m is defined recursively as:

$$\mathcal{V}^{(0)} = \{\boldsymbol{v}_0\},$$
$$\mathcal{V}^{(j)} = \left\{\boldsymbol{v}, \boldsymbol{G}_1\boldsymbol{v}, \ldots, \boldsymbol{G}_L\boldsymbol{v} \ : \ \boldsymbol{v} \in \mathcal{V}^{(j-1)}\right\}, j = 1, \ldots, m,$$

where $\boldsymbol{v}_0$ is an arbitrarily chosen nonzero $n(m)$-dimensional vector, and the matrices $\{G_\ell\}$ are $n(m) \times n(m)$ diagonal matrices as constructed in (3.7). Now we exploit the fact that the matrices $\{G_\ell\}$ are diagonal and that their products commute, to rewrite $\mathcal{V}^{(m)}$ as:

$$\mathcal{V}^{(m)} = \left\{\boldsymbol{G}_1^{j_1}\boldsymbol{G}_2^{j_2}\cdots\boldsymbol{G}_L^{j_L}\boldsymbol{v}_0 : j_1 + j_2 + \cdots + j_L \leq m\right\}. \tag{3.9}$$

By Assumption 3.1, the vectors in $\mathcal{V}^{(m)}$ are linearly independent (with probability one), and therefore the dimension of the signal space at each transmitter is equal to the number of vectors in $\mathcal{V}^{(m)}$, which can be obtained by determining the number of combinations of nonnegative integers that sum to a value no greater than m. A simple counting argument yields that

$$\left|\mathcal{V}^{(m)}\right| = \binom{m+L}{m}.$$

Now let us consider the spaces occupied by the desired signal and interference at receiver k. Since the vectors corresponding to the desired signal are obtained by pre-multiplying each vector in $\mathcal{V}^{(m)}$ by $\boldsymbol{H}_{kk}$, the dimension of the desired signal space at the receiver is given by

$$\left|\boldsymbol{H}_{kk}\mathcal{V}^{(m)}\right| = \left|\mathcal{V}^{(m)}\right| = \binom{m+L}{m}.$$

The vectors corresponding to the interference at receiver k are obtained by pre-multiplying each vector in $\mathcal{V}^{(m)}$ by each vector in $\mathcal{G}$. It is then easy to see from (3.9) that the interference space at the receiver is the span of the vectors in $\mathcal{V}^{(m+1)}$. Therefore, the dimension of the interference space at receiver k is given by

$$\left|\mathcal{V}^{(m+1)}\right| = \binom{m+1+L}{m+1}.$$

Also, by Assumption 3.1, there is no overlap between the desired signal and interference subspaces at the receiver, and hence one can obtain the desired signal without interference at the receiver through zero-forcing linear processing $\boldsymbol{U}_k$ at receiver k (at high SNR).

We now calculate the DoF that results from choosing the signal set $\mathcal{V}^{(m)}$ at each transmitter. The DoF achieved for each user is given by the ratio

$$\frac{|\mathcal{V}^{(m)}|}{|\mathcal{V}^{(m)}| + |\mathcal{V}^{(m+1)}|} = \frac{m+1}{2(m+1)+L}. \tag{3.10}$$

As $m \to \infty$, this ratio converges to $\frac{1}{2}$, and therefore the sum DoF converges to $\frac{K}{2}$, thus achieving the upper bound on sum DoF in (3.5) asymptotically.

3.3.1 Discussion

To conclude this section, we make some remarks about the asymptotic interference alignment scheme that achieves the sum DoF of $\frac{K}{2}$:

- The scheme requires the knowledge of all the cross channel gain matrices, i.e., the set $\mathcal{G}$ defined in (3.7), at all the transmitters and receivers, in addition to requiring knowledge of the desired signal channel gain matrix at each receiver. The training required to obtain this global channel knowledge can become prohibitive even for moderate values of K. Iterative schemes for alignment as discussed in Chapter 4 can get around this requirement to some extent. Also, when we allow for cooperation among the transmitters, interference avoidance schemes such as those discussed in Chapter 6 can be used to achieve sum degrees of freedom that scale linearly with K without the burden of global channel knowledge.
- The scheme requires the channel to change from symbol to symbol over the symbol extension period so that Assumption 3.1 holds. Also, the fact that the channel matrices are diagonal is crucial to the construction of the achievable scheme. We discuss this issue further in Section 3.4 (see also [47]).
- The number of symbol extensions required by the scheme to approach a sum DoF of $\frac{K}{2}$ grows exponentially in $L = K(K-1)$, and becomes impractical even for small values of K. For example, with $K = 4$, in order to achieve a sum DoF of 1.8, which is 90 percent of $\frac{K}{2}$, we see from (3.10) that the required value of m equals 6, and the corresponding value for the number of symbol extensions n as given in (3.8) exceeds 6×10^4 (see also [48]). Also, see Section 3.4.2 below for a more detailed analysis of the DoF achieved as a function of the number of symbol extensions (channel diversity).

3.4 Interference Alignment Bounds with Finite Diversity

In the previous section we studied an interference alignment scheme based on symbol extensions that achieves the sum DoF of $\frac{K}{2}$ asymptotically, as the number of symbol extensions, or equivalently the channel diversity, goes to infinity. In practice, we would have to work with a finite amount of diversity through symbol extensions in time/frequency, or through the use of multiple antennas at the transmitters and receivers. In this section, we investigate the best achievable DoF for a given finite level of channel diversity. We look at the cases of spatial diversity (obtained through multiple antennas) and time/frequency diversity separately, since in the latter case the channel matrices are diagonal, as we saw in the previous section, whereas in the former case they may not be.

3.4.1 Constant MIMO Channel

Consider first the case of spatial diversity for which we have the MIMO interference channel model depicted in Figure 2.1, where the output $\boldsymbol{y}_k \in \mathbb{C}^{N_\mathrm{r}}$ at receiver k is given by

$$\boldsymbol{y}_k = \sum_{j=1}^{K} \boldsymbol{H}_{kj}\boldsymbol{x}_j + \boldsymbol{z}_k, \ \forall k \in [K], \tag{3.11}$$

where $\boldsymbol{x}_j \in \mathbb{C}^{N_\mathrm{t}\times 1}$ denotes the signal of transmitter j, $\boldsymbol{z}_i \in \mathcal{CN}\left(\boldsymbol{0}, \boldsymbol{I}_{N_\mathrm{r}}\right)$ denotes the additive white Gaussian noise at receiver i, and $\boldsymbol{H}_{kj} \in \mathbb{C}^{N_\mathrm{r}\times N_\mathrm{t}}$ denotes the channel transfer matrix from transmitter j to receiver k. Each transmitter is assumed to have N_t transmit antennas, and each receiver is assumed to have N_r receive antennas. Note that this model could be generalized to allow for different numbers of antennas at the various transmitters and receivers, but we avoid this generalization to simplify the presentation. We assume that the channels $\{\boldsymbol{H}_{kj}\}$ are drawn once from a continuous joint distribution, without any symbol extensions, hence the name *constant MIMO channel.*

As in the previous section, we restrict to linear transmit and receive strategies where we pre-multiply the channel input $\boldsymbol{x}_j$ by an $N_\mathrm{r} \times d_j$ pre-coding matrix $\boldsymbol{V}_j$ and pre-multiply the channel output $\boldsymbol{y}_k$ by $\boldsymbol{U}_k^\dagger$, where $\boldsymbol{U}_k$ is an $N_\mathrm{t} \times d_k$ matrix, to obtain:

$$\boldsymbol{U}_k^\dagger \boldsymbol{y}_k = \boldsymbol{U}_k^\dagger \sum_{j=1}^{K} \boldsymbol{H}_{kj}\boldsymbol{V}_j\boldsymbol{x}_j + \boldsymbol{U}_k^\dagger \boldsymbol{z}_k.$$

The goal of interference alignment is to construct matrices $\{\boldsymbol{V}_k\}$ and $\{\boldsymbol{U}_k\}$ such that the receivers can zero-force the interference terms, yielding interference-free effective channels between each transmit/receive pair, with d_k interference-free streams for the kth transmit–receive pair.

To zero-force the interference at receiver k, we require

$$\boldsymbol{U}_k^\dagger \boldsymbol{H}_{kj}\boldsymbol{V}_j = \boldsymbol{0}, \ \forall k \neq j. \tag{3.12}$$

After zero-forcing, the effective channel seen by the kth transmit–receive pair is given by

$$\boldsymbol{U}_k^\dagger \boldsymbol{y}_k = \boldsymbol{U}_k^\dagger \boldsymbol{H}_{kk}\boldsymbol{V}_k\boldsymbol{x}_k + \boldsymbol{U}_k^\dagger \boldsymbol{z}_k.$$

This is a (single-user) MIMO channel with channel matrix $\boldsymbol{U}_k^\dagger \boldsymbol{H}_{kk}\boldsymbol{V}_k$, and therefore in order to create d_k interference-free channels, we require

$$\operatorname{rank}(\boldsymbol{U}_k^\dagger \boldsymbol{H}_{kk}\boldsymbol{V}_k) = d_k, \ \forall k. \tag{3.13}$$

Note that the transmit and receive matrices $\{\boldsymbol{V}_k\}$ and $\{\boldsymbol{U}_k\}$ are to be chosen to satisfy (3.12) using only the cross channel matrices $\{\boldsymbol{H}_{kj}\}_{k\neq j}$. If indeed (3.12) can be satisfied, this fact combined with our assumption that the channel matrices $\{\boldsymbol{H}_{kj}\}$ are drawn from a valid (non-degenerate) joint distribution, yields that (3.13) is automatically satisfied (see [38, 39] for a formal argument).

Therefore, the problem of interference alignment design boils down to constructing matrices $\{\boldsymbol{V}_k\}$ and $\{\boldsymbol{U}_k\}$ such that (3.12) is satisfied. Ideally one would also like to find $\{\boldsymbol{V}_k\}$ and $\{\boldsymbol{U}_k\}$ such that the sum DoF, which is equivalent to $\sum_{k\in[K]} d_k$, is maximized. There have been a number of papers (see, e.g., [49–53]) on finding necessary and sufficient conditions for the feasibility of solving the interference alignment equations in (3.12). It is relatively easy to derive necessary conditions for solving the interference alignment equations using variable and equation counting. For some specific system configurations, it is possible to derive sufficient conditions using tools from algebraic geometry [54–56].

To illustrate the necessary conditions, consider a symmetric constant MIMO interference channel with $N_\mathrm{t} = N_\mathrm{r} = N$ and $d_k = d$. In order to solve the interference alignment equations in (3.12), the number of variables should be no fewer than the number of equations; this intuitive result can be proved rigorously using algebraic geometry (see, e.g., [55]). The zero interference equation for transmitter j at receiver k is given by

$$\boldsymbol{U}_k^\dagger \boldsymbol{H}_{kj} \boldsymbol{V}_j = \boldsymbol{0},$$

and consists of d^2 equations. Therefore, the total number of equations to be satisfied to ensure zero interference at all receivers is:

$$\#\,\text{equations} = K(K-1)d^2. \tag{3.14}$$

Counting the number of variables is a little bit trickier. A direct count yields Nd variables in each of $\boldsymbol{U}_k$ and $\boldsymbol{V}_k$, and therefore a total of $2KNd$ variables. However, the conditions for satisfying the interference alignment equations only depend on the subspaces spanned by the columns of $\{\boldsymbol{U}_k\}$ and $\{\boldsymbol{V}_k\}$, and not the particular choices of $\{\boldsymbol{U}_k\}$ and $\{\boldsymbol{V}_k\}$. To illustrate this point, suppose that $\boldsymbol{U}_k^\dagger \boldsymbol{H}_{kj} \boldsymbol{V}_j = \boldsymbol{0}, \forall k \neq j$. For each $k \in [K]$, if we represent the columns of $\boldsymbol{U}_k$ and $\boldsymbol{V}_k$ using other bases, then the resulting matrices $\{\tilde{\boldsymbol{V}}_k\}$ and $\{\tilde{\boldsymbol{U}}_k\}$ satisfy

$$\tilde{\boldsymbol{V}}_k = \boldsymbol{V}_k \boldsymbol{Q}_k, \tag{3.15}$$

$$\tilde{\boldsymbol{U}}_k = \boldsymbol{U}_k \boldsymbol{P}_k, \tag{3.16}$$

where $\{\boldsymbol{Q}_k\}$ and $\{\boldsymbol{P}_k\}$ are $d \times d$ full rank matrices. Then we have

$$\tilde{\boldsymbol{U}}_k^\dagger \boldsymbol{H}_{kj} \tilde{\boldsymbol{V}}_j = \boldsymbol{P}_k^\dagger \underbrace{\boldsymbol{U}_k^\dagger \boldsymbol{H}_{kj} \boldsymbol{V}_j}_{=\boldsymbol{0}} \boldsymbol{Q}_j = \boldsymbol{0}.$$

Since satisfying the interference alignment equations only depends on subspaces spanned by $\{\boldsymbol{V}_k\}$ and $\{\boldsymbol{U}_k\}$, the number of unknown variables that we need to specify for alignment is less than the direct count of $2KNd$. By the preceding argument, due to freedom in choosing $\boldsymbol{P}_k$ and $\mathbf{Q}_k$, in fact we have

$$\dim(\mathrm{span}(\boldsymbol{V}_k)) = \dim(\mathrm{span}(\boldsymbol{U}_k)) = Nd - d^2 = (N-d)d.$$

This implies that

$$\# \text{ variables} = 2K(N-d)d. \tag{3.17}$$

Setting the number of variables in (3.17) to be no fewer than the number of equations in (3.14), we get that the necessary condition for interference alignment in a symmetric constant MIMO interference channel becomes

$$2K(N-d)d \geq K(K-1)d^2,$$

which can be simplified to

$$d \leq \frac{2N}{K+1}. \tag{3.18}$$

This implies that the achievable sum DoF across the K users is bounded by

$$\text{Sum DoF} \leq \frac{2KN}{K+1} \leq 2N. \tag{3.19}$$

This necessary condition (3.18) is not always sufficient; as shown in [49], with $N=3$, $d=2$, and $K=2$, the system satisfies (3.18) but the corresponding DoF is not achievable. Interestingly, it was shown in [53] that for the symmetric constant MIMO channel with $K \geq 3$, the necessary condition (3.18) is also sufficient. Necessary and sufficient conditions for feasibility of interference alignment for a number of other configurations are given in [50–52]. However, it should be noted that the conditions for the feasibility of solving the interference alignment equations for the constant MIMO interference channel are not known in general.

3.4.2 Finite Time/Frequency Diversity

When channel diversity is obtained in time/frequency through n symbol extensions, as we saw earlier, the channel matrix between transmitter j and receiver k is given by an $n \times n$ diagonal matrix of the form

$$\boldsymbol{H}_{kj} = \begin{bmatrix} H_{kj}(1) & & 0 \\ & \ddots & \\ 0 & & H_{kj}(n) \end{bmatrix}.$$

It can be shown that the asymptotic interference alignment scheme discussed in Section 3.3 requires n to be of the order of K^{K^2} (see the example in Section 3.3.1) in order to approach the sum DoF of $\frac{K}{2}$ (this scaling can be improved slightly to 2^{K^2} by a modification of the scheme [57]). In particular, the achievable DoF of the scheme described in Section 3.3 can be shown to scale as [58]

$$\text{Sum DoF} \geq \frac{K}{2}\left(1 - \frac{c\ell(K)}{\left(\frac{n}{2}\right)^{\frac{1}{\ell(K)}}}\right), \tag{3.20}$$

where $c > 0$ is a constant and $\ell(K) = (K-1)(K-2) - 1$.

It is of interest to find upper bounds on the achievable sum DoF as a function of channel diversity n and the number of users K, in order to understand whether the scaling obtained in (3.20) can be improved. Interestingly, for the special case where $K = 3$, matching upper and lower bounds for the scaling were obtained in [59], where it is shown that

$$\text{Sum DoF} = \frac{3}{2}\left(1 - \frac{1}{4n - 2\lfloor n/2 \rfloor - 1}\right) = \mathcal{O}\left(\frac{3}{2}\left(1 - \frac{1}{n}\right)\right).$$

As $n \to \infty$, we converge to the upper bound of $\frac{3}{2}$, with a gap that goes to zero as $\frac{1}{n}$.

For $K \geq 4$, it is shown in [58] that

$$\text{Sum DoF} \leq \frac{K}{2}\left(1 - b\min\left\{\frac{1}{n^{\frac{1}{4}}}, \frac{2^{g(K)}}{\sqrt{n}}\right\}\right),$$

where $b > 0$ is a constant and $g(K) = (K-2)(K-3)/4$. If we compare this upper bound with the lower bound in (3.20), we see that even for $K = 4$, for which $\ell(K) = 5$, there is a gap between the rates at which the upper and lower bounds decrease with n (i.e., $\frac{1}{n^{\frac{1}{5}}}$ versus $\frac{1}{n^{\frac{1}{4}}}$ or $\frac{1}{\sqrt{n}}$), and this gap only increases for larger values of K.

3.4.3 Discussion

From the preceding discussion we see that for the case of time and frequency diversity, when the amount of diversity (n) is large, the achievable sum DoF approaches $\frac{K}{2}$. But while the sum DoF scales linearly with K, for $K \geq 4$, we may need unreasonably large values of n in order to get close to a sum DoF of $\frac{K}{2}$.

For spatial diversity, in a symmetric constant MIMO channel with N antennas at every node, the maximum achievable sum DoF is $2N$, and it does not scale with K. Finally, we note that if we allow for symbol extensions in a symmetric MIMO channel, then asymptotically, as the number of symbol extensions goes to infinity, one can achieve the maximum sum DoF of $\frac{KN}{2}$ (see [47]).

4 Iterative Algorithms for Interference Management

In Chapter 3, we examined conditions under which the interference alignment equations can be solved. In particular, we showed that conditions (3.12) and (3.13) need to be satisfied, and we further used these conditions to obtain bounds on the achievable degrees of freedom – e.g., (3.18). However, this analysis did not provide a recipe for solving the interference alignment equations. In this chapter, we study iterative algorithms for approaching interference alignment solutions. The emphasis will be on time-division duplex (TDD) wireless systems, where channel reciprocity holds, i.e., the channel from a given transmitter to a receiver remains the same when the roles of the transmitter and receiver are interchanged.

There are two key ideas from interference alignment that drive the development of the iterative algorithms: (i) linear transmit strategies are optimal at high SNR from a DoF viewpoint, and therefore it may be desirable to design linear transmit strategies that mimic interference alignment at any finite SNR; (ii) the sum degrees of freedom of the interference channel provides guidelines for choosing the number of transmit streams at finite SNRs.

4.1 Channel Model and Preliminaries

We consider the channel model in (3.11), with each transmitter having N_t antennas, and each receiver having N_r antennas, with the understanding that the model can easily be generalized to allow for different numbers of antennas at each of the transmitters and receivers. We change notation slightly, adding tildes to the channel inputs and outputs before linear pre-coding and processing, to obtain the following model from the perspective of receiver k:

$$\tilde{\boldsymbol{y}}_k = \sum_{j=1}^{K} \boldsymbol{H}_{kj}\tilde{\boldsymbol{x}}_j + \tilde{\boldsymbol{z}}_k, \tag{4.1}$$

where $\tilde{\boldsymbol{y}}_k$ is the $N_r \times 1$ receive vector, $\tilde{\boldsymbol{z}}_k$ is the $N_r \times 1$ additive white Gaussian noise (AWGN) vector normalized so that $\tilde{\boldsymbol{z}}_k \sim \mathcal{CN}(0, \boldsymbol{I}_{N_r})$, $\boldsymbol{H}_{kj}$ is the $N_r \times N_t$ matrix of channel coefficients between transmitter j and receiver k, and $\tilde{\boldsymbol{x}}_j$ is the $N_t \times 1$ transmitted signal vector from user j.

User k transmits a total of d_k (complex) symbols for each use of the MIMO channel, with d_k chosen based on the DoF bounds discussed in Chapter 3. The vector of d_k symbols of user k is denoted by $\boldsymbol{x}_k$. The quantity $\rho_{k\ell}$ is the power allocated to symbol ℓ of user k, $\boldsymbol{x}_k(\ell)$. Then, with complex Gaussian signaling, we have

$$\boldsymbol{x}_k \sim \mathcal{CN}\left(\mathbf{0}, \text{diag}(\rho_{k1}, \ldots, \rho_{kd_k}),\right),$$

with $\rho_{k\ell}$ chosen to satisfy a power constraint P_k, i.e.,

$$\sum_{\ell=1}^{d_k} \rho_{k\ell} \leq P_k. \tag{4.2}$$

We concatenate all the stream powers across the users into a $\sum_{j=1}^{K} d_j \times 1$ vector:

$$\boldsymbol{\rho} = [\rho_{11}, \ldots, \rho_{1d_1}, \ldots, \rho_{K1}, \ldots, \rho_{Kd_K}]^\top. \tag{4.3}$$

The channel input $\tilde{\boldsymbol{x}}_k$ is formed as

$$\tilde{\boldsymbol{x}}_k = \boldsymbol{V}_k \boldsymbol{x}_k, \tag{4.4}$$

where the matrix $\boldsymbol{V}_k$ is a matrix of unit norm transmit (beamforming) vectors of size $N_\text{t} \times d_k$. Similarly, at the receiver, the channel outputs $\tilde{\boldsymbol{y}}_k$ are formed as

$$\boldsymbol{y}_k = \boldsymbol{U}_k^\dagger \tilde{\boldsymbol{y}}_k, \tag{4.5}$$

where $\boldsymbol{U}_k$ is the $N_\text{r} \times d_k$ matrix of unit norm receive (beamforming) vectors used by receiver k. This yields the effective system model

$$\boldsymbol{y}_k = \sum_{j=1}^{K} \boldsymbol{U}_k^\dagger \boldsymbol{H}_{kj} \boldsymbol{V}_j \boldsymbol{x}_j + \boldsymbol{z}_k, \tag{4.6}$$

with $\boldsymbol{z}_k \sim \mathcal{CN}(0, \boldsymbol{U}_k^\dagger \boldsymbol{U}_k)$.

The number of data streams per user d_k is chosen based on the interference alignment conditions given in Chapter 3. In particular, a slight generalization of (3.18), allowing for the number of streams to vary across the users, yields the bound:

$$\sum_{k=1}^{K} d_k(N_\text{t} + N_\text{r} - 2d_k) \geq \sum_{j \neq k} d_j d_k, \tag{4.7}$$

which we use as a guideline to choose d_k, $k = 1, 2, \ldots, K$.

The effective gain from stream s of user j to stream ℓ of user k is defined as

$$G_{k\ell}^{js} = \left| \boldsymbol{U}_k(\ell)^\dagger \boldsymbol{H}_{kj} \boldsymbol{V}_j(s) \right|^2. \tag{4.8}$$

The noise-plus-interference covariance matrices are defined as:

$$\boldsymbol{B}_{k\ell} = \boldsymbol{I}_{N_\text{r}} + \sum_{(j,s) \neq (k,\ell)} \rho_{js} \boldsymbol{H}_{kj} \boldsymbol{V}_j(s) \boldsymbol{V}_j(s)^\dagger \boldsymbol{H}_{kj}^\dagger. \tag{4.9}$$

This yields that the sum-rate for linear transmit and receive strategies is given by:

$$R_{\text{sum}} = \sum_{k=1}^{K} \log\det\left(\boldsymbol{I}_{N_\text{r}} + \boldsymbol{H}_{kk}\boldsymbol{V}_k\boldsymbol{D}_k\boldsymbol{V}_k^\dagger\boldsymbol{H}_{kk}^\dagger\boldsymbol{B}_k^{-1}\right), \tag{4.10}$$

with $\boldsymbol{D}_k = \text{diag}(\rho_{k1}, \dots, \rho_{kd_k})$, and

$$\boldsymbol{B}_k = \boldsymbol{I}_{N_\text{r}} + \sum_{j \neq k} \boldsymbol{H}_{kj}\boldsymbol{V}_j\boldsymbol{D}_j\boldsymbol{V}_j^\dagger\boldsymbol{H}_{kj}^\dagger.$$

The SINR for stream (k,ℓ) is defined as:

$$\text{SINR}_{k\ell} = \frac{\rho_{k\ell}G_{k\ell}^{k\ell}}{1 + \sum_{(j,s)\neq(k,\ell)} \rho_{js}G_{k\ell}^{js}}. \tag{4.11}$$

4.1.1 Reciprocal Channel

The algorithms presented in this chapter all assume that channel reciprocity holds, i.e., the channel from a given transmitter to a receiver remains the same when the roles of the transmitter and receiver are interchanged. In the reciprocal channel, the roles of $\boldsymbol{V}_k$ and $\boldsymbol{U}_k$ are reversed, while d_k is preserved to be the same as in the original channel. For all variables used in describing the channel, an arrow above a quantity will denote the direction of communication. Thus, for the reciprocal channel:

$$\overleftarrow{\boldsymbol{y}}_k = \sum_{j=1}^{K} \overleftarrow{\boldsymbol{U}}_k^\dagger \overleftarrow{\boldsymbol{H}}_{kj} \overleftarrow{\boldsymbol{V}}_j \overleftarrow{\boldsymbol{x}}_j + \overleftarrow{\boldsymbol{z}}_k, \tag{4.12}$$

where $\overleftarrow{\boldsymbol{H}}_{kj} = \overrightarrow{\boldsymbol{H}}_{jk}^\dagger$. The original channel will also be referred to as the forward channel and the reciprocal channel as the reverse channel when appropriate.

4.2 Iterative Algorithms Based On Interference Alignment

The Min Leakage algorithm [38, 39] was one of the first iterative algorithms proposed for interference management. The goal is to minimize the leaked interference power at each receiver after linear processing at the receiver, with the motivation that for interference alignment the leaked interference power equals zero at each receiver – see (3.12) and (3.13). This algorithm works by repeatedly reversing the direction of communication, and designing receive vectors to minimize the leaked interference power in each direction.

The receive vectors are designed by computing an *interference covariance matrix* $\boldsymbol{Q}_k$ at each receiver:

$$\boldsymbol{Q}_k = \sum_{\substack{j=1 \\ j\neq k}}^{K} \frac{P}{d_j}\boldsymbol{H}_{kj}\boldsymbol{V}_j\boldsymbol{V}_j^\dagger\boldsymbol{H}_{kj}^\dagger. \tag{4.13}$$

The leaked interference power at receiver k is given in terms of $\boldsymbol{Q}_k$ as

$$\operatorname{Tr}\left(\boldsymbol{U}_k^{\dagger}\boldsymbol{Q}_k\boldsymbol{U}_k\right). \tag{4.14}$$

To minimize the leaked interference power at receiver k, Gomadam, Cadambe, and Jafar [38,39] suggested choosing $\boldsymbol{U}_k$ to contain the d_k least dominant eigenvectors of $\boldsymbol{Q}_k$. An attractive feature of the Min Leakage algorithm is that it makes a metric known as *weighted leakage interference* decrease monotonically at each receiver, and therefore the algorithm is guaranteed to converge [38, 39]. However, there is no guarantee that the convergence will be to a global minimum of the leaked interference power; only convergence to a local optimum is guaranteed. The Min Leakage algorithm is summarized in Algorithm 4.1.

Algorithm 4.1 Min Leakage Algorithm

1. For each $k \in [K]$, set the forward transmit precoding matrix $\overrightarrow{\boldsymbol{V}}_k^{(1)}$ such that its columns $\{\overrightarrow{\boldsymbol{V}}_k^{(1)}(\ell)\}_{\ell=1}^{d_k}$ are in arbitrary directions with the only constraint that they have unit norm. Furthermore, set
$$\rho_{k\ell} = \frac{P}{d_k}, \quad \forall k \in [K],\ \ell \in \{1,\dots,d_k\}.$$
2. Compute $\overrightarrow{\boldsymbol{Q}}_k$ – see (4.13) – at each receiver k, and choose the receive vectors to minimize the interference leakage, i.e., set $\overrightarrow{\boldsymbol{U}}_k^{(1)}$ to be the d_k least dominant eigenvectors $\forall k \in [K]$.
3. Reverse the direction of communication, and set
$$\overleftarrow{\boldsymbol{V}}_k^{(1)}(\ell) = \overrightarrow{\boldsymbol{U}}_k^{(1)}(\ell), \quad \forall k \in [K],\ \ell \in \{1,\dots,d_k\}.$$
4. Compute $\overleftarrow{\boldsymbol{Q}}_k$ at each receiver k, and set $\overleftarrow{\boldsymbol{U}}_k^{(1)}$ to be the d_k least dominant eigenvectors $\forall k \in [K]$.
5. Reverse the direction of communication, and set
$$\overrightarrow{\boldsymbol{V}}_k^{(2)}(\ell) = \overleftarrow{\boldsymbol{U}}_k^{(1)}(\ell), \quad \forall k \in [K],\ \ell \in \{1,\dots,d_k\}.$$
6. Repeat steps 2 through 5 until the desired convergence accuracy is met.

4.2.1 Other Algorithms Based on Interference Alignment

A number of other algorithms have been proposed in the literature, with the same goal as the Min Leakage algorithm of approaching an interference alignment solution, i.e., one that satisfies (3.12) and (3.13). The Peters–Heath algorithm [60] can be considered as an extension of the Min Leakage algorithm that gets around the requirement of reciprocity and can be applied in frequency-division duplex (FDD) cellular communication systems. A modification of the Peters–Heath algorithm to achieve a better sum-rate

Algorithm 4.2 Max SINR Algorithm

1. For each $k \in [K]$, set the forward transmit precoding matrix $\overrightarrow{\boldsymbol{V}}_k^{(1)}$ such that its columns $\{\overrightarrow{\boldsymbol{V}}_k^{(1)}(\ell)\}_{\ell=1}^{d_k}$ are in arbitrary directions with the only constraint that they have unit norm. Furthermore, set
$$\rho_{k\ell} = \frac{P}{d_k}, \quad \forall k \in [K],\ \ell \in \{1,\ldots,d_k\}.$$
2. Compute the receive vectors $\overrightarrow{\boldsymbol{U}}_k^{(1)}$ $\forall k \in [K]$ that maximize the SINR (equivalently, minimize the mean squared error) for each data stream as defined in (4.11).
3. Reverse the direction of communication, and set
$$\overleftarrow{\boldsymbol{V}}_k^{(1)}(\ell) = \frac{\overrightarrow{\boldsymbol{U}}_k^{(1)}(\ell)}{\|\overrightarrow{\boldsymbol{U}}_k^{(1)}(\ell)\|}, \quad \forall k \in [K],\ \ell \in \{1,\ldots,d_k\}.$$
4. Compute the receive vectors $\overleftarrow{\boldsymbol{U}}_k^{(1)}$ $\forall k \in [K]$ to maximize the SINR.
5. Reverse the direction of communication, and set
$$\overrightarrow{\boldsymbol{V}}_k^{(2)}(\ell) = \frac{\overleftarrow{\boldsymbol{U}}_k^{(1)}(\ell)}{\|\overleftarrow{\boldsymbol{U}}_k^{(1)}(\ell)\|}, \quad \forall k \in [K],\ \ell \in \{1,\ldots,d_k\}.$$
6. Repeat steps 2 through 5.

is presented in the work of Kumar and Xue [61]. Other algorithms that are designed to mimic interference alignment include [62] and [63].

4.3 Max SINR Algorithm

The algorithms described in Section 4.2 are all inspired by interference alignment and attempt to indirectly solve (3.12) and (3.13). However, there may be many (or even infinite) solutions to (3.12) and (3.13) that all achieve different sum-rate performance. For example, with $N_\mathrm{t} = N_\mathrm{r} = N$, $K < 2N - 1$, and $d_k = 1, \forall k \in [K]$, Yetis et al. [64] established that there are an infinite number of solutions to the alignment problem. Therefore, rather than focusing on finding alignment solutions, it may be better to develop communication strategies that maximize the sum-rate.

In the Max SINR algorithm proposed by Gomadam et al. [38, 39], the goal is to maximize the SINR for each data stream as defined in (4.11), with the understanding that maximizing the SINR is related to maximizing the sum-rate as defined in (4.10). Since the processing at the transmitters and receivers is linear, maximizing the SINR is equivalent to minimizing the mean squared error (MSE) [65]. The Max SINR algorithm is summarized in Algorithm 4.2. Simulations show that the Max SINR algorithm results in a better sum-rate at all values of SNR (P), as compared to the algorithms that are based on interference alignment. For example, as seen in Figure 1 in [66], the performance of the Peters–Heath algorithm is nearly identical to that of the Min Leakage algorithm, and the Max SINR algorithm outperforms both of these algorithms at all SNRs. The performance gap between the algorithms gets smaller with larger SNR. One reason for the gap getting smaller at high SNRs might be that interference alignment is optimal from a sum-rate viewpoint as the SNR goes to infinity, and therefore algorithms that attempt to achieve alignment inherit this optimality at high SNR.

4.3.1 Convergence Behavior of Max SINR Algorithm

In simulations of the Max SINR algorithm [38,39,66], the sum rate appears to converge as the number of iterations goes to infinity. But unlike the other algorithms based on interference alignment, for which an appropriate performance metric (e.g., interference leakage for the Min Leakage algorithm) decreases (or increases) monotonically with iterations, the Max SINR algorithm does not exhibit such monotonic behavior and is hence not provably convergent.

To explore the convergence behavior of the Max SINR, we introduce the following notation, with reference to the steps in Algorithm 4.2. Within round n of the iteration, we denote the sum rate at the end of Step 2 of the algorithm by $\overrightarrow{R}^{(n)}_{\text{sum}}$, at the end of Step 3 by $\overleftarrow{R}^{(n)}_{\text{sum-switch}}$, at the end of Step 4 by $\overleftarrow{R}^{(n)}_{\text{sum}}$, and at the end of Step 5 by $\overrightarrow{R}^{(n)}_{\text{sum-switch}}$.

In simulations of the Max SINR algorithm, we observe the following behavior. First, within round n of the iteration, it is sometimes the case that $\overrightarrow{R}^{(n)}_{\text{sum}} > \overleftarrow{R}^{(n)}_{\text{sum-switch}}$, but after optimization $\overrightarrow{R}^{(n)}_{\text{sum}} \leq \overleftarrow{R}^{(n)}_{\text{sum}}$. A similar process occurs in the forward channel, with $\overrightarrow{R}^{(n)}_{\text{sum}} \leq \overrightarrow{R}^{(n+1)}_{\text{sum}}$. Despite this nonmonotonic behavior within each round of the iteration, in most simulations, the Max SINR algorithm does appear to converge at each SNR, even if the convergence is not monotonic across rounds of the iteration, as shown in [66]. Due to this nonmonotonic behavior, it is not easy to establish the convergence of the Max SINR algorithm. In the next section, we describe a variant of the Max SINR algorithm, introduced in [66], which has sum-rate performance that is nearly identical to that of Max SINR, while at the same time is guaranteed to converge.

4.4 Convergent Variant of Max SINR Algorithm

The metric that we use in establishing the convergence of a variant of the Max SINR algorithm is the *sum stream rate*:

$$R_{\text{sum stream}} = \sum_{k=1}^{K}\sum_{\ell=1}^{d_k} \log(1+\text{SINR}_{k\ell}).$$

Note that $R_{\text{sum stream}}$ is the sum of the rates achieved by decoding each stream separately, while treating all the other streams as noise. This is in contrast with the sum-rate given in (4.10), for which we allow for joint decoding of all the d_k streams at receiver k.

An important component of the new algorithm is a power control step, in which we impose a sum power constraint across all the users:

$$\sum_{k=1}^{K}\sum_{\ell=1}^{d_k}\rho_{k\ell} \leq KP, \tag{4.15}$$

which can also be written as $\mathbf{1}^\top\boldsymbol{\rho} \leq KP$, with $\boldsymbol{\rho}$ defined in (4.3). Imposing a sum power constraint, as opposed to the individual power constraint given in (4.2), allows for a more flexible design of linear transmit and receive strategies, albeit with the requirement that the individual powers are not constrained by P anymore.

4.4.1 Achieving Equal SINRs on Forward and Reverse Channels

We now show that by selecting the power appropriately we can achieve the same SINRs on the forward and reciprocal channels. This result is based on extending the ideas from the work of Song et al. [67] to allow for successive interference cancellation.

Let $\boldsymbol{\gamma}_{k\ell}$ be the target SINR for stream ℓ of user k, and define

$$\boldsymbol{D} = \text{diag}\left\{\frac{\boldsymbol{\gamma}_{11}}{\overrightarrow{G}_{11}^{11}}, \ldots, \frac{\boldsymbol{\gamma}_{Kd_K}}{\overrightarrow{G}_{Kd_K}^{Kd_K}}\right\} \tag{4.16}$$

and

$$\boldsymbol{G}\left(\sum_{m=1}^{k-1} d_m + \ell, \sum_{n=1}^{j-1} d_n + s\right) = \begin{cases} 0 & \text{if } (k,\ell) = (j,s), \\ \overrightarrow{G}_{k\ell}^{js} & \text{else}, \end{cases} \tag{4.17}$$

with $G_{k\ell}^{js}$ given in (4.8). Set

$$\boldsymbol{A} = \boldsymbol{D}^{-1} - \boldsymbol{G},$$

and use the power allocations

$$\overrightarrow{\boldsymbol{\rho}}^* = \boldsymbol{A}^{-1}\mathbf{1}, \tag{4.18}$$

$$\overleftarrow{\boldsymbol{\rho}}^* = \boldsymbol{A}^{-\top}\mathbf{1} \tag{4.19}$$

for the forward and reciprocal directions, respectively.

The following lemma establishes the duality of the SINRs on the forward and reciprocal channels. The result implies that as long as we meet the power constraint initially, $\overrightarrow{\boldsymbol{\rho}}^*$ and $\overleftarrow{\boldsymbol{\rho}}^*$ will meet it too.

LEMMA 4.1 *Suppose the initial power allocation $\boldsymbol{\rho}$ is such that $\mathbf{1}^\top\boldsymbol{\rho} \leq KP$. Let $\boldsymbol{\gamma}_{k\ell}$ be SINRs resulting from the choice of transmit and receive vectors $\{\boldsymbol{V}_k\}$ and $\{\boldsymbol{U}_k\}$, respectively. Then*

$$\mathbf{1}^\top\overrightarrow{\boldsymbol{\rho}}^* = \mathbf{1}^\top\overleftarrow{\boldsymbol{\rho}}^* \leq KP,$$

and

$$\overrightarrow{\text{SINR}}_{k\ell}(\overrightarrow{\boldsymbol{\rho}}^*) = \overleftarrow{\text{SINR}}_{k\ell}(\overleftarrow{\boldsymbol{\rho}}^*) = \boldsymbol{\gamma}_{k\ell}.$$

Proof The proof follows the arguments given in [67] and is developed in the appendix of [66].

While equations (4.18) and (4.19) provide a way to achieve equal SINRs in both the forward and reverse directions, this may be impractical due to the fact that it requires a centralized controller. Fortunately, the framework introduced by Yates [68] can be applied to our problem to compute the required powers in a distributed way. Define the vector-valued function $\boldsymbol{I}(\boldsymbol{\rho})$ by

$$\boldsymbol{I}_{k\ell}(\boldsymbol{\rho}) = \frac{\gamma_{k\ell}\rho_{k\ell}}{\mathrm{SINR}_{k\ell}(\boldsymbol{\rho})}.$$

Then it is easy to show that $\boldsymbol{I}$ is a *standard interference function* as defined in [68], and therefore the asynchronous power control algorithm developed in [68] can be used to develop a distributed algorithm (see Algorithm 4.3) that converges to $\boldsymbol{\rho}^*$.

Algorithm 4.3 Distributed Power Control Algorithm

1. Choose an initial power vector $\boldsymbol{\rho}^{(1)}$ such that $\mathbf{1}^\top \boldsymbol{\rho}^{(1)} \leq KP$.
2. Compute

$$\boldsymbol{\rho}^{(2)} = \boldsymbol{I}(\boldsymbol{\rho}^{(1)}).$$

3. Repeat step 2 until convergence.

4.4.2 Convergent Max SINR

We are now ready to formally define the convergent variant of the Max SINR algorithm, which we call the Convergent Max SINR algorithm; it is summarized in Algorithm 4.4. As in the analysis of the Max SINR algorithm, we introduce the following notation, with reference to the steps of Algorithm 4.4. Within round n of the iteration, we denote the sum stream rate at the end of Step 2 by $\overrightarrow{R}^{(n)}_{\text{sum stream}}$, at the end of Step 3 by $\overleftarrow{R}^{(n)}_{\text{sum stream-switch}}$, at the end of Step 4 by $\overleftarrow{R}^{(n)}_{\text{sum stream}}$, and at the end of Step 5 by $\overrightarrow{R}^{(n)}_{\text{sum stream-switch}}$.

The convergence of the Convergent Max SINR algorithm is established in the following result.

THEOREM 4.2 *In the Convergent Max SINR algorithm,* $\overrightarrow{R}^{(n)}_{\text{sum stream}}$ *converges. Also, the sum power constraint is met at every step.*

Proof The initial power allocation $\overrightarrow{\boldsymbol{\rho}}^{(1)}$ meets the power constraint, and therefore the constraint will continue to be met due to Lemma 4.1. Now,

$$\overrightarrow{R}^{(n)}_{\text{sum stream}} \overset{\text{(a)}}{=} \overleftarrow{R}^{(n)}_{\text{sum stream-switch}}$$
$$\overset{\text{(b)}}{\leq} \overleftarrow{R}^{(n)}_{\text{sum stream}}$$

Algorithm 4.4 Convergent Max SINR Algorithm

1. Choose the initial transmit vectors $\{\overrightarrow{V}_k^{(1)}\}_{k=1}^K$ and $\overrightarrow{\rho}^{(1)}$ to satisfy the power constraint.
2. Compute the receive vectors $\overrightarrow{U}_k^{(1)}$ $\forall k \in [K]$ to minimize the MSE.
3. Reverse the direction of communication, and calculate the power allocation $\overleftarrow{\rho}^{(1)}$ to achieve the same SINR in the reverse direction as in the forward direction. Set

$$\overleftarrow{V}_k^{(1)}(\ell) = \frac{\overrightarrow{U}_k^{(1)}(\ell)}{\|\overrightarrow{U}_k^{(1)}(\ell)\|},$$

$$\overleftarrow{U}_k^{(1)}(\ell) = \overrightarrow{V}_k^{(1)}(\ell),$$

$\forall k \in [K],\ \ell \in \{1,\ldots,d_k\}$.

4. Compute the MMSE receive vectors $\overleftarrow{U}_k^{(1)}$ $\forall k \in [K]$.
5. Reverse the direction of communication. Calculate the power allocation $\overrightarrow{\rho}^{(2)}$ to achieve the same SINR as in the previous direction. Set

$$\overrightarrow{V}_k^{(2)}(\ell) = \frac{\overleftarrow{U}_k^{(1)}(\ell)}{\|\overleftarrow{U}_k^{(1)}(\ell)\|},$$

$$\overrightarrow{U}_k^{(2)}(\ell) = \overleftarrow{V}_k^{(1)}(\ell),$$

$\forall k \in [K],\ \ell \in \{1,\ldots,d_k\}$.

6. Repeat steps 2 through 5 until convergence.

$$\overset{(c)}{=} \overrightarrow{R}^{(n)}_{\text{sum stream-switch}}$$

$$\overset{(d)}{\leq} \overrightarrow{R}^{(n+1)}_{\text{sum stream}},$$

where (a) and (c) follow by the fact that the SINR is the same in both directions, and (b) and (d) follow since the receive vectors are chosen to maximize the SINR. Therefore, the sum stream rate increases monotonically and the algorithm converges.

4.5 Performance of Convergent Max SINR Algorithm

As shown in simulation results reported in [66], the Max SINR and Convergent Max SINR algorithms have nearly identical sum-rate performance, and both of these algorithms outperform the other convergent algorithms, such as the Min Leakage [38, 39] and Peters–Heath [60] algorithms. The conclusion that we can draw from these results is that the Convergent Max SINR algorithm has performance comparable to the Max SINR algorithm, with the added advantage of having guaranteed convergence.

4.6 Discussion

In this chapter, we have studied iterative algorithms for designing linear transmitter and receiver processing to maximize the sum-rate for communication on interference channels. We first discussed a number of convergent iterative algorithms that were inspired by interference alignment, and then discussed the more ad hoc Max SINR algorithm, in which the linear processing is chosen to maximize the SINR (equivalently, minimize the MSE) at the receivers in each iteration. The Max SINR algorithm was shown to outperform the IA-based solutions, especially at low SNR, in simulations. However, a disadvantage of the Max SINR algorithm relative to the IA-based solutions is that it cannot be shown to be convergent.

We then studied a modification to the Max SINR algorithm called the Convergent Max SINR algorithm, for which a metric called the sum stream rate converges. A key step to making the algorithm provably convergent is to choose the power allocation appropriately every time we reverse the direction of communication in the iterations to equate the SINR in both directions of communication.

In Chapter 5, we will extend the study of interference alignment and degrees of freedom analysis to the case where subsets of the transmitters in the network are allowed to cooperate in transmitting messages to the receivers. For constant MIMO channels, we can use iterative schemes such as the ones described in the present chapter, while incorporating transmitter cooperation. See Section 5.7 for details.

5 Degrees of Freedom with Coordinated Multi-Point Transmission

In this chapter, we study cooperative communication in large interference networks, a theme that will be followed through the remainder of the book. The cooperative channel models we consider are based on the coordinated multi-point (CoMP) [69] transmission technology, where one message can be jointly transmitted using multiple transmitters. We aim to analyze the sum degrees of freedom, which we introduced in Chapter 3, under the cooperative communication framework. More specifically, in this chapter we focus on fully connected interference networks, where all links between transmitters and receivers are present. We model a limited capacity backhaul by allowing each message to only be available at a pre-specified number M of transmitters, where M is typically much smaller than the number of users K. One of our goals in this chapter is to study if the asymptotic per-user degrees of freedom as K goes to infinity can be improved through cooperation.

5.1 Interference Alignment with CoMP

In this section, we present an example where CoMP transmission aided by asymptotic interference alignment leads to DoF gains in a K-user fully connected interference channel. We use $\eta(K,M)$ to denote the best achievable DoF for a K-user fully connected interference channel, where each message can be available at M transmitters. We know from Chapter 3 that the sum DoF of a fully connected interference channel without cooperation is $\frac{K}{2}$, i.e., $\eta(K,1) = \frac{K}{2}$. We now show that $\eta(K,M) > \frac{K}{2}$, for $M > 1$, by using the following spiral message assignment for each K-user channel:

$$\mathcal{T}_k = \{k, k+1, \ldots, k+M-1\} \; \forall k \in [K], \tag{5.1}$$

with the indices taken modulo K. We use $\mathcal{T}_k$ to denote the transmit set for message W_k. For example, $\mathcal{T}_1 = \{1,2\}$ means W_1 is available at both transmitter 1 and transmitter 2.

Using this message assignment and an asymptotic interference alignment scheme, we prove the following result.

THEOREM 5.1

$$\eta(K,M) \geq \frac{K+M-1}{2}, \; \forall M \leq K < 10. \tag{5.2}$$

We provide the formal proof in Section 5.2. Here, we discuss the key ideas that enable the theorem statement. To achieve the stated lower bound, the M transmitters carrying each message are used to cancel the interference introduced by this message at the first $M-1$ receivers, thereby allowing each of these receivers to enjoy one degree of freedom. By coding over multiple parallel channels corresponding to different time slots, we use an interference alignment scheme to align the interfering signals at each other receiver to occupy half the signal space as the number of parallel channels goes to infinity.

The achievable scheme is based on transmit beam-forming. The beam design process is broken into two steps, as illustrated in Figure 5.1. First, we transform each parallel CoMP channel into a derived channel. Then, we design an asymptotic interference alignment scheme over the derived channel, achieving the required DoF in an asymptotic fashion as the number of parallel channels goes to infinity. Figure 5.2 provides a description of the derived channel for the special case of $K=4$ and $M=2$. In order to use asymptotic interference alignment in the achievable scheme, we need to show that at each receiver, polynomial transformations defining a set of derived channel coefficients determined by the receiver index are algebraically independent as functions of the original channel coefficients. We could verify in MATLAB that this is true for all the values of K and M that we checked. Specifically, we checked until $K \leq 9$, but we conjecture that the result holds true for any K and M.

The maximum achievable DoF without cooperation is $\frac{K}{2}$. We see from Theorem 5.1 that CoMP transmission allows for a gain in achievable DoF (relative to $\frac{K}{2}$), but this gain does not scale with the number of users K. If we define the asymptotic per-user DoF as a function of the cooperation constraint by

$$\tau(M) = \lim_{K\to\infty} \frac{\eta(K,M)}{K}, \tag{5.3}$$

then for the achievable scheme in Theorem 5.1, $\tau(M) = \frac{1}{2}$ for all fixed M. The question of whether there exist achievable schemes such that $\tau(M) > \frac{1}{2}$ for $M > 1$ remains open. We also note that the spiral message assignment that is used here relies only on local cooperation. In other words, the maximum difference between a message index and the index of any transmitter it is assigned to does not scale with the number of users K. In Section 5.4, we show that no gain in the asymptotic per-user DoF can be achieved through any message assignment that satisfies a local cooperation constraint.

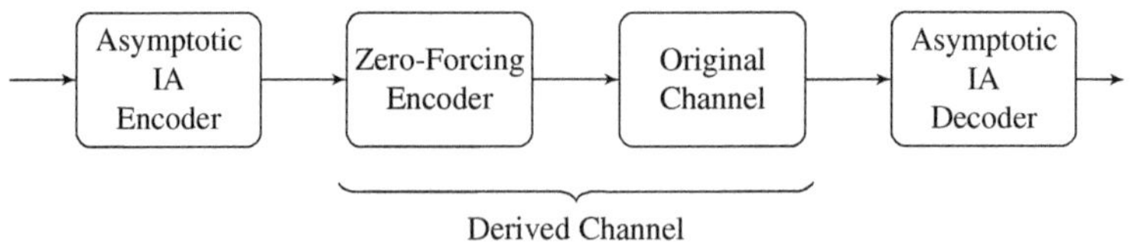

Fig. 5.1 Summary of the achievable scheme of Theorem 5.1. ©[2017] IEEE. Reprinted, with permission, from [70].

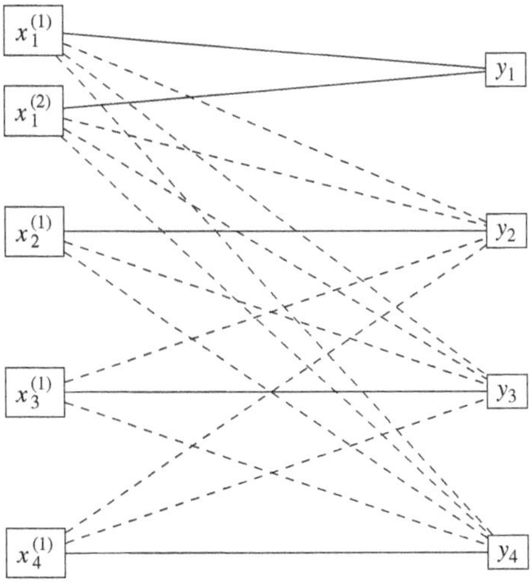

Fig. 5.2 The derived channel with $K = 4$ and $M = 2$. The thick lines indicate the links carrying signal. The dashed lines indicate the links carrying interference. ©[2017] IEEE. Reprinted, with permission, from [70].

5.2 Proving that Interference Alignment Achieves CoMP Transmission Gains

In this section, we prove Theorem 5.1. We show that the DoF of a K-user fully connected interference channel with a maximum transmit set size constraint of M is lower-bounded by

$$\eta(K,M) \geq \frac{K+M-1}{2}.$$

We prove this by assigning each message to the transmitter with the same index, as well as the $M-1$ succeeding transmitters, and arguing that the DoF vector with components

$$d_i = \begin{cases} 1 & 1 \leq i \leq M-1, \\ 0.5 & M \leq i \leq K \end{cases}$$

is achievable; i.e., the first $M-1$ users benefit from cooperation and achieve 1 degree of freedom, whereas the remaining $K-M+1$ users achieve half a degree of freedom, just like in the interference channel without cooperation. Conceptually, the achievable scheme in this section is based on converting the CoMP channel into a derived channel, and then employing the asymptotic interference alignment scheme on the derived channel, as summarized in Figure 5.1. We now provide a detailed description of the design steps summarized in Figure 5.1.

5.2.1 Derived Channel

Since our objective is to achieve a DoF vector that is asymmetric, the derived channel is also chosen to be asymmetric. The derived channel we consider in this section has two antennas for each of the first $M-1$ transmitters, and one antenna for each of the remaining $K-M+1$ transmitters. The received signal at receiver i is given by

$$y_i = \sum_{j=1}^{K} g_{ij}^{(1)} x_j^{(1)} + \sum_{j=1}^{M-1} g_{ij}^{(2)} x_j^{(2)} + z_i, \tag{5.4}$$

where $g_{ij}^{(m)}$ is the derived channel coefficient between the mth antenna at transmitter j and receiver i, and $x_j^{(m)}$ is the transmit signal of the mth antenna at transmitter j in the derived channel. We assume that the channel inputs of the CoMP channel are related to the channel inputs of the derived channel through a linear transformation. The contribution of the derived channel input $x_j^{(m)}$ in the real transmit signals $x_j, x_{j+1}, \ldots, x_{j+M_t-1}$ is defined by an $M \times 1$ beam-forming vector, i.e.,

$$\begin{bmatrix} x_j \\ x_{j+1} \\ \vdots \\ x_{j+M-1} \end{bmatrix} = (*) + \boldsymbol{v}_j^{(m)} x_j^{(m)},$$

where $(*)$ represents the contribution from other derived channel inputs. It is easy to see that the derived channel coefficients are related to the original channel coefficients as

$$g_{ij}^{(m)} = \boldsymbol{H}_{\{i\},\mathcal{T}_j} \boldsymbol{v}_j^{(m)},$$

for all $i,j \in [K]$ and appropriate m, where we use the notation $\boldsymbol{H}_{\{i\},\mathcal{S}}$ to denote the matrix of channel coefficients between receiver i and transmitters with indices in $\mathcal{S}$. Since we are designing the transmit beam-forming scheme to achieve one degree of freedom for the first $M-1$ users, it must be that the first $M-1$ receivers in the derived channel do not see any interference.

5.2.2 Zero-Forcing Step

We now explain our choice of the beam-forming vectors that ensures that the first $M-1$ receivers do not see any interference.

Zero-Forcing Beam Design

We first describe the general idea of constructing a zero-forcing beam. Consider the problem of designing a zero-forcing beam $\boldsymbol{v}$ to be transmitted by n transmit antennas indexed by the set $\mathcal{T} \subseteq [K]$ such that it does not cause interference at $n-1$ receive antennas indexed by the set $\mathcal{I} \subseteq [K]$, i.e.,

$$\boldsymbol{H}_{\mathcal{I},\mathcal{T}} \boldsymbol{v} = \boldsymbol{0},$$

where we use the notation $\boldsymbol{H}_{\mathcal{S}_1,\mathcal{S}_2}$ to denote the matrix of channel coefficients between receivers with indices in $\mathcal{S}_1$ and transmitters with indices in $\mathcal{S}_2$. Since $\boldsymbol{H}_{\mathcal{I},\mathcal{T}}$ is an

$(n-1) \times n$ matrix, the choice for $\boldsymbol{v}$ is unique up to a scaling factor. For any arbitrary row vector $\boldsymbol{a}$ of length n, we can use the Laplace expansion to expand the determinant

$$\det \begin{bmatrix} \boldsymbol{H}_{\mathcal{I},\mathcal{T}} \\ \boldsymbol{a} \end{bmatrix} = \sum_{j=1}^{n} a_j c_j,$$

where c_j is the cofactor of a_j, which depends only on the channel coefficients in $\boldsymbol{H}_{\mathcal{I},\mathcal{T}}$, and is independent of $\boldsymbol{a}$. By setting the beam-forming vector $\boldsymbol{v}$ as $\boldsymbol{v} = [c_1 \; c_2 \; \cdots \; c_n]$, we see that an arbitrary receiver i sees the signal transmitted along the beam $\boldsymbol{v}$ with a strength equal to

$$\boldsymbol{H}_{\{i\},\mathcal{T}}\boldsymbol{v} = \det \begin{bmatrix} \boldsymbol{H}_{\mathcal{I},\mathcal{T}} \\ \boldsymbol{H}_{\{i\},\mathcal{T}} \end{bmatrix} = \det \boldsymbol{H}_{\mathcal{I}\cup\{i\},\mathcal{T}}.$$

Clearly, this satisfies the zero-forcing condition $\boldsymbol{H}_{\{i\},\mathcal{T}}\boldsymbol{v} = 0$ for all $i \in \mathcal{I}$.

Design of Transmit Beam $\boldsymbol{v}_j^{(1)}$ for $j \geq M$

The signal $x_j^{(1)}$ is transmitted by the M transmitters from the transmit set $\mathcal{T}_j = \{j, j+1, \ldots, j+M-1\}$. The corresponding beam $\boldsymbol{v}_j^{(1)}$ is designed to avoid the interference at the first $M-1$ receivers $\mathcal{I} = [M-1]$. Therefore, we see that the contribution of $x_j^{(1)}$ at receiver i is given by

$$g_{ij}^{(1)} = \det \boldsymbol{H}_{\mathcal{A},\mathcal{B}}, \tag{5.5}$$

where

$$\mathcal{A} = \{1, 2, \ldots, M-1, i\},$$
$$\mathcal{B} = \{j, j+1, \ldots, j+M-1\}.$$

Design of Transmit Beams $\boldsymbol{v}_j^{(1)}$ and $\boldsymbol{v}_j^{(2)}$ for $j < M$

The signals $x_j^{(1)}$ and $x_j^{(2)}$ are transmitted by the M transmitters from the transmit set $\mathcal{T}_j = [M]$. They must avoid interference at the $M-2$ receivers

$$\mathcal{I} = \{1, 2, \ldots, j-1, j+1, \ldots, M-1\}.$$

Since we only need to avoid interference at $M-2$ receivers, it is sufficient to transmit each signal from $M-1$ transmitters. We use the first $M-1$ antennas of the transmit set $\mathcal{T}_j$ to transmit $x_j^{(1)}$, and the last $M-1$ antennas of the transmit set $\mathcal{T}_j$ to transmit $x_j^{(2)}$. Thus, we obtain

$$\begin{aligned} g_{ij}^{(1)} &= \det \boldsymbol{H}_{\mathcal{A},\mathcal{B}_1}, \\ g_{ij}^{(2)} &= \det \boldsymbol{H}_{\mathcal{A},\mathcal{B}_2}, \end{aligned} \tag{5.6}$$

where

$$\mathcal{A} = \{1, 2, \ldots, j-1, j+1, M-1, i\},$$
$$\mathcal{B}_1 = \{j, j+1, \ldots, j+M-2\},$$
$$\mathcal{B}_2 = \{j+1, j+1, \ldots, j+M-1\}.$$

Thus, the derived channel (5.4) can be simplified as

$$\begin{aligned} y_i &= g_{ii}^{(1)} x_j^{(1)} + g_{ii}^{(2)} x_j^{(2)} + z_i, \ 1 \le i < M, \\ y_i &= \sum_{j=1}^{K} g_{ij}^{(1)} x_j^{(1)} + \sum_{j=1}^{M-1} g_{ij}^{(2)} x_j^{(2)} + z_i, \ M \le i \le K, \end{aligned} \tag{5.7}$$

where the derived channel coefficients are as described in (5.5) and (5.6).

5.2.3 Asymptotic Interference Alignment

In this section, we consider L parallel derived channels, and propose a scheme achieving a DoF arbitrary close to $(K+M-1)/2$ in the limit $L \to \infty$. We can combine L parallel derived channels (5.7) and express them together as

$$\begin{aligned} \underline{\boldsymbol{y}}_i &= \boldsymbol{G}_{ii}^{(1)} \underline{\boldsymbol{x}}_j^{(1)} + \boldsymbol{G}_{ii}^{(2)} \underline{\boldsymbol{x}}_j^{(2)} + \underline{\boldsymbol{z}}_i, \ 1 \le i < M, \\ \underline{\boldsymbol{y}}_i &= \sum_{j=1}^{K} \boldsymbol{G}_{ij}^{(1)} \underline{\boldsymbol{x}}_j^{(1)} + \sum_{j=1}^{M_t-1} \boldsymbol{G}_{ij}^{(2)} \underline{\boldsymbol{x}}_j^{(2)} + \underline{\boldsymbol{z}}_i, \ M \le i \le K, \end{aligned}$$

where $\underline{\boldsymbol{x}}_j^{(m)}$, $\underline{\boldsymbol{y}}_i$, and $\underline{\boldsymbol{z}}_i$ are $L \times 1$ column vectors and $\boldsymbol{G}_{ij}^{(m)}$ is the $L \times L$ diagonal channel transfer matrix given by

$$\boldsymbol{G}_{ij}^{(m)} = \begin{bmatrix} g_{ij}^{(m)}(1) & & & \\ & g_{ij}^{(m)}(2) & & \\ & & \ddots & \\ & & & g_{ij}^{(m)}(L) \end{bmatrix}.$$

The achievable scheme that we propose is based on the asymptotic alignment scheme introduced by Cadambe and Jafar in [46] (also see Section 3.3).

DEFINITION 5.1 **(Cadambe–Jafar (CJ) subspace)** The order-n CJ subspace generated by the diagonal matrices $\boldsymbol{G}_1, \boldsymbol{G}_2, \cdots, \boldsymbol{G}_N$ is defined as the linear subspace spanned by the vectors

$$\{\boldsymbol{G}_1^{a_1} \boldsymbol{G}_2^{a_2} \cdots \boldsymbol{G}_N^{a_N} \mathbf{1} : \mathbf{a} \in \mathbf{Z}_+^{\mathbf{N}} \text{ and } \sum_{\mathbf{i}} \mathbf{a_i} \le \mathbf{n}\},$$

where $\mathbf{1}$ is the $L \times 1$ column vector of all ones. The matrix containing these $\binom{N+n}{n}$ vectors as columns is said to be the order-n CJ matrix.

Let $\boldsymbol{V}$ denote the order-n CJ subspace (and the corresponding matrix) generated by the nontrivial channel matrices carrying interference:

$$\{\boldsymbol{G}_{ij}^{(1)}, \boldsymbol{G}_{ij}^{(2)} : i \ge M, j < M\} \cup \{\boldsymbol{G}_{ij}^{(1)} : i \ne j \ge M\}. \tag{5.8}$$

We use $\boldsymbol{V}$, defined as the transmit beam-forming matrix at every transmitter of the derived channel. The first $M-1$ receivers do not see any interference. Therefore, for each $k < M$, the receiver k can decode all the desired streams free of interference if the matrix

$$\boldsymbol{M}_k = \begin{bmatrix} \boldsymbol{G}_{kk}^{(1)}\boldsymbol{V} & \boldsymbol{G}_{kk}^{(2)}\boldsymbol{V} \end{bmatrix}$$

has full column rank. Assuming that the number of rows in $\boldsymbol{M}_k$, equal to the number of parallel channels L, is greater than or equal to the number of columns, i.e., $L \geq 2|\boldsymbol{V}|$, the matrix $\boldsymbol{M}_k$ has full column rank for generic channel coefficients $\{h_{ij}\}$ (which are realized with probability one) if the following claim is true. See Corollary B.1 in Appendix B for an explanation.

CLAIM 5.1 For each $k < M$, the polynomials denoted by the variables

$$\begin{aligned} &\{g_{kk}^{(1)}, g_{kk}^{(2)}\} \cup \{g_{ij}^{(1)}, g_{ij}^{(2)} : i \geq M, j < M\} \\ &\qquad \cup \{g_{ij}^{(1)} : i \neq j \geq M\} \end{aligned} \tag{5.9}$$

are algebraically independent.

For each $k \geq M$, the interference seen at receiver k is limited to the order-$(n+1)$ CJ subspace, denoted by **INT**. Therefore, receiver k can decode all the desired streams free of interference if the matrix

$$\boldsymbol{M}_k = \begin{bmatrix} \boldsymbol{G}_{kk}^{(1)}\boldsymbol{V} & \mathbf{INT} \end{bmatrix}$$

has full column rank. Assuming that the number of rows is greater than or equal to the number of columns, i.e., $L \geq |\boldsymbol{V}| + |\mathbf{INT}|$, the matrix $\boldsymbol{M}_k$ has full column rank for generic (original) channel coefficients $\{h_{ij}\}$ if the following claim is true.

CLAIM 5.2 For each $k \geq M$, the polynomials denoted by the variables

$$\begin{aligned} &\{g_{kk}^{(1)}\} \cup \{g_{ij}^{(1)}, g_{ij}^{(2)} : i \geq M, j < M\} \\ &\qquad \cup \{g_{ij}^{(1)} : i \geq M, j \geq M, i \neq j\} \end{aligned} \tag{5.10}$$

are algebraically independent.

To satisfy the requirements on L, we choose L as

$$L = \max(2|\boldsymbol{V}|, |\boldsymbol{V}| + |\mathbf{INT}|) = |\boldsymbol{V}| + |\mathbf{INT}|.$$

Observe that

$$|\boldsymbol{V}| = \binom{N+n}{n} \text{ and } |\mathbf{INT}| = \binom{N+n+1}{n+1},$$

where N is the number of matrices (5.8) used to generate the CJ subspace, and is given by

$$\begin{aligned} N &= 2(K-M+1)(M-1) + (K-M+1)(K-M) \\ &= (K-M+1)(K+M-2). \end{aligned} \tag{5.11}$$

Therefore, if we let $\eta(K,M,L)$ be the maximum achievable DoF by coding over at most L parallel channels, then we obtain

$$\begin{aligned}\eta(K,M,L) &\geq \frac{2(M-1)|V|+(K-M+1)|V|}{L}\\ &= \frac{(K+M-1)|V|}{|V|+|\mathbf{INT}|}\\ &= \frac{K+M-1}{2+\frac{N}{n+1}}.\end{aligned}$$

Therefore, we obtain that

$$\begin{aligned}\eta(K,M) &= \limsup_{L\to\infty} \eta(K,M,L)\\ &\geq \lim_{n\to\infty} \frac{K+M-1}{2+\frac{N}{n+1}}\\ &= \frac{K+M-1}{2}.\end{aligned}$$

5.2.4 Proof of Algebraic Independence

We use the Jacobian criterion of Lemma B.1 in Appendix B to prove Claims 5.1 and 5.2. Recall that each derived channel coefficient is a polynomial in the K^2 variables $\{h_{ij} : 1 \leq i,j, \leq K\}$. Let $\boldsymbol{g}$ denote the vector consisting of the polynomials specified by the derived channel coefficients in the respective claims. The exact description of the polynomials can be obtained from (5.5) and (5.6) in Section 5.2.2. The number of polynomials in Claims 5.1 and 5.2 is equal to $N+2$ and $N+1$, respectively, where N is given by (5.11). From Lemma B.1 in Appendix B, we see that a collection of polynomials is algebraically independent if and only if the corresponding Jacobian matrix has full row rank. It can be easily verified that $N+2 \leq K^2$, and hence $N+1 \leq K^2$, for any K and M, which is a necessary condition for the corresponding Jacobian matrices to have full row rank. It is easy to verify that the Jacobian matrices corresponding to the polynomials in Claims 5.1 and 5.2 have full row rank using the symbolic toolbox of MATLAB for any fixed K and M. In particular, we verified that the Jacobian matrices have full row rank for all values of $M < K \leq 9$.

5.3 Degrees of Freedom Upper Bound for CoMP

In this section, we present an information-theoretic upper bound for the DoF of fully connected interference channels with CoMP transmission. We first need to discuss a general lemma that is used to obtain DoF upper bounds in any interference network, where each message could be available at an arbitrary set of transmitters. We need the following definition: For any set of receiver indices $\mathcal{A} \subseteq [K]$, define $U_{\mathcal{A}}$ as the set of indices of transmitters that exclusively carry the messages for the receivers in $\mathcal{A}$, where the complement set $\bar{U}_{\mathcal{A}}$ is the set of indices of transmitters that carry messages for receivers outside $\mathcal{A}$. More precisely, $U_{\mathcal{A}} = [K] \setminus \cup_{i \notin \mathcal{A}} \mathcal{T}_i$. Then, we have the following result.

LEMMA 5.2 *If there exists a set $\mathcal{A} \subseteq [K]$, a function f_1, and a function f_2 whose definition does not depend on the transmit power constraint P, and $f_1\left(y_{\mathcal{A}}, x_{U_{\mathcal{A}}}\right) = x_{\bar{U}_{\mathcal{A}}} + f_2(z_{\mathcal{A}})$, then $\eta \leq |\mathcal{A}|$.*

Proof The formal proof is available in [71]. We provide an overview here. Recall that $y_{\mathcal{A}} = \{y_i, i \in \mathcal{A}\}$, and $W_{\mathcal{A}} = \{W_i, i \in \mathcal{A}\}$, and note that $x_{U_{\mathcal{A}}}$ is the set of transmit signals that do not carry messages outside $W_{\mathcal{A}}$. Fix a reliable communication scheme for the considered K-user channel, and assume that there is only one centralized decoder that has access to the received signals $y_{\mathcal{A}}$. We show that using the centralized decoder, the only uncertainty in recovering all the messages $W_{[K]}$ is due to the Gaussian noise signals. In this case, the sum DoF is bounded by $|\mathcal{A}|$, as it is the number of received signals used for decoding.

Using $y_{\mathcal{A}}$, the messages $W_{\mathcal{A}}$ can be recovered reliably, and hence the signals $x_{U_{\mathcal{A}}}$ can be reconstructed. Using $y_{\mathcal{A}}$ and $x_{U_{\mathcal{A}}}$, the remaining transmit signals can be approximately reconstructed using the function f_1 of the hypothesis. Finally, using all transmit signals, the received signals $y_{\bar{\mathcal{A}}}$ can be approximately reconstructed, and the messages $W_{\bar{\mathcal{A}}}$ can then be recovered.

In order to characterize the asymptotic per-user DoF of the fully connected channel $\tau(M)$, we need to consider all possible message assignments satisfying the maximum transmit set size constraint. Through the following corollary of Lemma 5.2, we provide a way to bound the sum DoF η of a K-user fully connected channel with a fixed message assignment, thereby introducing a criterion for comparing different message assignments using the special cases where this bound holds tightly.

For a set of transmitter indices $\mathcal{S}$, we define the set $C_{\mathcal{S}}$ as the set of messages carried by transmitters in $\mathcal{S}$.

COROLLARY 5.1 *For any pair of positive integers $m, \bar{m} : m + \bar{m} \geq K$, if there exists a set $\mathcal{S}$ of indices for transmitters carrying no more than m messages, and $|\mathcal{S}| = \bar{m}$, then $\eta \leq m$, or more precisely,*

$$\eta \leq \min_{\mathcal{S} \subseteq [K]} \max(|C_{\mathcal{S}}|, K - |\mathcal{S}|). \tag{5.12}$$

Proof For each subset of transmitter indices $\mathcal{S} \subseteq [K]$, we apply Lemma 5.2 with the set $\mathcal{A}$ defined as follows.

Initially, set $\mathcal{A}$ as the set of indices for messages carried by transmitters with indices in $\mathcal{S}$. That is, $\mathcal{A} = C_{\mathcal{S}}$. Now, if $|\mathcal{A}| < K - |\mathcal{S}|$, then augment the set $\mathcal{A}$ with arbitrary message indices such that $|\mathcal{A}| = K - |\mathcal{S}|$.

We now note that the above construction guarantees that $|\mathcal{A}| + |\mathcal{S}| \geq K$ and that $U_{\mathcal{A}} \subseteq \bar{\mathcal{S}}$. Hence, using Lemma 5.2, it suffices to show the existence of functions f_1 and f_2 such that $f_1(y_{\mathcal{A}}, x_{\mathcal{S}}) = x_{\bar{\mathcal{S}}} + f_2(z_{\mathcal{A}})$, where f_2 is a linear function that does not depend on the transmit power.

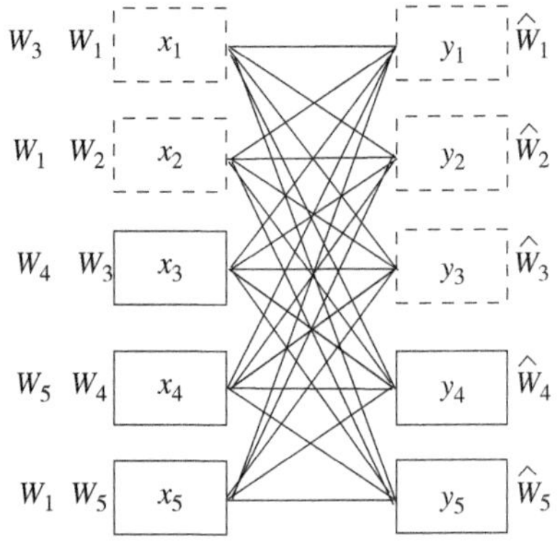

Fig. 5.3 Example application of Corollary 5.1, with $\mathcal{S} = \{1,2\}$ and $C_{\mathcal{S}} = \{1,2,3\}$. Transmit signals with indices in $\mathcal{S}$, and messages as well as receive signals with indices in $C_{\mathcal{S}}$, are shown in italics and dashed boxes. The DoF $\eta \leq |C_{\mathcal{S}}| = K - |\mathcal{S}| = 3$.

Consider the following argument. Given $y_{\mathcal{A}}$, $z_{\mathcal{A}}$, and $x_{\mathcal{S}}$, we can construct the set of signals $\tilde{y}_{\mathcal{A}}$ as follows:

$$\begin{aligned}\tilde{y}_i &= y_i - \left(\sum_{j \in \mathcal{S}} h_{ij} x_j + z_i\right) \\ &= \sum_{j \in \bar{\mathcal{S}}} h_{ij} x_j, \forall i \in \mathcal{A}. \end{aligned} \tag{5.13}$$

Since the channel is fully connected, by removing the Gaussian noise signals $z_{\mathcal{A}}$ and transmit signals in $x_{\mathcal{S}}$ from received signals in $y_{\mathcal{A}}$, we obtain the set of signals $\{\tilde{y}_i : i \in \mathcal{A}\}$, which has at least $K - |\mathcal{S}| = |\bar{\mathcal{S}}|$ linear equations in the transmit signals in $x_{\bar{\mathcal{S}}}$. Moreover, since the channel coefficients are generic, these equations will be linearly independent with high probability. Now, if we do not remove the noise signals $z_{\mathcal{A}}$ from (5.13), then by using $y_{\mathcal{A}}$ and $x_{\mathcal{S}}$, we can reconstruct $x_{\bar{\mathcal{S}}} + f_2(z_{\mathcal{A}})$, where f_2 depends on the inverse transformation of $|\bar{\mathcal{S}}|$ linearly independent equations in $x_{\bar{\mathcal{S}}}$, and the coefficients of the linear equations depend only on the channel coefficients.

See Figure 5.3 for an example illustration of Corollary 5.1. Using the corollary, we can identify the optimal DoF using spiral message assignments.

COROLLARY 5.2 *Let $\eta_{\text{sp}}(K,M)$ be the DoF of a fully connected K-user channel with each message assigned to $M \leq K$ transmitters according to the spiral message assignment*

$$\mathcal{T}_k = \{k, k+1, \ldots, k+M-1\}, \forall k \in [K], \tag{5.14}$$

with the indices taken modulo K; then the following bound holds:

$$\eta_{\text{sp}}(K,M) \leq \left\lceil \frac{K+M-1}{2} \right\rceil. \tag{5.15}$$

Proof We apply Corollary 5.1 with the set $\mathcal{S} = \left\{1,2,\ldots,\left\lceil\frac{K-(M-1)}{2}\right\rceil\right\}$. Note that transmitters with indices in $\mathcal{S}$ are carrying messages with indices in the set $C_{\mathcal{S}} = \left\{1,2,\ldots,\left\lceil\frac{K+M-1}{2}\right\rceil\right\}$, and it is true that $|\mathcal{S}| + |C_{\mathcal{S}}| \geq K$. Hence, the bound in (5.15) follows from (5.12).

Note that combining the above result with the lower bound in Theorem 5.1 means we obtain an approximate characterization of the DoF with spiral message assignments.

5.4 Degrees of Freedom Cooperation Gain

In the previous section we considered a specific choice of message assignments to transmitters, namely, the spiral message assignments where each message is available at the transmitter having the same index as well as $M-1$ following transmitters. In this section, we consider a general *cooperation order* constraint that only imposes a limit on the maximum transmit set size. More precisely,

$$|\mathcal{T}_i| \leq M, \forall i \in [K]. \tag{5.16}$$

We define $\tau(M)$ to denote the asymptotic per-user DoF with a cooperation order constraint of M. We now use Corollary 5.1 to prove upper bounds on the asymptotic per-user DoF $\tau(M)$.

In an attempt to reduce the complexity of the problem of finding an optimal message assignment, we begin by considering message assignments that rely only on a local cooperation constraint, i.e., each message can only be assigned within a neighborhood of transmitters whose size does not scale with the size of the network. We will formally define local cooperation below. We show that a scalable cooperation DoF gain cannot be achieved using local cooperation.

5.4.1 Local Cooperation

In order to formally define what we mean by local cooperation in large networks, we need to capture the constraint that the indices of neighboring transmitters that can carry a message cannot differ from a message index by a value that scales with the number of users. In order to do so, we need to describe message assignments for families of network topologies with increasing numbers of users K. We call this a message assignment strategy, defined formally as follows.

DEFINITION 5.2 A message assignment strategy $\{\mathcal{T}_{K,i}, K \in \mathbf{Z}^+, i \in [K]\}$ defines a message assignment for each network with K users. For the network having K_x users, the message assignment for user i is the set $\mathcal{T}_{K_x,i}$.

We define the local cooperation constraint for message assignment strategies.

DEFINITION 5.3 We say that a message assignment strategy $\{\mathcal{T}_{K,i}, K \in \mathbf{Z}^+, i \in [K]\}$ satisfies the local cooperation constraint if there exists a local neighborhood radius $r(K)$ such that

$$\lim_{K\to\infty} \frac{r(K)}{K} = 0, \tag{5.17}$$

and the following holds:

$$\mathcal{T}_{K,i} \subseteq \{i - r(K), i - r(K) + 1, \ldots, i + r(K)\}, \ \forall K, i. \tag{5.18}$$

We define $\tau^{(\text{loc})}(M)$ as the asymptotic per-user DoF with a cooperation order constraint of M, and the additional local cooperation constraint. Then we have following result (see also [71]):

THEOREM 5.3 *Any message assignment strategy satisfying the local cooperation constraint of* (5.18) *cannot be used to achieve an asymptotic per-user DoF greater than that achieved without cooperation. More precisely,*

$$\tau^{(\text{loc})}(M) = \frac{1}{2}, \text{ for all } M. \tag{5.19}$$

Proof Fix $M \in \mathbf{Z}^+$. For any value of $K \in \mathbf{Z}^+$, we use Corollary 5.1 with the set $\mathcal{S} = \{1, 2, \ldots, \lceil \frac{K}{2} \rceil\}$. Note that $C_{\mathcal{S}} \subseteq \{1, 2, \ldots, \lceil \frac{K}{2} \rceil + r(K)\}$, where $\lim_{K\to\infty} \frac{r(K)}{K} = 0$, and hence it follows that $\eta^{(\text{loc})}(K, M) \le \lceil \frac{K}{2} \rceil + r(K)$. Finally, $\tau^{(\text{loc})}(M) = \lim_{K\to\infty} \frac{\eta^{(\text{loc})}(K,M)}{K} \le \frac{1}{2}$. The lower bound follows from [46] without cooperation.

5.4.2 General Upper Bounds

We now investigate if it is possible for the cooperation gain to scale linearly with K for fixed M. It was shown in Theorem 5.3 that such a gain is not possible for message assignment strategies that satisfy the local cooperation constraint. Here, we only impose the maximum transmit set size constraint and show an upper bound on $\tau(M)$ that is tight enough for finding $\tau(2)$.

THEOREM 5.4 *For any cooperation order constraint $M \ge 2$, the following upper bound holds for the asymptotic per-user DoF:*

$$\tau(M) \le \frac{M-1}{M}. \tag{5.20}$$

Proof For any value of M and K, we show that $\eta(K, M) \le \frac{K(M-1)}{M} + o(K)$. For every value of K such that $\frac{K-1}{M}$ is an integer, we show that $\eta(K, M) \le \frac{K(M-1)+1}{M}$. When $\frac{K-1}{M}$ is not an integer, we add $x = o(K)$ extra users such that $\frac{K+x-1}{M}$ is an integer, and bound the DoF as follows:

$$\begin{aligned}
\eta(K, M) &\le \eta(K + x, M) && (5.21)\\
&\le \frac{(K + x)(M - 1) + 1}{M} && (5.22)\\
&= \frac{K(M-1)}{M} + o(K). && (5.23)
\end{aligned}$$

It then suffices to consider the case where $\frac{K-1}{M}$ is an integer. The idea is to show that for any assignment of messages satisfying the cooperation order constraint, there exists a set of indices $\mathcal{S} \subseteq [K]$ for $\frac{K-1}{M}$ transmitters that do not carry more than $K - \frac{K-1}{M}$ messages, and then the DoF upper bound follows by applying Corollary 5.1. More precisely, it suffices to show that the following holds: For all K such that $\frac{K-1}{M} \in \mathbf{Z}^+$, there exists $\mathcal{S} \subseteq [K]$ such that

$$|\mathcal{S}| = \frac{K-1}{M}, \; |\mathcal{C}_{\mathcal{S}}| = \frac{K(M-1)+1}{M} = K - |\mathcal{S}|. \tag{5.24}$$

We first illustrate simple examples that demonstrate the validity of (5.24). Consider the case where $K = 3, M = 2$. We need to show in this case that there exists a transmitter that does not carry more than two messages, which follows by the pigeonhole principle since each message can only be available at a maximum of two transmitters. Now, consider the slightly more complex example of $K = 5$, $M = 2$. We need to show in this case that there exists a set of two transmitters that do not carry more than three messages. We know that there is a transmitter carrying at most two messages, and we select this transmitter as the first element of the desired set. Without loss of generality, let the two messages available at the selected transmitter be W_1 and W_2. Now, we need to find another transmitter that carries at most one message among the messages in the set $\{W_3, W_4, W_5\}$. Since each of these three messages can be available at a maximum of two transmitters, and we have four transmitters to choose from, one of these transmitters has to carry at most one of these messages. By adding the transmitter satisfying this condition as the second element of the set, we obtain a set of two transmitters carrying no more than three messages, and (5.24) holds.

We extend the argument used in the above examples through Lemmas 5.6 and 5.7, provided in Section 5.5. We know by induction using these lemmas that (5.24) holds, and the theorem statement follows.

Together with the asymptotic interference alignment achievability result in [46], the statement in Theorem 5.4 implies the following corollary.

COROLLARY 5.3 *For any message assignment strategy such that each message is available at a maximum of two transmitters, the asymptotic per-user DoF is the same as that achieved without cooperation. More precisely,*

$$\tau(2) = \frac{1}{2}. \tag{5.25}$$

The characterization of $\tau(M)$ for values of $M > 2$ remains an open question, as Theorem 5.4 is only an upper bound. Moreover, the following result shows that the upper bound in Theorem 5.4 is loose for $M = 3$.

THEOREM 5.5 *For any message assignment strategy such that each message is available at a maximum of three transmitters, the following bound holds for the asymptotic per-user DoF:*

$$\tau(3) \leq \frac{5}{8}. \tag{5.26}$$

Proof In a similar fashion to the proof of Theorem 5.4, we prove the statement by induction. The idea is to prove the existence of a set $\mathcal{S}$ with approximately $\frac{3K}{8}$ transmitter indices, with these transmitters carrying no more than approximately $\frac{5K}{8} = K - |\mathcal{S}| + o(K)$ messages, and then use Corollary 5.1 to derive the DoF outer bound. In the proof of Theorem 5.4, we used Lemmas 5.6 and 5.7 in Section 5.5 to provide the basis and induction step of the proof, respectively. Here, we follow the same path until we show that there exists a set $\mathcal{S}$ such that $|\mathcal{S}| = \frac{K+1}{4}$ and $|C_{\mathcal{S}}| \leq (M-1)|\mathcal{S}| + 1$, and then we use Lemma 5.9 in Section 5.5 to provide a stronger induction step that establishes a tighter bound on the size of the set $C_{\mathcal{S}}$.

We note that it suffices to show that $\eta(K,3) \leq \frac{5K}{8} + o(K)$ for all values of K such that $\frac{K+1}{4}$ is an even positive integer, and hence we make that assumption for K. Define the following:

$$n_1 = \frac{K+1}{4}, \tag{5.27}$$

$$n_2 = \frac{K-7}{8}, \tag{5.28}$$

$$n_3 = 2n_1 + 1 + n_2. \tag{5.29}$$

Now, we note that

$$n_3 = K - (n_1 + n_2), \tag{5.30}$$

and by induction, it follows from Lemmas 5.6 and 5.7 that there exists $\mathcal{S}_1 \subset [K]$ such that $|\mathcal{S}_1| = n_1$ and $|C_{\mathcal{S}_1}| \leq 2n_1 + 1$. We now apply induction again with the set $\mathcal{S}_1$ as a basis, and use Lemma 5.9 for the induction step to show that there exists $\mathcal{S}_2 \subset [K]$ such that $|\mathcal{S}_2| = n_1 + n_2$ and $|C_{\mathcal{S}_2}| \leq n_3 = K - |\mathcal{S}_2|$. Hence, we get the following upper bound using Corollary 5.1:

$$\begin{aligned}\eta(K,3) &\leq n_3\\ &= \frac{5(K+1)}{8},\end{aligned} \tag{5.31}$$

from which (5.26) holds.

5.5 Auxiliary Lemmas for Large Network Upper Bounds

LEMMA 5.6 *For a K-user channel where each message is available at a maximum of M transmitters, there exists a transmitter carrying at most M messages, i.e., there exists* $i \in [K]$ *such that* $|C_{\{i\}}| \leq M$.

Proof The statement follows by the pigeonhole principle, since the following holds:

$$\sum_{i=1}^{K} |C_{\{i\}}| = \sum_{i=1}^{K} |\mathcal{T}_i| \leq MK. \tag{5.32}$$

LEMMA 5.7 *Consider a K-user channel, where each message is available at a maximum of M transmitters, and* $M \geq 2$. *If there exists a set* $\mathcal{A}$ *of n transmitters carrying at most* $(M-1)n+1$ *messages, then there exists a set* $\mathcal{B}$ *of* $n+1$ *transmitters carrying at most* $(M-1)(n+1)+1$ *messages. More precisely, if there exists* $\mathcal{A} \subset [K]$ *such that* $|\mathcal{A}| = n < K$, *and* $|C_{\mathcal{A}}| \leq (M-1)n+1$, *then there exists* $\mathcal{B} \subseteq [K]$ *such that* $|\mathcal{B}| = n+1$, *and* $|C_{\mathcal{B}}| \leq (M-1)(n+1)+1$.

Proof We only consider the case where $K > (M-1)(n+1)+1$, as otherwise the statement holds trivially. In this case, we can show that

$$M(K-|C_{\mathcal{A}}|) < (K-n)((M-1)(n+1)+2-|C_{\mathcal{A}}|). \tag{5.33}$$

The proof of (5.33) is available in Lemma 5.8 below. Note that the left-hand side in (5.33) is the maximum number of message instances for messages outside the set $C_{\mathcal{A}}$, i.e.,

$$\begin{aligned}\sum_{i\in[K],i\notin\mathcal{A}} |C_{\{i\}}\backslash C_{\mathcal{A}}| &\leq M(K-|C_{\mathcal{A}}|)\\ &< (K-n)((M-1)(n+1)+2-|C_{\mathcal{A}}|).\end{aligned} \tag{5.34}$$

Since the number of transmitters outside the set $\mathcal{A}$ is $K-n$, it follows by the pigeonhole principle that there exists a transmitter whose index is outside $\mathcal{A}$ and carries at most $(M-1)(n+1)+1-|C_{\mathcal{A}}|$ messages whose indices are outside $C_{\mathcal{A}}$. More precisely,

$$\exists i \in [K]\backslash\mathcal{A} : |C_{\{i\}}\backslash C_{\mathcal{A}}| \leq (M-1)(n+1)+1-|C_{\mathcal{A}}|. \tag{5.35}$$

It follows that there exists a transmitter whose index is outside the set $\mathcal{A}$ and can be added to the set $\mathcal{A}$ to form the set $\mathcal{B}$ that satisfies the statement.

LEMMA 5.8 *For any positive integers* $K, M, n \in \mathbf{Z}^+$ *such that* $M \geq 2$ *and* $K \geq (M-1)(n+1)+1$, *the following holds for any set* $\mathcal{S} \subseteq [K]$ *such that* $|\mathcal{S}| \leq (M-1)n+1$:

$$M(K-|\mathcal{S}|) < (K-n)\left((M-1)(n+1)+2-|\mathcal{S}|\right). \tag{5.36}$$

Proof We first prove the statement for the case where $|\mathcal{S}| = (M-1)n+1$. This directly follows, as

$$\begin{aligned}M(K-|\mathcal{S}|) &= M(K-((M-1)n+1))\\ &\leq M(K-(n+1))\\ &< M(K-n)\\ &= (K-n)\left((M-1)(n+1)+2-|\mathcal{S}|\right).\end{aligned} \tag{5.37}$$

In order to complete the proof, we note that each decrement of $|\mathcal{S}|$ leads to an increase in the left-hand side by M, and in the right-hand side by $K-n$, and

$$\begin{aligned}K-n &\geq (M-1)(n+1)+1-n\\ &= (M-2)n+M\\ &\geq M.\end{aligned} \tag{5.38}$$

LEMMA 5.9 *For a K-user channel, if each message is available at a maximum of $M = 3$ transmitters, and there exists a set $\mathcal{A}$ of n transmitters carrying at most $n + \frac{K+1}{4} + 1$ messages, and $\frac{K+1}{4} \leq n < K$, then there exists a set $\mathcal{B}$ of $n+1$ transmitters carrying at most $n + \frac{K+1}{4} + 2$ messages. More precisely, if $\exists \mathcal{A} \subset [K]$ such that $|\mathcal{A}| = n$, $\frac{K+1}{4} \leq n < K$, and $|C_{\mathcal{A}}| \leq n + \frac{K+1}{4} + 1$, then $\exists \mathcal{B} \subset [K]$ such that $|\mathcal{B}| = n+1$, $|C_{\mathcal{B}}| \leq n + \frac{K+1}{4} + 2$.*

Proof The proof follows in a similar fashion to that of Lemma 5.7. Let $x = n + \frac{K+1}{4} + 1$. We only consider the case where $K > x + 1$, as otherwise the proof is trivial. We first assume the following:

$$3(K - |C_{\mathcal{A}}|) < (K - n)\left(n + \frac{K+1}{4} + 3 - |C_{\mathcal{A}}|\right). \tag{5.39}$$

Now, it follows that

$$\begin{aligned}\sum_{i \in [K], i \notin \mathcal{A}} |C_{\{i\}} \backslash C_{\mathcal{A}}| &\leq M(K - |C_{\mathcal{A}}|) \\ &< (K - n)\left(n + \frac{K+1}{4} + 3 - |C_{\mathcal{A}}|\right),\end{aligned} \tag{5.40}$$

and hence

$$\exists i \in [K] \backslash \mathcal{A} : |C_{\{i\}} \backslash C_{\mathcal{A}}| \leq n + \frac{K+1}{4} + 2 - |C_{\mathcal{A}}|, \tag{5.41}$$

and then the set $\mathcal{B} = \mathcal{A} \cup \{i\}$ satisfies the statement of the lemma. Finally, we need to show that (5.39) is true. For the case where $|C_{\mathcal{A}}| = x$,

$$\begin{aligned}3x &= \frac{3K}{4} + \frac{15}{4} + 3n \\ &= (2n + K) + \left(n - \frac{K}{4} + \frac{15}{4}\right) \\ &> 2n + K,\end{aligned} \tag{5.42}$$

and hence $3(K - x) < 2(K - n)$, which implies (5.39) for the case where $|C_{\mathcal{A}}| = x$. Moreover, we note that each decrement of $|C_{\mathcal{A}}|$ increases the left-hand side of (5.39) by 3 and the right-hand side by $(K - n)$, and we know that

$$\begin{aligned}K &> x + 1 \\ &= n + \frac{K+1}{4} + 2 \\ &\geq n + 2,\end{aligned} \tag{5.43}$$

and hence $K - n \geq 3$. Therefore, there is no loss of generality in assuming that $|C_{\mathcal{A}}| = x$ in the proof of (5.39), and the proof is complete.

5.6 Connection between DoF Upper Bound and Bipartite Vertex Expanders

We note that all the DoF upper-bounding proofs used so far employ Corollary 5.1. We now show that under the hypothesis that the upper bound in Corollary 5.1 is tight for any K-user fully connected interference channel with a cooperation order constraint M, then scalable DoF cooperation gains are achievable for any value of $M \geq 3$. Hence, a solution to the general problem necessitates the discovery of either new upper-bounding techniques or new coding schemes.

In this section, we restrict our attention to upper bounds on $\tau(M)$ that follow by a direct application of Corollary 5.1. More precisely, for a K-user fully connected channel with an assignment of the transmit sets $\{\mathcal{T}_i\}_{i\in[K]}$, define $B(K,\{\mathcal{T}_i\})$ as the upper bound that follows by Corollary 5.1 for this channel, i.e.,

$$B(K,\{\mathcal{T}_i\}) = \min_{\mathcal{S}\subseteq[K]} \max(|C_{\mathcal{S}}|, K-|\mathcal{S}|). \tag{5.44}$$

Now, let $\eta_{\text{out}}(K,M)$ and $\tau_{\text{out}}(M)$ be the corresponding upper bounds that apply on $\eta(K,M)$ and $\tau(M)$, respectively:

$$\eta_{\text{out}}(K,M) = \max_{\{\mathcal{T}_i\}_{i\in[K]}:\mathcal{T}_i\subseteq[K],|\mathcal{T}_i|\leq M,\forall i\in[K]} B(K,\{\mathcal{T}_i\}), \tag{5.45}$$

$$\tau_{\text{out}}(M) = \lim_{K\to\infty} \frac{\eta_{\text{out}}(K,M)}{K}. \tag{5.46}$$

All the facts that we stated above about $\tau(M)$ hold for $\tau_{\text{out}}(M)$, since all the discussed upper-bounding proofs follow by a direct application of Corollary 5.1. We now identify a property for message assignment strategies that leads us to prove that $\tau_{\text{out}}(M) > \frac{1}{2}, \forall M > 2$. Note that this does not necessarily imply that $\tau(M) > \frac{1}{2}, \forall M > 2$, but it provides some insight into whether this statement might be true [72].

For each possible message assignment, define a bipartite graph with partite sets of size K. Vertices in one of the partite sets represent transmitters, and vertices in the other set represent messages. There exists an edge between two vertices if and only if the corresponding message is available at the designated transmitter. We note that the maximum transmit set size constraint implies that the maximum degree of nodes in one of the partite sets is bounded by M. We now observe that for any set $\mathcal{A}$ of transmitters, $C_{\mathcal{A}} = \{i : \mathcal{T}_i \cap \mathcal{A} \neq \emptyset\}$ is just the neighboring set $N_G(\mathcal{A})$ in the corresponding bipartite graph G. See Figure 5.4 for an illustration of the bipartite graph representation of message assignments.

Let $\mathcal{U}_G, \mathcal{V}_G$ denote the partite sets corresponding to transmitters and messages in graph G, with respect to order. For all values of $i \in [K]$, define the following:

$$e_G(i) = \min_{\mathcal{A}\subseteq\mathcal{U}_G:|\mathcal{A}|=i} |N_G(\mathcal{A})|. \tag{5.47}$$

Then we can readily see that

$$\eta_{\text{out}}(K,M) = \max_{G\in\mathcal{G}_M(K)} \min_{i\in[K]} \max(K-i, e_G(i)), \tag{5.48}$$

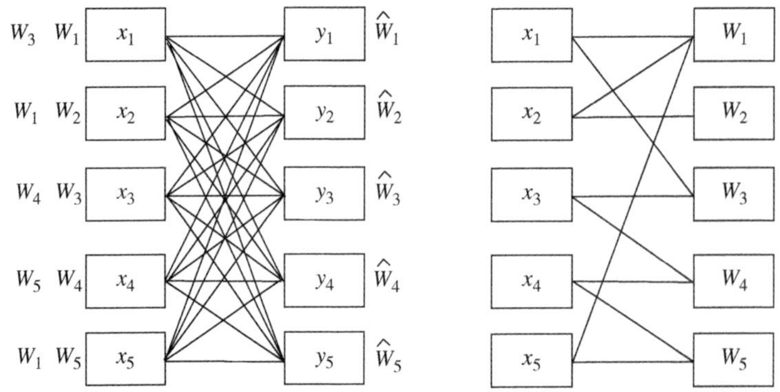

Fig. 5.4 The bipartite graph on the right side represents the message assignment for the five-user channel shown on the left side. ©[2017] IEEE. Reprinted, with permission, from [72].

where $\mathcal{G}_M(K)$ is the set of all bipartite graphs whose equi-sized partite sets have size K, and the maximum degree of the nodes in the partite set $\mathcal{V}_G$ is M.

For values of $M > 2$, Pinsker proved the following result in 1973 [73].

THEOREM 5.10 *For any $M > 2$, there exists a constant $c > 1$ and a sequence of M-regular bipartite graphs $(G_{M,K})$ whose partite sets have K vertices such that the following is true:*

$$\lim_{K\to\infty} \frac{e_{G_{M,K}}(\alpha K)}{\alpha K} \geq c, \ \forall 0 < \alpha \leq \frac{1}{2}. \tag{5.49}$$

We next show that the above statement implies that $\tau_{\text{out}}(M) > 2, \forall M > 2$.

COROLLARY 5.4

$$\tau_{\text{out}}(M) > \frac{1}{2}, \ \forall M > 2. \tag{5.50}$$

Proof For each bipartite graph G with partite sets of size K, define $i_{\min}(G)$ as

$$i_{\min}(G) = \operatorname{argmin}_i \max(K - i, e_G(i)). \tag{5.51}$$

Now, assume that $\tau_{\text{out}}(M) \leq \frac{1}{2}$; then, for the sequence $(G_{M,K})$ chosen as in the statement of Theorem 5.10,

$$\lim_{K\to\infty} \frac{\max(K - i_{\min}(G_{M,K}), e_{G_{M,K}}(i_{\min}(G_{M,K})))}{K} \leq \frac{1}{2}. \tag{5.52}$$

It follows that

$$\lim_{K\to\infty} \frac{K - i_{\min}(G_{M,K})}{K} \leq \frac{1}{2}, \tag{5.53}$$

or

$$\lim_{K\to\infty} \frac{i_{\min}(G_{M,K})}{K} \geq \frac{1}{2}. \tag{5.54}$$

But then, since $e_G(i)$ is non-decreasing in i, (5.49) implies that

$$\lim_{K\to\infty} \frac{e_{G_{M,K}}(i_{\min}(G_{M,K}))}{K} > \frac{1}{2}. \tag{5.55}$$

Therefore, the result in (5.49) implies that $\tau_{\text{out}}(M) > \frac{1}{2}$, for all $M > 2$.

It is worth noting that a sequence of bipartite graphs satisfying (5.49) is said to define a *vertex expander* as $K \to \infty$. To summarize, we have shown that because of message assignment strategies corresponding to vertex expanders, one cannot apply the bound in (5.44) directly to show that $\tau(M) = \tau(1) = \frac{1}{2}$ for any $M > 2$. Finally, we show that in the case that the upper bound $\tau_{\text{out}}(M)$ is tight, then, using partial cooperation, the DoF gain can approach that achieved through assigning each message to all transmitters (full cooperation). More precisely, we show the following.

THEOREM 5.11

$$\lim_{M\to\infty} \tau_{\text{out}}(M) = 1. \tag{5.56}$$

Proof We show that

$$\forall \epsilon > 0, \exists M(\epsilon) : \forall M \geq M(\epsilon), \tau_{\text{out}}(M) > (1-\epsilon). \tag{5.57}$$

For each positive integer K, we construct a bipartite graph $G_M(K)$ whose partite sets are of order K by taking the union of M *random* perfect matchings between the two partite sets. That is, the matchings are probabilistically independent, and each is drawn uniformly from the set of all possible matchings. One can easily see that the maximum degree of nodes in $G_M(K)$ is bounded by M, i.e., $\Delta(G_M(K)) \leq M$, and hence $G_M(K) \in \mathcal{G}_M(K)$. We will prove that for any $\epsilon > 0$, there exists an $M(\epsilon)$ sufficiently large such that for any $M \geq M(\epsilon)$, the probability that each set of ϵK nodes in the partite set $\mathcal{U}_{G_M(K)}$ have more than $(1-\epsilon)K$ neighbors is bounded away from zero for large enough K. More precisely, we show that

$$\lim_{K\to\infty} \Pr[\forall \mathcal{A} \subset \mathcal{U}_{G_M(K)} : |\mathcal{A}| = \epsilon K, |N_{G_M(K)}(\mathcal{A})| > (1-\epsilon)K] > 0, \tag{5.58}$$

and hence, for large enough K, there exists a graph G in $\mathcal{G}_M(K)$ where all subsets of $\mathcal{U}_G$ of order ϵK have more than $(1-\epsilon)K$ neighbors in $\mathcal{V}_G$, i.e.,

$$e_G(i) > (1-\epsilon)K, \forall i \geq \epsilon K, \tag{5.59}$$

and it follows that $\eta_{\text{out}}(K,M) > (1-\epsilon)K$, and (5.57) holds.

We now show that (5.58) holds. Let $\mathcal{A} \subset \mathcal{U}_{G_M(K)}$, $\mathcal{B} \subset \mathcal{V}_{G_M(K)}$ such that $|\mathcal{A}| = \epsilon K$, $|\mathcal{B}| = (1-\epsilon)K$. For any random perfect matching, the probability that all the neighbors of $\mathcal{A}$ are in $\mathcal{B}$ is $\frac{\binom{(1-\epsilon)K}{\epsilon K}}{\binom{K}{\epsilon K}}$. By independence of the matchings, we get the following:

$$\begin{aligned} \Pr[N_{G_M(K)}(\mathcal{A}) \subseteq \mathcal{B}] &= \left(\frac{\binom{(1-\epsilon)K}{\epsilon K}}{\binom{K}{\epsilon K}}\right)^M \\ &\leq \left((1-\epsilon)^{\epsilon K}\right)^M. \end{aligned} \tag{5.60}$$

A direct application of the union bound results in the following:

$$\begin{aligned}\Pr[|\mathcal{N}_{G_M(K)}(\mathcal{A})| \le (1-\epsilon)K] \\ &\le \sum_{\mathcal{B}\subset\mathcal{V}_{G_M(K)}:|\mathcal{B}|=(1-\epsilon)K} \Pr[\mathcal{N}_{G_M(K)}(\mathcal{A}) \subseteq \mathcal{B}] \\ &\le \binom{K}{(1-\epsilon)K}(1-\epsilon)^{\epsilon MK},\end{aligned} \tag{5.61}$$

and

$$\begin{aligned}\Pr[\exists\mathcal{A}\subset\mathcal{U}_{G_M(K)} : |\mathcal{A}| = \epsilon K, |\mathcal{N}_{G_M(K)}(\mathcal{A})| \le (1-\epsilon)K] \\ &\le \sum_{\mathcal{A}\subset\mathcal{U}_{G_M(K)}:|\mathcal{A}|=\epsilon K} \Pr[|\mathcal{N}_{G_M(K)}(\mathcal{A})| \le (1-\epsilon)K] \\ &\le \binom{K}{\epsilon K}\binom{K}{(1-\epsilon)K}(1-\epsilon)^{\epsilon MK} \\ &= \binom{K}{\epsilon K}^2 (1-\epsilon)^{\epsilon MK} \\ &\overset{(a)}{\approx} 2^{2K\mathsf{H}(\epsilon)}(1-\epsilon)^{\epsilon MK} \\ &= 2^{(2\mathsf{H}(\epsilon)+\epsilon M\log(1-\epsilon))K},\end{aligned} \tag{5.62}$$

where $\mathsf{H}(\cdot)$ is the binary entropy function, and (a) follows as $\binom{n}{\epsilon n} \approx 2^{n\mathsf{H}(\epsilon)}$ for large enough n. Now, we choose $M(\epsilon) > \frac{2\mathsf{H}(\epsilon)}{-\epsilon\log(1-\epsilon)}$, to make the above exponent negative, and the above probability will be strictly less than unity, i.e., we showed that for any $M \ge M(\epsilon)$,

$$\lim_{K\to\infty} \Pr[\exists\mathcal{A}\subset\mathcal{U}_{G_M(K)} : |\mathcal{A}| = \epsilon K, |\mathcal{N}_{G_M(K)}(\mathcal{A})| \le (1-\epsilon)K] < 1, \tag{5.63}$$

which implies that (5.58) is true.

5.7 Iterative Algorithms for Constant MIMO Channel with CoMP

In Chapter 4, we studied iterative algorithms for designing linear transmitter and receiver processing to maximize the sum-rate for communication on interference channels. In particular, we developed a convergent version of the Max SINR algorithm proposed by Gomadam et al. [38, 39]. In this section, we extend the iterative approach to the case where we allow transmit cooperation through CoMP. The model that we use is the one given in (4.1), with the additional simplification that $N_\text{t} = N_\text{r} = N$. We focus on the spiral set cooperation model introduced in (5.1).

The feasibility conditions for interference alignment given in Section 3.4 can be extended to the case with transmit cooperation. This analysis is facilitated by the two-stage scheme described in Section 5.2, consisting of first applying zero-forcing and then IA, as shown in Figure 5.1. The derived channel depicted in Figure 5.2

corresponding to the constant MIMO channel has $2N$ transmit antennas for each of the first $M-1$ users, and N transmit antennas for each of the remaining users. Thus, the achievable DoF may not be the same for all users. If every user is constrained to have the same DoF, it is shown in [74] that the achievable sum DoF (assuming $M \leq K-2$) is given by

$$\eta_{\text{ach}}^{\text{sym}} = K \left\lfloor \frac{2N}{K-M+2} \right\rfloor. \tag{5.64}$$

In the asymmetric case, where the per-user DoF is variable, the achievable sum DoF is given by

$$\eta_{\text{ach}}^{\text{asym}} = \max\left\{(M-1)N + (K-M+1)\left\lfloor \frac{N}{K-M+1} \right\rfloor, K\left\lfloor \frac{2N}{K-M+2} \right\rfloor\right\}. \tag{5.65}$$

These achievable sum DoF values should be compared with that achievable without cooperation, given in (3.19), i.e.,

$$\eta_{\text{ach}}^{\text{no coop}} = K \left\lfloor \frac{2N}{K+1} \right\rfloor. \tag{5.66}$$

For large N, if we ignore the floor function in the degrees of freedom expressions, we have

$$\begin{aligned} \eta_{\text{ach}}^{\text{no coop}} &\approx \frac{2KN}{K+1}, \\ \eta_{\text{ach}}^{\text{sym}} &\approx \frac{2KN}{K-M+2}, \\ \eta_{\text{ach}}^{\text{asym}} &\approx \max\left\{MN, \frac{2KN}{K-M+2}\right\}. \end{aligned} \tag{5.67}$$

Therefore, if $M \geq 2$, we have $\eta_{\text{ach}}^{\text{sym}} > \eta_{\text{ach}}^{\text{no coop}}$ and $\eta_{\text{ach}}^{\text{asym}} > \eta_{\text{ach}}^{\text{no coop}}$. We use the degrees of freedom guidelines of (5.67) to choose the number of beams in the iterative algorithms for interference management with CoMP.

In particular, suppose user k wants to send d_k symbols denoted by $\boldsymbol{x}_k$. We assume that

$$\boldsymbol{x}_k \sim \mathcal{CN}\left(\mathbf{0}, \text{diag}(\rho_{k1}, \ldots, \rho_{kd_k})\right) \tag{5.68}$$

with $\rho_{k\ell}$ chosen to satisfy a per-transmitter power constraint P, i.e.,

$$\sum_{j=0}^{M_t-1} \sum_{\ell=1}^{d_{k-j}} \rho_{k-j,\ell} \leq P. \tag{5.69}$$

Let $\boldsymbol{\Sigma}_k$ be a diagonal matrix with the powers used to send the d_k streams of user k on the diagonal, i.e.,

$$\boldsymbol{\Sigma}_k = \text{diag}(\rho_{k1}, \ldots, \rho_{kd_k}). \tag{5.70}$$

To construct its $\boldsymbol{x}_j$, as in (4.4) user j uses a linear transmit strategy denoted $\boldsymbol{V}_j$, so $\tilde{\boldsymbol{x}}_j = \boldsymbol{V}_j \boldsymbol{x}_j$, where $\boldsymbol{V}_j$ is an $MN \times d_j$ matrix. At receiver k we zero-force the interfering signals using an $N \times d_k$ matrix of zero-forcing vectors $\boldsymbol{U}_k$, as in (4.5). Note that we do not need to construct the derived channels first; we can simply design the transmit and receive vectors directly for the original channel.

As in Section 4.1.1, the algorithms for iterative interference management involve working with a reciprocal channel corresponding to reversing the direction of communication. The iterative algorithms presented in Chapter 4 can readily incorporate CoMP transmission, by replacing the channels in those algorithms with effective channels defined by

$$\hat{\boldsymbol{H}}_{kj} = \left[\boldsymbol{H}_{k,j} \cdots \boldsymbol{H}_{k,j+M-1}\right],$$

where the indices $j,\ldots,j+M-1$ are taken modulo K.

It is shown in numerical results given in [74] that iterative algorithms for interference management (Max SINR and Min Leakage) show significant improvements in sum-rate with the inclusion of CoMP transmission, even with $M = 2$. The gains are most significant in regimes where interference alignment is not feasible without CoMP transmission but is feasible with CoMP transmission. More importantly, with the inclusion of CoMP transmission, the iterative algorithms converge much faster, with as little as one iteration being enough to achieve good sum-rate performance for some system configurations.

5.8 Discussion: Justifying Model Choices

In this chapter, we have drawn conclusions about the value of CoMP transmission in dense fully connected networks. In the following chapters, we observe that some of these conclusions are specific to the assumption of full connectivity. In Chapter 6, we alter that assumption to investigate DoF gains in large locally connected networks, where each receiver observes only the transmit signals that originate in a local neighborhood. We will see that conclusions regarding the value of CoMP transmission and the right choices of assigning messages to transmitters as well as the joint transmission scheme will differ dramatically between the study of fully connected networks in this chapter and the study of locally connected networks in Chapter 6. In Chapter 9, we introduce *dynamic interference networks* where the topology can change due to deep fading conditions, and it is desired to find the assignment of backhaul resources that results in the maximum DoF averaged over all possible network realizations. We now discuss the choices of the system model considered in both this chapter and the rest of this book. Namely, the cooperation constraint that is used to model a backhaul network with limited rate resources, the choice of the degrees of freedom criterion for studying the information-theoretic capacity, and the choice of studying how the DoF scales with the number of users in the network.

5.8.1 Cooperation Constraint

We only consider the sharing of whole messages over a backhaul network that connects basestations with a central controller. We also started from Section 5.3 considering a maximum transmit set size constraint, where each message can only be available at a maximum number of transmitters. The soundness of these assumptions originates from two aspects. First, the knowledge of solutions using the considered cooperation constraint under different settings can significantly reduce the difficulty of obtaining solutions under more complex assumptions that are closer to practice. For example, we will see in Chapter 7 that optimal solutions obtained using the per-message maximum transmit set size constraint can be used to construct optimal solutions under an overall backhaul load constraint that allows for distributing one message to more transmitters at the cost of distributing other messages to fewer transmitters. The second aspect is the fitness of this model in the emerging cloud-based communication paradigm with a digital backhaul infrastructure. Understanding the structural impact of the network topology on the choice of assigning messages to basestation transmitters – or more generally, associating mobile terminals with basestations – could be very useful for future wireless networks. The tractability of the considered cooperation constraint will be demonstrated by characterizing the optimal message assignment and per-user DoF for multiple network topologies. Having a constraint that is a function of the number of message instances distributed through the backhaul, rather than imposing a direct backhaul capacity constraint, simplifies the statistical nature of the problem. Further, imposing a per-message constraint, rather than an overall backhaul constraint, simplifies the combinatorial nature of the problem.

5.8.2 Large Networks

We discussed in Section 5.1 an example where CoMP transmission can be used to achieve DoF gains through asymptotic interference alignment, by having each message available at two transmitters. However, we saw in the rest of the chapter that these DoF gains are not significant in large networks. In particular, assigning each message to two transmitters cannot lead to DoF gains that scale with the size of the network, regardless of the choices of message assignment and coding schemes. Our choice to discuss analysis of large networks allows us to avoid difficult problems that arise when analyzing exact DoF gains, which do not lead to insights that scale with the size of the network. Understanding large wireless networks with centralized designs is expected to be particularly relevant to future network designs and applications.

5.8.3 Degrees of Freedom

Degrees of freedom is the pre-log factor of the information-theoretic capacity at high signal-to-noise ratio. When studied in a network with more than one transmitter–receiver pair, it roughly captures the number of interference-free communication links in an equivalent network that has the same sum capacity. While the capacity of the simple two-user interference network is still a hard open problem, much progress has been made in characterizing the DoF of wireless networks in multiple settings over the past decade. Most notably, as we discussed in Chapter 3, the DoF of a K-user interference network with each message being assigned to the transmitter carrying the same index was identified as $\frac{K}{2}$, and the achieving scheme relies on asymptotic interference alignment. The main reason for considering DoF analysis is tractability, due to the many simplifications offered by this analysis, which can be summarized as follows:

- **Ignoring all sublinear capacity terms:** By definition, the DoF analysis captures only the pre-log factor of the information-theoretic capacity. Any term that is an offset, or scales sublinearly with the logarithm of the transmit power, is not accounted for. This simplifies the analysis as we only look for factors that scale with the point-to-point capacity. For example, the DoF of any two-user interference channel with an interfering link is unity.
- **Ignoring transmit power:** All considered DoF analyses assume that all users have the same transmit power, and the DoF is measured at the infinite limit of that power. Hence, any difficulty associated with the impact of different transmit powers for different users is neglected.
- **Ignoring Gaussian noise:** Since the power of Gaussian noise does not scale with the transmit power, no consideration is given to combating the Gaussian noise in DoF analysis. This shifts the focus to understanding the best schemes for mitigating interference.
- **Ignoring different channel strengths:** The DoF analysis we consider here is oblivious to differences in channel strengths. However, there exists a generalization in the literature that is known as the generalized degrees of freedom (GDoF) that takes this factor into account. Ignoring variations in channel strengths allows us to focus our study on obtaining insights for resource allocation and coding schemes based solely on the network topology.

The soundness of DoF analyses can also be derived from the above reasons. In particular, through the insights we obtain from DoF analysis, we are able to ignore the point-to-point code analysis and focus on network interference management. Therefore, regardless of the mechanism used to alleviate the effects mentioned in the above points, and most notably the Gaussian noise, we could incorporate these mechanisms as building blocks in the obtained framework for network-level interference management.

6 Locally Connected Channels with CoMP

In the previous chapter, we highlighted an important negative conclusion regarding the potential of CoMP transmission in dense wireless networks. That is, if sharing of messages is restricted to local neighborhoods whose size does not increase linearly with the size of the network, then CoMP transmission cannot be used to increase the per-user DoF in large fully connected networks. We now question the practical relevance of the full connectivity assumption. Understanding the analysis of fully connected networks can enable theoretical insights for extreme scenarios where all the interfering channel links are strong enough, such that it is not appropriate to treat any interfering signal as noise. However, in practice, path loss effects result in a weaker received interference power as the distance between the interfering transmitter and receiver increases. For this reason, it is the case that the number of dominant interfering links in most cellular networks ranges from three to seven. By "dominant," here, we mean links for which the received interference power is considerably larger than the noise power. We therefore study the DoF in locally connected interference channels with CoMP transmission in this chapter.

6.1 Channel Model

We introduce the locally connected channel model by first defining a parameter L to denote the number of dominant interfering signals per receiver. Each transmitter is connected to $\left\lfloor \frac{L}{2} \right\rfloor$ preceding receivers and $\left\lceil \frac{L}{2} \right\rceil$ succeeding receivers. We use $\eta_L(K,M)$ as the best achievable DoF η over all choices of transmit sets satisfying a maximum transmit set size constraint of M, and $\tau_L(M)$ to denote the asymptotic per-user DoF under the same constraint.

We may use a more convenient alternative model for the locally connected channel without affecting the conclusions regarding the asymptotic per-user DoF. Instead of assuming that the set of receivers connected to a transmitter is approximately split between receivers that have indices preceding the transmitter index and those with succeeding indices, we assume that each transmitter with index i is connected to receivers with indices $\{i, i+1, \ldots, i+L\}$. We can see that the two models are equivalent for the purposes of analyzing the per-user DoF from the following argument. Let $x = \left\lfloor \frac{L}{2} \right\rfloor$. We silence the first x transmitters, deactivate the last x receivers, and relabel the transmit signals to obtain a $(K-x)$-user channel, where transmitter j is connected to

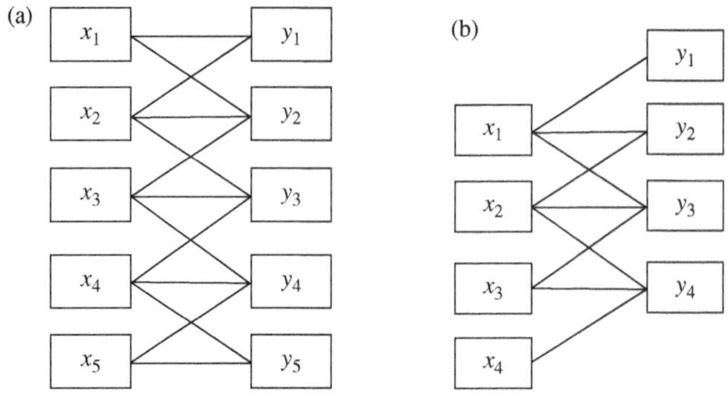

Fig. 6.1 Construction of the equivalent locally connected channel model with the number of users $K = 5$ and connectivity parameter $L = 2$. (a): the original model. (b): the new model. ©[2017] IEEE. Reprinted, with permission, from [71].

receivers in the set $\{y_i : i \in \{j, j+1, \ldots, j+L\}\}$. We note that the new channel model gives the same value of $\tau_L(M)$ as the original one, since $x = o(K)$. Unless explicitly stated otherwise, we will be using this equivalent model in the rest of this chapter. Finally, we assume that the set of values for all existing communication links is drawn from a joint continuous distribution. We show an example construction of the equivalent channel model in Figure 6.1.

For the case when $L = 1$, each transmitter is connected to the receiver carrying the same index, as well as one succeeding receiver. This network is known as Wyner's asymmetric interference network ($L = 2$ refers to the symmetric case) [75]. We will also refer to this network as the linear interference network due to its connectivity pattern. The rationale for analyzing this network is that it allows us to obtain rigorous conclusions that yield insights into understanding more complex networks that may have practical significance.

6.2 Interference-Aware Cell Association

We start our discussion of locally connected channels by studying the simpler scenario where no cooperation between transmitters is allowed. Suppose that each message can be available at one transmitter. For the case where $L = 1$, i.e., Wyner's asymmetric linear interference network, if message i is available at transmitter i, for all $i \in [K]$, then the asymptotic per-user DoF is exactly $\frac{1}{2}$. The information-theoretic converse can be proved in this case by arguing that a centralized decoder that has access to all received signals with even indices can decode all messages (up to the uncertainty in the noise) because of the simple connectivity of the channel. Given y_2, we can determine W_2, and hence we can determine x_2, and hence we can determine x_1, and then W_1. What we argue for next is that by allowing for a more flexible pattern of assigning messages (or receivers) to single basestation transmitters, or what we call a *cell association* decision that is *interference aware*, we could achieve gains in the asymptotic per-user DoF even for this simple scenario.

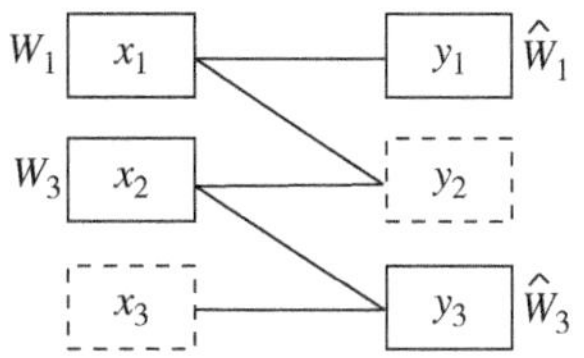

Fig. 6.2 Achieving $\frac{2}{3}$ per-user DoF without cooperative transmission. Only signals corresponding to the first subnetwork in a general K-user network are shown. The signals in the dashed boxes are deactivated.

Consider the following cell association pattern. We designate the transmit signal x_1 to serve the first receiver, and the transmit signal x_2 to serve the third receiver; this is the same as saying that we assign message W_1 to the first transmitter and message W_3 to the second transmitter. Further, we deactivate the third transmitter and the second receiver. By deactivating the third transmitter, we disconnect the subnetwork consisting of the first three transmitters and receivers from the rest of the network. Also, in that subnetwork consisting of the first three users, two degrees of freedom can be achieved, since messages W_1 and W_3 can be communicated without interference. The same cell association pattern can be repeated for every subsequent subnetwork that consists of three consecutive transmitter–receiver pairs. For example, assigning message W_4 to the fourth transmitter and message W_6 to the fifth transmitter, and deactivating the sixth transmitter and the fifth receiver. Since we achieve two degrees of freedom for every set of three users, the asymptotic per-user DoF achieved through this cell association pattern is $\frac{2}{3}$. We illustrate the scheme in Figure 6.2.

This example highlights the key idea that we wish to present in this chapter. By taking into account knowledge of the locally connected network topology, we can make smarter message assignments that lead to schemes that are simpler than those based on interference alignment (see Chapter 3), and still achieve unprecedented gains. For example, the coding scheme in the example presented relies only on interference avoidance. In the coming sections, we extend this insight to the case where cooperative transmission is possible by allowing each message to be available at more than one transmitter, and illustrate how an extension of the scheme in this simple example is in fact optimal.

6.3 One-Shot Linear Beamforming Schemes

A crucial advantage of interference avoidance coding schemes, besides their simple analysis, is that they require no symbol extensions. This is in contrast to interference alignment schemes that deliver their promised gains in DoF only asymptotically as the length of the symbol extension used goes to infinity (see Chapter 3). However, basic interference avoidance that relies on a time-division multiple-access (TDMA) strategy does not exploit the availability of a message at more than one transmitter. In order to exploit cooperative transmission, while maintaining the no-delay advantage of interference avoidance, we discuss here a generalization of basic interference avoidance schemes to incorporate cooperation by designing zero-forcing transmit beams to cancel undesired interference. Consider a simple linear pre-coding scheme, where a zero-forcing transmit beamforming strategy is employed. The transmit signal at the jth transmitter is given by

$$x_j = \sum_{i:j\in\mathcal{T}_i} x_{j,i}, \tag{6.1}$$

where $x_{j,i}$ depends only on message W_i. Let us also only consider message assignments where each message can only be assigned to a set of M successive transmitters. Intuitively, restricting the transmit set of each message to a set of transmitters with successive indices should not hurt performance. This follows from the connectivity of the channel, since any two transmitters whose indices differ by more than one cannot both be connected to a single receiver, and we would not assign a message to a transmitter unless this transmitter will either help in delivering the message to its intended destination or help to cancel the interference caused by this message at an unintended receiver. Canceling or reducing interference through a transmit signal can only happen if the corresponding transmitter is connected to a receiver that is also connected to another transmitter carrying the same message. We formalize this concept in the next section by characterizing a class of *useful* message assignments for any network topology.

Like basic interference avoidance, the zero-forcing transmission strategy involves deactivating some receivers in each communication session and achieving interference-free delivery of each active message to its intended receiver. First, a unique transmit signal is designated for the delivery of each active message to its destination. Then, each remaining transmit signal in the transmit set is used to cancel the interference caused by the message at an active receiver. Completely eliminating interference *over the air* is valid when we assume that perfect information about the channel state is available at all transmitters, and also assume perfect synchronization between the transmitters. It is important to keep in mind that both of these assumptions only hold approximately in practice. However, they remain very useful for theoretical analysis as they allow us to make progress toward understanding the fundamental limitations of interference management regardless of the state of the available technology for combating the issues of synchronization and channel estimation.

6.3.1 Linear Network with Two Transmitters per Message

Consider a linear network scenario where we can make each message available at two transmitters. We split the network into subnetworks, where each subnetwork has five transmitter–receiver pairs, and the last transmitter in each subnetwork is inactive. Because of the connectivity of the linear network, if the last transmitter in a subnetwork is inactive, there is no interference caused by this subnetwork to the next subnetwork. We then know that the per-user DoF that we can achieve in each subnetwork is simply the overall asymptotic per-user DoF, assuming we repeat the same scheme for all subnetworks. We therefore describe a message assignment and coding scheme for the first subnetwork. Message W_1 is assigned to the first transmitter and delivered through x_1 to the first receiver without interference; the same message is also assigned to the second transmitter and its interference at the second receiver is canceled through a careful design of the transmit signal $x_{2,1}$. In particular, we set $x_{2,1}$ as

$$x_{2,1} = \frac{-h_{21}x_1}{h_{22}}, \tag{6.2}$$

and hence the interference due to W_1 at y_2 is completely eliminated, because

$$h_{21}x_{1,1} + h_{22}x_{2,1} = 0.$$

Message W_2 is assigned to the second transmitter and delivered through $x_{2,2}$ to the second receiver without interference. Message W_3 is not transmitted. Here it is worth mentioning that fairness across message rates can be achieved through a fractional reuse mechanism where user indices are shuffled across communication sessions (or resource units). Message W_5 is assigned to the fourth transmitter and delivered through $x_{3,4}$ to the fourth receiver without interference. Finally, we assign both messages W_4 and W_5 to the third transmitter. The transmit signal $x_{3,5}$ is used to eliminate the interference of W_5 at y_4, and the transmit signal $x_{3,4}$ is used to deliver message W_4 to the fourth receiver without interference.

As we discussed, in order to analyze the achieved asymptotic per-user DoF, when using the scheme described above in a large network where we repeat the same pattern of message assignment and coding scheme for each subnetwork of five users, it suffices to analyze the achieved per-user DoF in each subnetwork. In this case, we have four messages delivered to their intended destinations without interference, and hence, by using capacity achieving point-to-point codes, the achieved DoF is 4 within the subnetwork. Since the subnetwork has five users, the achieved per-user DoF is $\frac{4}{5}$. We show in Section 6.5 that this in fact is the optimal per-user DoF value if each message can be available at any two transmitters. This strategy is illustrated in Figure 6.3.

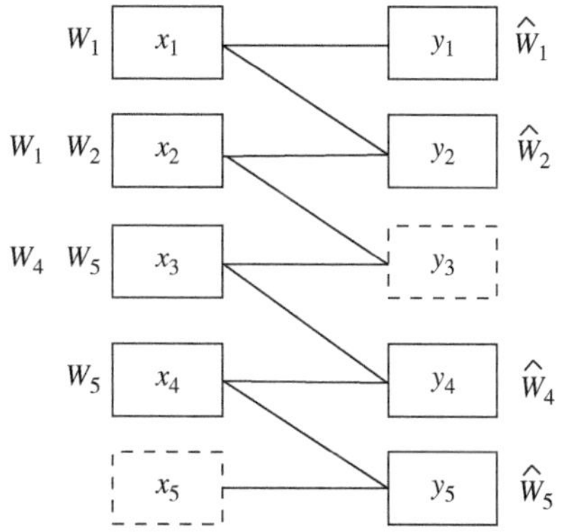

Fig. 6.3 Achieving a per-user DoF of $\frac{4}{5}$ by assigning each message to two transmitters. Dashed boxes signify inactive signals.

We now make the following observations that will pave the way for the rest of this chapter:

- It is not the case that each message W_i is available at the transmitter with the same index, i.e., the ith transmitter. For example, the two transmitters selected for assigning message W_5 are the third and fourth transmitters. This is different from traditional approaches where having each message available at the transmitter with the same index is assumed to be a constraint (see, e.g., [76]). It is worth highlighting the potential advantages of this flexibility in a practical setting. For example, in the cellular downlink, it can be beneficial to allow for a *flexible* association of receiving mobile terminals to basestation transmitters in a fashion that is *asymmetric* across the different mobile terminals, with the goal of maximizing the overall rate. We further illustrate the benefits of asymmetry in message assignment in Chapter 7, where the cooperation constraint is an overall one rather than a per-message constraint.
- As we will see in Section 6.5, if each message can be available at only one transmitter, then the per-user DoF is $\frac{2}{3}$. Hence, using CoMP results in a significant DoF gain that scales with the size of the network. Further, as we will show in Section 6.3.2, the asymptotic per-user DoF achieved through CoMP approaches unity as the number of transmitters allowed per message increases.
- As we split the network into small subnetworks of fixed size, we notice that the scheme relies on assigning all the messages with indices in the subnetwork within only transmitters inside the subnetwork. For example, messages W_1 through W_5 are assigned only within the first five transmitters. Hence, cooperation can be carried over within neighborhoods of a fixed small size that does not increase with the number of users in the network. We formalize this point in Section 6.4, as we show that local cooperation suffices for any locally connected network, not only from a degrees of freedom perspective.

6.3.2 Linear Network with Multiple Transmitters per Message

Consider the case where each message can be available at M transmitters, where M can be any positive integer. We illustrate an extension of the scheme presented in the previous section that achieves $\frac{2M}{2M+1}$ asymptotic per-user DoF. The network is again split into subnetworks, and the last transmitter in each subnetwork is deactivated to avoid inter-subnetwork interference. Each subnetwork consists of $2M+1$ transmitter–receiver pairs with consecutive indices. The case where $M = 2$ is the case discussed in the previous section; in that scenario, the third receiver in each subnetwork was deactivated. In general, the middle receiver, i.e., the one with index $(M+1)$, is inactive. We discuss next the message assignment and coding scheme for the first subnetwork.

As in the scheme for the case where $M = 2$, message W_1 is delivered through x_1 to y_1 without interference, messages W_1 and W_2 are assigned to the second transmitter, and x_2 is used to both eliminate the interference due to W_1 at y_2 and to deliver W_2 to y_2 without interference. Further, if $M > 2$, then messages W_1, W_2, and W_3 are assigned to the third transmitter, which will use its transmit signals $x_{3,1}$ and $x_{3,2}$ to eliminate the interference due to W_1 and W_2 at y_3, respectively. Also, $x_{3,3}$ can now be used to deliver W_3 to y_3 without interference. We keep assigning messages according to this pattern for each of the

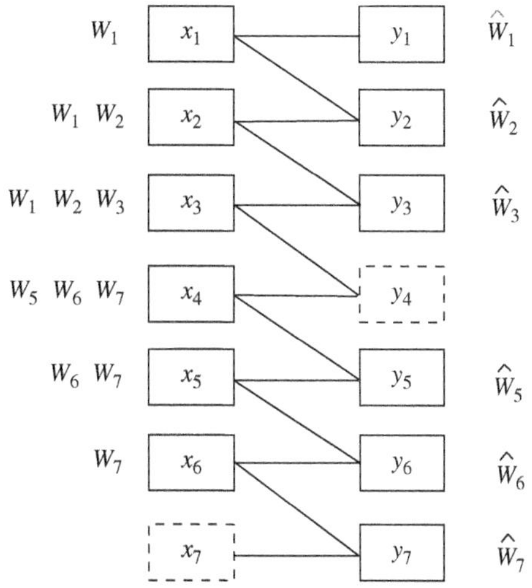

Fig. 6.4 Achieving a per-user DoF of $\frac{6}{7}$ by assigning each message to three transmitters. Dashed boxes signify inactive signals.

first M messages. More precisely, the ith message is assigned to the first i transmitters, for each $i \in \{1,2,\ldots,M\}$, and the ith transmit signal is used to cancel the interference due to $W_1, W_2, \ldots, W_{i-1}$ at y_i, as well as to deliver W_i to y_i without interference.

Recall that message W_{M+1} is not transmitted. For the remaining M messages in the subnetwork, from W_{M+2} up to W_{2M+1}, the same pattern that was used for the assignment of the first M messages is applied in reverse with the M transmitters from $M+1$ up to $2M$. More specifically, message W_{2M+1} is delivered through x_{2M} to y_{2M+1} without interference (since x_{2M+1} is inactive). Messages W_{2M+1} and W_{2M} are available for encoding x_{2M-1} to cancel the interference of W_{2M+1} at y_{2M} and deliver W_{2M} to y_{2M} without interference. We continue with this pattern of assigning messages $W_{2M+1}, W_{2M}, \ldots, W_{i+1}$ to the ith transmitter for every $i \in \{M+1, M+2, \ldots, 2M\}$, and the transmit signal x_i is used to cancel the interference due to $W_{2M+1}, W_{2M}, \ldots, W_{i+2}$ at y_{i+1} and deliver the message W_{i+1} to y_{i+1} without interference. Since we have $2M$ interference-free message deliveries for every subnetwork of $2M+1$ users, the achieved per-user DoF is $\frac{2M}{2M+1}$.

Observe how fast we can approach the interference-free per-user DoF of unity while assigning each message to few transmitters. For example, in a large linear network where we are allowed to make each message available at three transmitters, an asymptotic per-user DoF of $\frac{6}{7}$ can be achieved. Figure 6.4 depicts the message assignment for the case where $M = 3$.

6.3.3 Locally Connected Network with Multiple Transmitters per Message

Consider the extension of the scheme presented in Section 6.3.2 to general locally connected networks, where each transmitter i is connected to receivers $i, i+1, \ldots, i+L$, and L can be greater than one. We again split the network into subnetworks, but instead of deactivating the last transmitter in each subnetwork, we need to activate

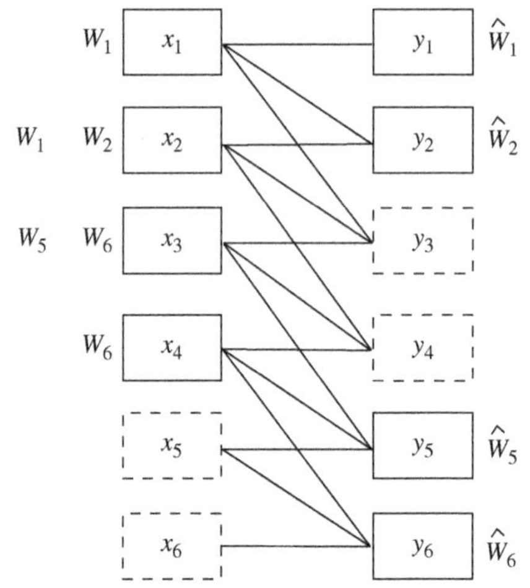

Fig. 6.5 Achieving a per-user DoF of $\frac{2}{3}$ by assigning each message to two transmitters in a locally connected network, with three transmitters connected to each receiver. Dashed boxes signify inactive signals. ©[2017] IEEE. Reprinted, with permission, from [71].

more than one transmitter. In order to avoid inter-subnetwork interference, we need to deactivate the last L transmitters in each subnetwork. Each subnetwork has $2M+L$ transmitter–receiver pairs, and the middle L receivers are inactive. The goal is to create $2M$ interference-free point-to-point communication links in each subnetwork, so that the achieved per-user DoF is $\frac{2M}{2M+L}$.

We outline the message assignment and interference cancellation scheme in the first subnetwork below:

- Message W_1 is assigned to the first transmitter and delivered through x_1 to y_1 without interference.
- For each $i \in \{1,2,\ldots,M\}$, messages $W_1, W_2, \ldots, W_i$ are assigned to the ith transmitter. The transmit signal x_i is used to eliminate the interference due to $W_1, W_2, \ldots, W_{i-1}$ at y_i, as well as to deliver W_i to y_i without interference.
- Message W_{2M+L} is assigned to the transmitter with index $2M$, and delivered through x_{2M} to y_{2M+L} without interference.
- For each $i \in \{2M, 2M-1, \ldots, M+1\}$, messages $W_{i+L}, W_{i+L+1}, \ldots, W_{2M+L}$ are assigned to the ith transmitter. The transmit signal x_i is used to eliminate the interference due to $W_{i+L+1}, \ldots, W_{2M+L}$ at y_{i+L}, as well as to deliver W_{i+L} to y_{i+L} without interference.

Note that here, the L messages $W_{M+1}, W_{M+2}, \ldots, W_{M+L}$ are inactive. We show the assignment for the first subnetwork for the case where $M=2$ and $L=2$ in Figure 6.5.

6.3.4 Optimal Zero-Forcing Schemes

The one-shot linear beamforming schemes that we have presented so far are in fact the optimal zero-forcing schemes for locally connected channels. What we mean by *zero-forcing schemes* is that interference is canceled *over the air*. In other words, each receiver is either inactive or receives only its desired signal, so each active receiver can directly employ a point-to-point decoder that is oblivious to the potential presence of interference. Formally, if the jth receiver is active, then the mutual information between its received signal and any message other than W_j is zero:

$$\mathrm{I}\left(W_i; y_j\right) = 0, \ \forall i \neq j, \text{ for every active receiver with index } j. \tag{6.3}$$

In order to prove that the schemes presented are the optimal zero-forcing schemes, we show that for any locally connected network with connectivity parameter L, the maximum achievable asymptotic per-user DoF is $\frac{2M}{2M+L}$. We prove this by showing that for any consecutive set of $2M+L$ users, at most $2M$ of these users can be active in any zero-forcing scheme. We first state the following two facts, which are formally defined and proved in [71]:

How many times can we cancel a message's interference? Suppose that we have a scheme where message W_i causes interference at N undesired destinations; then it must be the case that W_i is assigned to at least $N+1$ transmitters, i.e., $|\mathcal{T}_i| \geq N+1$. The reason is that we need one transmitter to deliver the message W_i to its desired destination, as well as N other transmitters to cancel the interference caused by W_i at the N undesired destinations. In other words, what we are saying is that a single transmitter can be used to cancel the interference due to any given message at no more than one receiver. The reason is that for a generic channel, using one transmitter to cancel a message's interference at more than one receiver will necessitate that the set of involved channel coefficients is drawn from a realization that can only take place with zero probability. It follows that the number of active receivers connected to transmitters carrying any given message cannot exceed the number of transmitters carrying this message (i.e., at most M).

A message's interference affects a block of consecutive receivers: If message W_i causes interference at receiver j, then the same message *must* also cause interference at all active receivers with an index between i and j. The reason for this fact will become clearer as we discuss the concept of *useful message assignments* in Section 6.4, but the intuition is that we do not assign a message to a transmitter unless it is used either for delivering the message to its intended destination or for canceling the interference caused by this message at an undesired receiver. We also know that only the L transmitters that are connected to the ith receiver can be used to deliver message W_i to its destination. If W_i causes interference at y_j, that means that W_i is assigned to one of the transmitters connected to the jth receiver, and then we know that there has to be a list of transmitters starting at a transmitter connected to the ith receiver and ending with the transmitter that is carrying W_i and connected to the jth receiver, where each pair of succeeding transmitters in the list share at least one receiver to which they are both connected. This chain of transmitters is the reason why W_i causes interference at every active receiver with an index between i and j.

Consider any set $\mathcal{S}$ of $2M+L$ messages with consecutive indices. If there are at least $2M+1$ active messages in $\mathcal{S}$, then there has to be a message with index i in the middle with at least M active messages in $\mathcal{S}$ with lower indices than i, and at least M active messages in $\mathcal{S}$ with higher indices than i. Let $s_{\min}$ and $s_{\max}$ be the minimum and maximum indices in $\mathcal{S}$, respectively. Then we have the following cases:

1. Message W_i causes interference at $y_{s_{\min}}$. In this case, W_i has to cause interference at every receiver with an index x such that $s_{\min} \leq x \leq i$, but we know that the number of active receivers with indices in that range is $M+1$, and W_i can only be available at M transmitters. Therefore, one active receiver will have interference from W_i that is not canceled.
2. Message W_i causes interference at $y_{s_{\max}}$. In this case, W_i causes interference at the $M+1$ active receivers with indices between i and $s_{\max}$, resulting in at least one active receiver with undesired interference from W_i that is not canceled.
3. For the remaining case, all the interference caused by W_i is at receivers with indices in $\mathcal{S}$. It follows that the number of active receivers connected to transmitters carrying W_i is $|\mathcal{T}_i|+L$. Further, since $\mathcal{S}$ has $2M+L$ receivers, among which at least $2M+1$ are active, any subset of $\mathcal{S}$ that has $|\mathcal{T}_i|+L$ receivers has to have at least $|\mathcal{T}_i|+1$ active receivers. Hence, we have shown that the size of the set of active receivers connected to transmitters carrying W_i is at least $|\mathcal{T}_i|+1$. In summary, there is at least one active receiver that suffers from interference of W_i, and it is not canceled.

Thus far we have characterized the optimal interference avoidance schemes using cooperative transmission in any locally connected network. The message assignment exploits only local cooperation by splitting the network into subnetworks and assigning the messages of each subnetwork within only the transmitters in the subnetwork. In Section 6.4, we formally prove the optimality of local cooperation in any locally connected network, even without the restriction to zero-forcing schemes. Finally, we note that the presented coding scheme is easy to implement in practice, not only because it relies on only local cooperation but also because unlike interference alignment schemes it can be implemented without symbol extensions that introduce decoding delay.

6.4 Optimality of Local Cooperation

A key idea that allows us to make significant progress toward characterizing the DoF of locally connected networks with cooperative transmission is that of understanding which assignments of messages to transmitters are required and which are not. In other words, we would like to understand when can we assert that assigning a message to a certain transmitter cannot increase the rate of communication regardless of the coding scheme employed. For that purpose, we introduce the following definition of *useful* message assignments.

6.4.1 Useful Message Assignments

Intuitively, we say that an assignment of a message W_i to a transmitter T_j is useful if it serves at least one of the following two purposes:

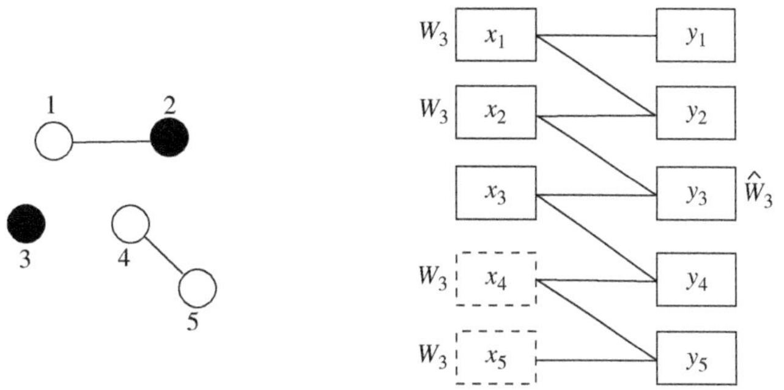

Fig. 6.6 Construction of $G_{W_3,\mathcal{T}_3}$ in a five-user channel with $L = 1$. Marked vertices are represented with filled circles. W_3 can be removed at both x_4 and x_5 without decreasing the sum-rate, as the corresponding vertices lie in a component that does not contain a marked vertex.

Signal delivery: The assignment can be useful for delivering W_i to receiver R_i, if T_j is connected to R_i.

Interference mitigation: The assignment can be useful for reducing the interference caused by W_i at a receiver R_k, if T_j is connected to R_k, and there is already another transmitter connected to R_k to which the message W_i is already assigned.

We now formalize the definition of useful message assignments. We first introduce a graph-theoretic representation that simplifies the presentation of the necessary conditions on useful message assignments. For message W_i, and a fixed transmit set $\mathcal{T}_i$, we construct the following graph $G_{W_i,\mathcal{T}_i}$ that has $\{1,2,\ldots,K\}$ as its set of vertices, and an edge exists between any given pair of vertices $v,w \in [K]$ if and only if

$$v,w \in \mathcal{T}_i,$$
$$|v-w| \leq L.$$

Vertices corresponding to transmitters connected to y_i are given a special mark, i.e., vertices with labels in the set $\{i,i-1,\ldots,i-L\}$ are marked for the considered channel model. See Figure 6.6 for an example illustration of $G_{W_i,\mathcal{T}_i}$. A necessary property for any useful message assignment is stated in the following lemma.

LEMMA 6.1 *For any $k \in \mathcal{T}_i$ such that the vertex k in $G_{W_i,\mathcal{T}_i}$ is not connected to a marked vertex, removing k from $\mathcal{T}_i$ does not decrease the sum-rate.*

The formal information-theoretic proof of Lemma 6.1 is available in [71]. We explain the insight behind it here. Let $\mathcal{S}$ denote the set of indices of vertices in a component of $G_{W_i,\mathcal{T}_i}$ that has no marked vertices. No vertex in $\mathcal{S}$ is connected to a marked vertex. We can then argue that replacing W_i at any transmitter in $\mathcal{S}$ with an independent random variable that has the same distribution will give the same rate, regardless of the coding scheme.

As a result of Lemma 6.1, we can conclude that if message W_i is available at transmitter T_j in a useful message assignment, then there has to be a chain of transmitters starting at T_j and ending at a transmitter connected to the receiver R_i, where every two consecutive transmitters in the chain are both carrying W_i and there is at least one receiver that is connected to both. We can describe this corollary more succinctly in terms of the graph $G_{W_i,\mathcal{T}_i}$ as follows.

COROLLARY 6.1 *Let $\mathcal{T}_i$ be a useful message assignment and $|\mathcal{T}_i| \leq M$. Then, for all $k \in [K]$, $k \in \mathcal{T}_i$ only if the vertex k in $G_{W_i,\mathcal{T}_i}$ lies at a distance that is less than or equal to $M-1$ from a marked vertex.*

We now interpret what the above corollary implies for locally connected networks where each transmitter i is connected to receivers $\{i, i+1, \ldots, i+L\}$. Suppose we wish to compute the distance in $G_{W_i,\mathcal{T}_i}$ between transmitters k and $k+2L$. We know that this distance would depend on the message assignment $\mathcal{T}_i$. Therefore, if we want to have an answer that relies only on the network topology and not on the message assignment, we change our question to ask rather about the least possible distance between transmitters k and $k+2L$ in any graph $G_{W_i,\mathcal{T}_i}$ for any $i \in [K]$. If transmitters k, $k+L$, and $k+2L$ all carry message W_i, i.e., $\{k, k+L, k+2L\} \subseteq \mathcal{T}_i$, then the distance in $G_{W_i,\mathcal{T}_i}$ between k and $k+2L$ is two, because k and $k+L$ would be connected, and so would $k+L$ and $k+2L$. In fact, this is the scenario that would give the minimum distance between k and $k+2L$, simply because in the considered network topology, transmitters k and $k+2L$ are not both connected to a common receiver, and hence the aforementioned distance cannot be one.

We can see that the corollary implies that

$$\mathcal{T}_i \subseteq \{i-ML, i-ML+1, \ldots, i+(M-1)L\}, \ \forall i \in [K]$$

is a necessary condition for any useful message assignment. This necessary condition does not depend on the number of users in the network, K. So, no matter how large the network is, we have a guarantee that we need only worry about assigning each message within a neighborhood of transmitters whose size depends only on the cooperation constraint M and the connectivity parameter L. We formalize this conclusion by defining local cooperation schemes.

6.4.2 Local Cooperation

Recall the definitions of a *message assignment strategy* and *local cooperation* in Section 5.4. From Lemma 6.1, we can conclude the following about local cooperation schemes:

THEOREM 6.2 *Local cooperation is optimal for locally connected channels.*

Proof We know from the corollary to Lemma 6.1 that

$$\mathcal{T}_i \subseteq \{i-ML, i-ML+1, \ldots, i+(M-1)L\}, \ \forall i \in [K]$$

is a necessary condition for useful message assignments, regardless of the number of users K. Hence, the constraint in (5.18) is satisfied with $r(K) = ML$, for all $K \in \mathbf{Z}^+$. Further, (5.17) is also satisfied with the same radius function.

We note how dramatically different the conclusions that we obtain are for different network topologies. For a fully connected network, local cooperation does not result in degrees of freedom gains that scale with the number of users. For any locally connected network, not only does local cooperation allow for scalable degrees of freedom gains, but restricting the message assignments to satisfy the local cooperation constraint does not hurt performance.

6.5 Degrees of Freedom in Linear Networks

We use the knowledge we have developed about useful message assignments to reach a tight characterization of the asymptotic degrees of freedom for linear networks. We have identified through our discussion of one-shot schemes that one can achieve $\frac{2M}{2M+1}$ per-user degrees of freedom in large linear networks by a clever message assignment and zero-forcing-based cooperative transmission scheme that allows each message to be available at M transmitters. We show in this section that under this cooperation constraint, this is in fact the optimal per-user degrees of freedom from an information-theoretic standpoint.

Consider the case where each message can be available at only one transmitter in a linear interference network, i.e., $M = 1$. We know that a per-user DoF of $\frac{2}{3}$ can be achieved by interference avoidance. The following argument establishes that this is the maximum achievable per-user DoF. Since this analysis is asymptotic in the number of users, assume without loss of generality that the number of users K is a multiple of 3. Now, we make an argument to show that any achievable per-user DoF is also achievable if we remove every receiver with an index i such that $i \bmod 3 = 2$, and connect all remaining receivers to a centralized decoder. Note that if we can do so, then we have proved that any achievable per-user DoF is at most $\frac{2}{3}$, since this is the fraction of remaining receivers.

We use Lemma 5.2 to capture the converse argument discussed above. We apply Lemma 5.2 with the set $\mathcal{A} = \{i \in [K] : i \bmod (2M+1) \neq M+1\}$. From the discussion of useful message assignments, we know that message W_{M+1} cannot be available at transmitter $2M+1$, and message W_{3M+2} cannot be available at transmitters $2M+1$ and $4M+2$. In general, for any positive integer x in a large network, message $W_{(2M+1)x+M+1}$ cannot be available at transmitters $(2M+1)x$ and $(2M+1)(x+1)$. It then follows that all transmit signals with an index that is a multiple of $2M+1$ belong to $x_{U_{\mathcal{A}}}$, i.e., $U_{\mathcal{A}} \subseteq \{i \in [K] : i \bmod (2M+1) = 0\}$. Armed with this fact about the set $U_{\mathcal{A}}$, we can briefly explain why Lemma 5.2 applies here. We will assume that we can ignore the effect of the Gaussian noise, and the reason for that assumption is that the construction of the function $f_2(z_{\mathcal{A}})$ will be clear once we explain how to obtain the transmit signals $x_{\bar{U}_{\mathcal{A}}}$ from $y_{\mathcal{A}}$ and $x_{U_{\mathcal{A}}}$. Since we know the received signal y_{2M+1} and the transmit signal x_{2M+1}, we can reconstruct the transmit signal x_{2M} by a simple linear computation. Now, we know x_{2M} and y_{2M}, and hence we can reconstruct x_{2M-1}. Proceeding in this manner, we can reconstruct all transmit signals $x_{\{1,2,\ldots,2M\}}$. Further, the same procedure can be used to reconstruct $x_{\{2M+2,2M+3,\ldots,4M+1\}}$ from $y_{\{2M+1,2M+2,\ldots,4M+2\}}$ and x_{4M+2}. We can thus reconstruct $x_{\bar{U}_{\mathcal{A}}}$ from $y_{\mathcal{A}}$ and $x_{U_{\mathcal{A}}}$, assuming that we can ignore the Gaussian noise. We only followed this assumption to simplify the presentation of the proof. If we are not ignoring the Gaussian noise, then the function $f_2(z_{\mathcal{A}})$ should be constructed in each step described above.

6.6 Discussion: SISO Interference Channels

There are two design parameters in the considered problem: the message assignment strategy satisfying the maximum transmit set size constraint, and the design of transmit beams. We characterized the asymptotic per-user DoF when one of the design parameters is restricted to a special choice, i.e., restricting message assignment strategies by a local cooperation constraint for fully connected networks or restricting the design of transmit beams to zero-forcing transmit beams for locally connected networks. The restriction of one of the design parameters can significantly simplify the problem because of the interdependence of the two design parameters. On one hand, the achievable scheme is enabled by the choice of the message assignment strategy, and on the other hand, the assignment of messages to transmitters is governed by the technique followed in the design of transmit beams, e.g., zero-forcing transmit beamforming or interference alignment. In the following, we discuss each of these design parameters.

6.6.1 Message Assignment Strategy

The assignment of each message to more than one transmitter (CoMP transmission) creates a virtual multiple-input single-output (MISO) network without cooperative transmission. A real MISO network, where multiple dedicated antennas are assigned to the transmission of each message (see, e.g., [77]), differs from the created virtual one in two aspects. First, in a CoMP transmission setting, the same transmit antenna can carry more than one message. Second, for locally connected channels, the number of receivers at which a message causes undesired interference depends on the number of transmit antennas carrying the message. We study MISO networks in Section 6.7.

For fully connected channels, the number of receivers at which a message causes undesired interference is the same regardless of the size of the transmit set, as long as it is non-empty. The only aspect that governs the assignment of messages to transmitters is the pattern of overlap between transmit sets corresponding to different messages. It is expected that the larger the sizes of the intersections between sets of messages carried by different transmit antennas, the more dependent the coefficients of the virtual MISO channel, and hence the lower the available DoF. For the spiral assignments of messages considered in Section 3.2, $|\mathcal{T}_i \cap \mathcal{T}_{i+1}| = M-1$, and the same value holds for the size of the intersection between sets of messages carried by successive transmitters. In general, local cooperation implies large intersections between sets of messages carried by different transmitters, resulting in the negative conclusion we reached for $\tau^{(\text{loc})}(M)$.

For the case where we are restricted to zero-forcing transmit beamforming as in Section 6.3, the number of receivers at which each message causes undesired interference governs the choice of transmit sets, and hence we saw that for locally connected channels, the message assignment strategy illustrated in Section 6.3.3 selects transmit sets that consist of successive transmitters, to minimize the number of receivers at which each message should be canceled. This strategy is optimal under the restriction to zero-forcing transmit beamforming schemes.

6.6.2 Design of Transmit Beams

While it was shown in Section 3.2 that CoMP transmission accompanied by both zero-forcing transmit beams and asymptotic interference alignment can achieve a DoF cooperation gain beyond what can be achieved using only transmit zero-forcing, this is not obvious for locally connected channels. Though not obvious, we believe the reason is that, unlike in the fully connected channel, the addition of a transmitter to a transmit set in a locally connected channel may result in an increase in the number of receivers at which the message causes undesired interference.

We note that, unlike the asymptotic interference alignment scheme, the zero-forcing transmit beamforming scheme illustrated in Section 6.3 does not need symbol extensions, since it achieves its target DoF in one channel realization. However, it is not clear whether asymptotic interference alignment can be used to show an asymptotic per-user DoF cooperation gain beyond that achieved through simple zero-forcing transmit beamforming in an arbitrary network topology. We believe that the answer to this question is closely related to the two problems that remain open at this point, namely fully characterizing $\tau(M)$ and $\tau_L(M)$.

6.7 Multiple-Antenna Transmitters: Shared versus Dedicated Antennas

In order to compare the cases of having dedicated versus shared antennas for the transmission of each message, we consider in this section the scenario where each transmitter is equipped with N antennas. We use the standard model for the K-user interference channel with N-antenna transmitters and single-antenna receivers:

$$y_i = \sum_{j=1}^{K}\sum_{n=1}^{N} h_{ij}^{(n)} x_j^{(n)} + z_i, \tag{6.4}$$

where $x_i^{(n)}(t)$ is the transmitted signal of the nth antenna at transmitter i, and $h_{ij}^{(n)}$ is the channel coefficient from the nth antenna at transmitter j to receiver i. The condition for the locally connected channel model is extended here to

$$h_{ij}^{(n)} \text{ is not identically 0 if and only if } i \in \left[j, j+1, \ldots, j+L\right],\ n \in [N], \tag{6.5}$$

and all channel coefficients that are not identically zero are generic.

In [78], the characterization of the asymptotic per-user DoF was extended under the restriction to zero-forcing transmit beamforming coding schemes to the considered locally connected channel with multiple-antenna transmitters. More precisely, let $\tau_L^{(\text{zf})}(M,N)$ be the asymptotic per-user DoF for locally connected channels with connectivity parameter L, N antennas at each transmitter, a maximum transmit set size constraint M, and under the restriction to the class of zero-forcing transmit beamforming coding schemes.

We show here that the optimality of the zero-forcing schemes defined in Section 6.3 extends to the considered setting where each transmitter has N antennas. We have the following constraint on each transmit signal: The transmit signal at the nth antenna of the jth transmitter is given by

$$x_j^{(n)} = \sum_{i:j\in\mathcal{T}_i} x_{j,i}^{(n)}, \tag{6.6}$$

where $x_{j,i}^{(n)}$ depends only on message W_i. The DoF characterization of locally connected networks with a maximum transmit set size constraint can be extended to obtain the following characterization.

THEOREM 6.3 *Under the restriction to zero-forcing transmit beamforming coding schemes (interference avoidance), the asymptotic per-user DoF of a locally connected channel with connectivity parameter L and N-antenna transmitters is given as follows:*

If $MN \geq M+L$, then

$$\tau_L^{(\mathrm{zf})}(M,N) = 1.$$

Otherwise,

$$\tau_L^{(\mathrm{zf})}(M,N) = \frac{2MN}{M(N+1)+L}. \tag{6.7}$$

Proof The proofs of the lower and upper bounds are given in Sections 6.7.1 and 6.7.2, respectively.

6.7.1 Coding Scheme

We first consider the case where $MN < L+M$ by treating the network as clusters, each consisting of consecutive $M(N+1)+L$ transceivers. The last L transmitters in each cluster are deactivated to eliminate inter-cluster interference, and hence it suffices to show that $2MN$ DoF can be achieved in each cluster. Without loss of generality, consider the cluster with users of indices in the set $[M(N+1)+L]$. Define the following subsets of $[M(N+1)+L]$:

$$\mathcal{S}_1 = [MN], \tag{6.8}$$

$$\mathcal{S}_2 = \{L+M+1, L+M+2, \ldots, L+M(N+1)\}, \tag{6.9}$$

where, in the proposed scheme, messages with indices in the set $[M(N+1)+L]\backslash(\mathcal{S}_1 \cup \mathcal{S}_2)$ are not transmitted and the corresponding receivers are deactivated. The remaining messages are assigned as follows:

$$\mathcal{T}_i = \begin{cases} \{1,2,\ldots,M\} & \forall i \in \mathcal{S}_1, \\ \{MN+1, MN+2, \ldots, M(N+1)\} & \forall i \in \mathcal{S}_2, \end{cases}$$

and all other transmitters in the cluster are deactivated. In other words, the first MN messages in the cluster are assigned to transmitters in the set $[M]$, and the last MN

messages in the cluster are assigned to the M transmitters with indices in the set $\{MN+1, MN+2, \ldots, M(N+1)\}$.

We note that messages with indices in $\mathcal{S}_1$ are not available outside transmitters with indices in $[M]$, and hence do not cause interference at receivers with indices in $\mathcal{S}_2$. Also, messages with indices in $\mathcal{S}_2$ are not available at transmitters with indices in $[MN]$, and hence do not cause interference at receivers with indices in $\mathcal{S}_1$.

In order to complete the proof by showing that each user in $\mathcal{S}_1 \cup \mathcal{S}_2$ achieves one degree of freedom, we next show that transmissions corresponding to messages with indices in $\mathcal{S}_1$ ($\mathcal{S}_2$) do not cause interference at receivers with indices in the same set. Let $\mathcal{S}_{1,j}$ denote the set $\{(j-1)N+1, (j-1)N+2, \ldots, jN\}$ where $j \in [M]$, and consider the design of transmit beams for messages W_i, $i \in \mathcal{S}_{1,j}$. Our aim is to create interference-free communication between the $(i-(j-1)N)$th antenna at the jth transmitter and the ith receiver. We prove this by showing the existence of a choice of transmit signals $\{x_{k,i}^{(n)} : n \in [N], k \in [M], (k,n) \neq (j, i-(j-1)N)\}$ to cancel the interference caused by W_i at the $MN-1$ receivers in $\mathcal{S}_1 \backslash \{i\}$. Consider the design of the transmit beam at the nth antenna of the kth transmitter $x_{k,i}^{(n)}$, where $n \in [N], k \in [M], (k,n) \neq (j, i-(j-1)N)$, and note that, given all other transmit signals carrying W_i, $x_{k,i}^{(n)}$ can be designed such that the interference caused by W_i at the $((k-1)N+n)$th receiver is canceled. Therefore, the interference cancellation constraints imply a system of $MN-1$ equations in $MN-1$ variables, where each equation is assigned a distinct variable that can be set to satisfy it, given any assignment of the other $MN-2$ variables. We now describe a simple algorithm that finds an assignment for the variables to satisfy the equations. Fix an order on the abovementioned equations, label them from 1 to $MN-1$, and recall that each equation is assigned a distinct variable that can be set to satisfy it given all other variables. For the xth equation, $x \in [MN+1]$, let that abovementioned distinct variable have the label x. In the first step of the algorithm, the first variable is removed by setting it as a function of all other variables to satisfy the first equation, and we have a reduced problem of $MN-2$ equations in $MN-2$ variables. Similarly, in the xth step of the algorithm, the xth variable is set as a function of all variables in the set $\{x+1, \ldots, MN+1\}$ to satisfy the xth equation. Once we reach the $(MN+1)$th and final step, the $(MN+1)$th variable will be set to satisfy the $(MN+1)$th equation, and recursively, all of the variables will be set to satisfy all of the equations. A solution is found using this algorithm for almost all channel realizations, as the assumption of a generic set of channel coefficients leads to linearly independent equations, almost surely.

Note that the validity of the above argument relies on the fact that $\forall j \in [M]$, the jth transmitter is connected to all receivers in the set $\{(j-1)N+1, (j-1)N+2, \ldots, jN\}$. This follows as we consider the case where $MN < L+M$, which implies that $jN < L+j \; \forall j \in [M]$, and the jth transmitter is connected to receivers with indices in the set $\{j, j+1, \ldots, L+j\}$.

We finally note that the channel between transmitters with indices in the sequence $(M(N+1), M(N+1)-1, \ldots, MN+1)$ and receivers with indices in the sequence $(L+M(N+1), L+M(N+1)-1, \ldots, L+M+1)$ has the same *connectivity pattern* as the channel between transmitters with indices in the sequence $(1,2,\ldots,M)$ and receivers with indices in the sequence $(1,2,\ldots,MN)$, and hence the argument in the previous

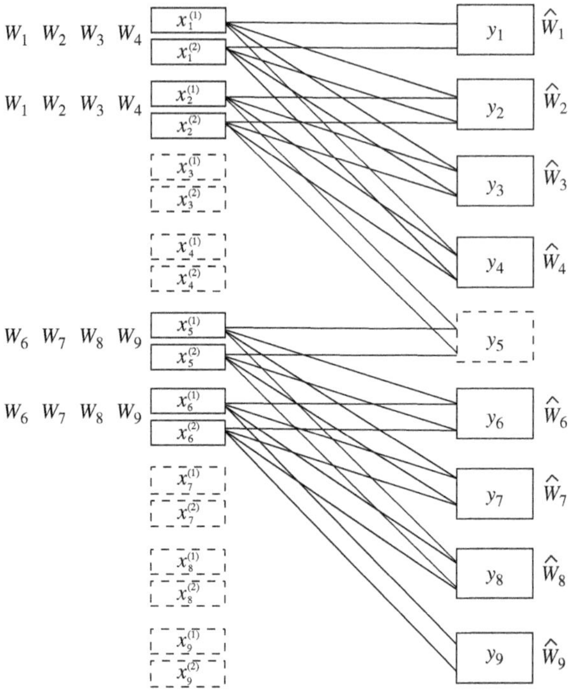

Fig. 6.7 Assignment of messages as explained in the proof of Theorem 6.3 for the case where $N = M = 2$ and $L = 3$. Only signals corresponding to the first cluster are shown. Signals in dashed boxes are deactivated. Note that the last L signals are deactivated to eliminate inter-cluster interference. Also, W_5 is not transmitted, while each other message with indices in $\{1,\ldots,9\}$ has one degree of freedom.

paragraph can be used to construct transmit beams for messages $W_i, i \in \mathcal{S}_2$ such that each user in $\mathcal{S}_2$ gets access to an interference-free transmission (one degree of freedom).

The proof is simpler for the case where $NM \geq L+M$. Let $x_{\min} = \min_x\{x \in [M] : Nx \geq L+x\}$; then $x_{\min} \leq M$, and the messages are assigned as $\mathcal{T}_i = \{i, i+1, \ldots, i+x_{\min}-1\}$. Consider the design of transmit beams for message W_i. Our aim is to allow for interference-free communication between the first antenna at the ith transmitter and the ith receiver, and eliminate the interference caused by W_i at all receivers in the set $\{i+1,\ldots,i+L+x_{\min}-1\}$. In a similar fashion to the proof described above, each receiver in $\{i+1,\ldots,i+L+x_{\min}-1\}$ is assigned a distinct transmit signal from the set $\{x_{j,i}^{(n)}, n \in [N], j \in \mathcal{T}_i, (j,n) \neq (i,1)\}$, where, given all other transmit signals, that transmit signal can be set to cancel the interference caused by W_i at that receiver, and hence there exists a setting for all transmit signals carrying W_i that cancels its interference at all receivers connected to transmitters in $\mathcal{T}_i$ other than its own receiver.

In Figure 6.7, we provide an illustration of this coding scheme.

6.7.2 Zero-Forcing Transmit Beamforming Upper Bound

In Section 6.3, we presented an intuitive argument to establish an upper bound on the DoF achieved by any zero-forcing scheme in a locally connected network with single-antenna transmitters. Here, we will present a formal argument that implies the upper bound for single-antenna transmitters as a special case when $N = 1$. In the following lemma, we present an upper bound on the number of active receivers connected to a transmit set in the considered multiple-antenna transmitters setting. For any set $\mathcal{S}$ of transmitter indices, we define $\mathcal{V}_\mathcal{S}$ as the set of active receivers connected to at least one transmitter in $\mathcal{S}$.

LEMMA 6.4 *For any message W_i, the number of active receivers connected to at least one transmitter carrying the message is no greater than the number of transmit antennas carrying the message,*

$$|\mathcal{V}_{\mathcal{T}_i}| \leq N|\mathcal{T}_i|. \tag{6.10}$$

Proof We only consider the non-trivial case where $\mathcal{T}_i \neq \emptyset$. For each receiver $j \in \mathcal{V}_{\mathcal{T}_i}$, there exists a transmit signal $x_{k,i}^{(n)}$, $k \in [K], n \in [N]$ such that, conditioned on all other transmit signals, the received signal y_j is correlated with the message W_i. More precisely,

$$\mathrm{I}\left(W_i; y_j | \{x_{v,i}^{(m)}, (v,m) \in [K] \times [N], (v,m) \neq (k,n)\}\right) > 0.$$

Since we impose the constraint $\mathrm{I}\left(W_i; y_j\right) = 0 \ \forall j \in \mathcal{V}_{\mathcal{T}_i}$, the interference seen at all receivers in $\mathcal{V}_{\mathcal{T}_i}$ has to be canceled. Finally, since the probability of a set of channel realizations with zero Lebesgue measure is zero, the $N|\mathcal{T}_i|$ transmit signals carrying W_i cannot be designed to cancel W_i at more than $N|\mathcal{T}_i| - 1$ receivers for almost all channel realizations.

We show that the sum DoF in each set $\mathcal{S} \subseteq [K]$ of consecutive $M(N+1)+L$ users is bounded by $2MN$. We focus on proving this statement by fixing a set $\mathcal{S}$ of consecutive $M(N+1)+L$ users, and make the following definitions. For a user $i \in [\mathcal{S}]$, let $\mathcal{U}_i$ be the set of active users in $\mathcal{S}$ with an index $j > i$, i.e.,

$$\mathcal{U}_i = \{j : j > i, j \in \mathcal{S}, \mathrm{I}\left(y_j; W_j\right) > 0\}.$$

Similarly, let $\mathcal{D}_i$ be the set of active users in $\mathcal{S}$ with an index $j < i$, i.e.,

$$\mathcal{D}_i = \{j : j < i, j \in \mathcal{S}, \mathrm{I}\left(y_j; W_j\right) > 0\}.$$

Assume that $\mathcal{S}$ has at least $2MN+1$ active users. Then there is an active user in $\mathcal{S}$ that lies in the middle of a subset of $2MN+1$ active users in $\mathcal{S}$. More precisely, $\exists i \in \mathcal{S} : |\mathcal{T}_i| > 0, |\mathcal{U}_i| \geq MN, |\mathcal{D}_i| \geq MN$; we let this middle user have the ith index for the rest of the proof.

Let $s_{\min}$ and $s_{\max}$ be the users in $\mathcal{S}$ with the minimum and maximum indices, respectively, i.e., $s_{\min} = \min_s\{s : s \in \mathcal{S}\}$ and $s_{\max} = \max_s\{s : s \in \mathcal{S}\}$. We then consider the following cases to complete the proof:

Case 1: W_i is being transmitted from a transmitter that is connected to the receiver with index $s_{\min}$, i.e., $\exists s \in \mathcal{T}_i : s \in \{s_{\min}, s_{\min}-1, \ldots, s_{\min}-L\}$. It follows from Lemma 6.1 that $\mathcal{V}_{\mathcal{T}_i} \supseteq \mathcal{D}_i \cup \{i\}$, and hence $|\mathcal{V}_{\mathcal{T}_i}| \geq MN+1$, which contradicts (6.10), since $|\mathcal{T}_i| \leq M$.

Case 2: W_i is being transmitted from a transmitter that is connected to the receiver with index $s_{\max}$, i.e., $\exists s \in \mathcal{T}_i : s \in \{s_{\max}, s_{\max}-1, \ldots, s_{\max}-L\}$. It follows from Lemma 6.1 that $\mathcal{V}_{\mathcal{T}_i} \supseteq \mathcal{U}_i \cup \{i\}$, and hence $|\mathcal{V}_{\mathcal{T}_i}| \geq MN+1$, which again contradicts (6.10).

Case 3: For the remaining case, there is no transmitter in $\mathcal{T}_i$ that is connected to any of the receivers with indices $s_{\min}$ and $s_{\max}$. In this case, it follows from Lemma 6.1 that $\mathcal{T}_i$ does not contain a transmitter that is connected to a receiver with an index less than $s_{\min}$ or greater than $s_{\max}$, and hence all the receivers connected to transmitters carrying W_i belong to $\mathcal{S}$. It follows that at least $L+|\mathcal{T}_i|$ receivers in $\mathcal{S}$ are connected to one or more transmitters in $\mathcal{T}_i$. Since $\mathcal{S}$ has at least $2MN+1$ active receivers, any subset of $L+|\mathcal{T}_i|$ receivers in $\mathcal{S}$ has to have at least $2MN+1-((M(N+1)+L)-(L+|\mathcal{T}_i|)) = MN+|\mathcal{T}_i|-(M-1)$ elements, and hence

$$\begin{aligned}|\mathcal{V}_{\mathcal{T}_i}| &\geq MN+|\mathcal{T}_i|-(M-1)\\ &= N|\mathcal{T}_i|+(M-|\mathcal{T}_i|)(N-1)+1\\ &\geq N|\mathcal{T}_i|+1.\end{aligned}$$

The statement is then proved by reaching a contradiction to (6.10) in the last case.

6.7.3 Successive Transmit Sets Upper Bound

We note that in the coding scheme used to prove Theorem 6.3, we used a message assignment that satisfies the useful message assignments condition in Corollary 6.1. Furthermore, each transmit set consists of a successive set of transmitter indices. More precisely,

$$\begin{aligned}&\mathcal{T}_i = \{s, s+1, \ldots, s+x-1\},\\ &s \in \{i-L-(x-1), i-L-(x-1)+1, \ldots, i\},\\ &x \in \{1, 2, \ldots, M\},\end{aligned} \tag{6.11}$$

and hence assigning each message to a successive set of transmitters is a property of the optimal message assignments with the restriction to zero-forcing transmit beamforming coding schemes. While we observe that Lemma 6.1 does not imply that transmit sets have to consist of successive transmitter indices without the restriction to zero-forcing schemes, it might be intuitive to consider that such a condition is necessary in general as it minimizes the number of receivers at which each message causes undesired interference.

Let $\tilde{\tau}_L(M,N)$ be the maximum achievable per-user DoF for the considered channel model with parameters M, N, and L, where only message assignments satisfying (6.11) are considered. We provide a DoF upper bound for general values of the system parameters.

THEOREM 6.5 *Under the restriction to successive transmit sets defined in* (6.11)*, the asymptotic per-user DoF of a locally connected channel with connectivity parameter L and N-antenna transmitters is given as follows:*

If $MN \geq M+L$, then

$$\tilde{\tau}_L(M,N) = 1.$$

Otherwise,

$$\frac{2MN}{M(N+1)+L} \leq \tilde{\tau}_L(M,N) \leq \frac{M(N+1)+L-1}{M(N+1)+L}. \tag{6.12}$$

Proof Since the coding scheme used to prove Theorem 6.3 is based on a message assignment that satisfies (6.11), the lower bound follows from the same coding scheme. We only need to show the upper bound for the case where $NM < L+M$. We apply Lemma 5.2 with the set $\mathcal{A}$ defined as follows. We view the network as clusters, each consisting of successive $M(N+1)+L$ users, and we exclude from $\mathcal{A}$ the $(L+M)$th receiver from each cluster. It then suffices to show that the condition in Lemma 5.2 holds for this choice of the set $\mathcal{A}$. More precisely, let the set $\mathcal{A}$ be defined as

$$\mathcal{A} = \{i, i \in [K], i \neq (M(N+1)+L)(j-1)+L+M, \forall j \in \mathbf{Z}^+\}. \tag{6.13}$$

We then need to show that there exist functions f_1 and f_2 such that $f_1\left(y_{\mathcal{A}}, x_{U_{\mathcal{A}}}\right) = x_{\bar{U}_{\mathcal{A}}} + f_2(z_{\mathcal{A}})$, where the definition of f_2 does not depend on the transmit power. The function f_2 that we construct is a linear function whose coefficients depend only on the channel coefficients.

We first show the existence of functions f_1 and f_2 for the case where each transmitter has a single antenna, i.e., $N=1$. Note that by using the condition in (6.11), we know that for any message with an index that lies at the intersection between the set $\bar{\mathcal{A}}$ and a given cluster, all members of its transmit set have indices that belong to the same cluster. We show how to reconstruct transmit signals in $x_{\bar{U}_{\mathcal{A}}}$ that lie in the first cluster. That is, transmit signals in the set $\{x_i : i \in \mathcal{T}_{L+M}\}$, and the rest of the proof for the remaining clusters will follow similarly. Note that because of (6.11), we know that $\mathcal{T}_{L+M} \subseteq [L+2M-1]$, and also $L+2M \notin \mathcal{T}_i, \forall i \notin \mathcal{A}$. Given y_1 and $\frac{z_1}{h_{11}^{(1)}}$, one can obtain $x_1^{(1)}$ as $x_1^{(1)} = \frac{y_1 - z_1}{h_{11}^{(1)}}$. Also, given $x_1^{(1)}$, y_2, and a linear function of z_2 whose coefficients depend only on the channel coefficients, one can obtain $x_2^{(1)}$. Similarly, transmit signals $x_3^{(1)}, \ldots, x_{L+M-1}^{(1)}$ can be reconstructed from $y_{[L+M-1]}$ and a linear function of the noise signals $z_{[L+M-1]}$. It remains to show how to obtain transmit signals in the set $\{x_{L+M}^{(1)}, x_{L+M+1}^{(1)}, \ldots, x_{L+2M-1}^{(1)}\}$. We note that the relation between those transmit signals and the signals $\{y_i : i \in \{L+M+1, \ldots, L+2M\}\}$ and $\{z_i : i \in \{L+M+1, \ldots, L+2M\}\}$ is given as follows:

$$\begin{bmatrix} \tilde{y}_{L+M+1} - z_{L+M+1} \\ \tilde{y}_{L+M+2} - z_{L+M+2} \\ \vdots \\ \tilde{y}_{L+2M} - z_{L+2M} \end{bmatrix} = \boldsymbol{M}_1 \begin{bmatrix} x_{L+M}^{(1)} \\ x_{L+M+1}^{(1)} \\ \vdots \\ x_{L+2M-1}^{(1)} \end{bmatrix}, \tag{6.14}$$

where $\forall i \in \{L+M+1, \ldots, L+2M\}$, $\tilde{y}_i = y_i - \sum_{j=1}^{L+M-1} h_{ij}^{(1)} x_j^{(1)} - h_{i,L+2M}^{(1)} x_{L+2M}^{(1)}$, and $\boldsymbol{M}_1$ is the $M \times M$ matrix defined as

$$\begin{bmatrix} h^{(1)}_{L+M+1,L+M} & h^{(1)}_{L+M+1,L+M+1} & 0 & 0 & \cdots & 0 & 0 \\ h^{(1)}_{L+M+2,L+M} & h^{(1)}_{L+M+2,L+M+1} & h^{(1)}_{L+M+2,L+M+2} & 0 & \cdots & 0 & 0 \\ \vdots & \vdots & \vdots & \vdots & \vdots & \vdots & \vdots \\ h^{(1)}_{L+2M,L+M} & h^{(1)}_{L+2M,L+M+1} & \cdots & \cdots & \cdots & h^{(1)}_{L+2M,L+2M-2} & h^{(1)}_{L+2M,L+2M-1} \end{bmatrix}. \tag{6.15}$$

If $\boldsymbol{M}_1$ is invertible, then all the transmit signals in $x_{[L+2M-1]}$ can be reconstructed, and it follows that all the transmit signals encoding the message W_{L+M} can be obtained. We show in Section 6.8.1 that the matrix $\boldsymbol{M}_1$ is full rank for almost all channel realizations. By constructing a similar proof for the remaining clusters, the upper bound proof for the case where $N=1$ is complete.

We next prove the statement for the case where $N>1$. As in the previous proof, we show how to obtain the transmit signals carrying W_{L+M} in the first cluster, and then the proof follows similarly for the remaining clusters. Let i be the smallest index in $\mathcal{T}_{L+M}$. Then we know from (6.11) that $i \in [L+M]$ and that $\mathcal{T}_{L+M} \subseteq \{i, i+1, \ldots, i+M-1\}$. Hence, it suffices to show how to obtain the transmit signals in the set $x_{\mathcal{S}}$ where $\mathcal{S} = \{i, \ldots, i+M-1\}$ from $y_{\mathcal{A}}$, $x_{U_{\mathcal{A}}} \backslash x_{\mathcal{S}}$, and a linear function of $z_{\mathcal{A}}$ whose coefficients depend only on the channel coefficients. Let $\tilde{y}_k = y_k - \sum_{j \in U_{\mathcal{A}} \backslash \mathcal{S}, n \in [N]} h^{(n)}_{kj} x^{(n)}_j, \forall k \in [K]$; then each of the signals $\tilde{y}_k - z_k, k \in [M(N+1)+L]$ is a (possibly zero) linear combination of the transmit signals in $x_{\mathcal{S}}$. As $|\mathcal{S}| = M$, and each transmitter has N antennas, then we need at least MN such linear combinations to be able to reconstruct $x_{\mathcal{S}}$. In order to do so, we pick $MN+1$ received signals, among which at most one is in the set $y_{\bar{\mathcal{A}}}$. We will also show that the linear equations corresponding to any MN signals of those picked are linearly independent, and hence suffice to reconstruct $x_{\mathcal{S}}$. By observing that we are considering the case where $NM < L+M$, or in particular that $L+1 \geq M(N-1)+2$, we pick the $MN+1$ received signals as the $M-1$ signals $\{y_i, \ldots, y_{i+M-2}\}$ together with the last $M(N-1)+2$ signals connected to the transmitter with index $i+M-1$, i.e., the set $\{i+M-1+x, i+M+x, \ldots, i+M-1+L\}$, where $x = L+1-(M(N-1)+2)$. The relation between those signals and $x_{\mathcal{S}}$ can be described as

$$\begin{bmatrix} \tilde{y}_i - z_i \\ \vdots \\ \tilde{y}_{i+M-2} - z_{i+M-2} \\ \tilde{y}_{i+M-1+x} - z_{i+M-1+x} \\ \vdots \\ \tilde{y}_{i+M-1+L} - z_{i+M-1+L} \end{bmatrix} = \boldsymbol{M}_N \begin{bmatrix} x^{(1)}_i \\ \vdots \\ x^{(N)}_i \\ x^{(1)}_{i+1} \\ \vdots \\ x^{(N)}_{i+M-1} \end{bmatrix}, \tag{6.16}$$

where $\boldsymbol{M}_N$ is the $MN+1 \times MN$ matrix

$$\begin{bmatrix} h_{ii}^{(1)} & \cdots & h_{ii}^{(N)} & 0 & \cdots & \cdots & \cdots & \cdots & 0 \\ h_{i+1,i}^{(1)} & \cdots & \cdots & \cdots & h_{i+1,i+1}^{(N)} & 0 & \cdots & \cdots & 0 \\ \vdots & & & & & & & & \vdots \\ h_{i+M-2,i}^{(1)} & \cdots & \cdots & \cdots & \cdots & h_{i+M-2,i+M-2}^{(N)} & 0 & \cdots & 0 \\ h_{i+M-1+x,i}^{(1)} & \cdots & \cdots & \cdots & \cdots & \cdots & \cdots & \cdots & h_{i+M-1+x,i+M-1}^{(N)} \\ \vdots & & & & & & & & \vdots \\ h_{i+L,i}^{(1)} & \cdots & \cdots & \cdots & \cdots & \cdots & \cdots & \cdots & h_{i+L,i+M-1}^{(N)} \\ 0 & \cdots & 0 & h_{i+L+1,i+1}^{(1)} & \cdots & \cdots & \cdots & \cdots & h_{i+L+1,i+M-1}^{(N)} \\ \vdots & & & & & & & & \vdots \\ 0 & \cdots & \cdots & \cdots & \cdots & 0 & h_{i+M-1+L,i+M-1}^{(1)} & \cdots & h_{i+M-1+L,i+M-1}^{(N)} \end{bmatrix}. \tag{6.17}$$

We note here that the missing received signal y_{L+M} can be one of the $MN+1$ received signals considered. However, we show in Section 6.8.2 that any $MN \times MN$ submatrix of M_N is full rank for all values of $N > 1$, hence proving that the transmit signals in x_S can be obtained from the remaining MN received signals and the corresponding linear combinations of the Gaussian noise signals, where the linear coefficients depend only on channel coefficients. This completes the proof for the case where $N > 1$.

By carefully inspecting the lower and upper bounds, we note that they coincide for the case where $NM = L+M-1$. However, we recall that even in this special case, the tight characterization of the asymptotic per-user DoF is available only under the restriction to successive transmit sets as defined in (6.11).

We also note that the lower bound in Theorem 6.5 is not optimal for general values of the parameters. This is because for fixed values of N and M, $\frac{2MN}{M(N+1)+L} \rightarrow 0$ as $L \rightarrow \infty$, while we know that the asymptotic interference alignment scheme can be applied in the considered channel model to achieve a per-user DoF number of $\frac{1}{2}$ without using multiple antennas or assigning any message to more than one transmitter. Furthermore, we do not expect the $\frac{L+M(N+1)-1}{L+M(N+1)}$ upper bound to be tight in general, since for fixed L, it has a lower value for the case where two antennas are dedicated to the transmission of each message ($N=2, M=1$) than the case where each message is allowed to be available at two antennas that may be carrying other messages as well ($N=1, M=2$).

6.7.4 Discussion: Dedicated versus Shared Antennas

Consider a comparison between two different scenarios. In the first, each message can be transmitted from a single transmitter that has x antennas, i.e., $N=x$ and $M=1$, while in the second scenario, each message can be transmitted from x single-antenna transmitters, i.e., $N=1$ and $M=x$. We note that the number of receivers at which a given message causes undesired interference is L in the first scenario, and is at least $L+x-1$ in the second. This leads to the result that $\tau_L^{(zf)}(M=1, N=x) > \tau_L^{(zf)}(M=x, N=1), \forall x > 1$. It is worth noting that the number of receivers at which each message causes undesired interference is not the only difference between the considered scenarios. In particular, other differences between the two scenarios affect the available DoF when considering general coding schemes beyond the simple zero-forcing transmit beamforming scheme. In the fully connected model, the number of receivers at which a given message causes undesired interference is the same for both considered scenarios. However, the per-user DoF number for the first scenario where x antennas are dedicated to each message is $\frac{x}{x+1}$, while for the case where each transmitter has a single antenna, and $M = x = 2$, the per-user DoF number $\tau(M = 2)$ is $\frac{1}{2}$, as discussed in Chapter 3.

6.8 Proof of Multiple-Antenna Transmitters Upper Bound

6.8.1 Proof of Non-Singularity of Matrix $\boldsymbol{M}_1$

The matrix $\boldsymbol{M}_1$ has the following form:

$$\begin{bmatrix} a_{11} & a_{12} & 0 & 0 & 0 & 0 & \cdots & \cdots & 0 \\ a_{21} & a_{22} & a_{23} & 0 & 0 & 0 & \cdots & \cdots & 0 \\ \vdots & & & & & & & & \vdots \\ 0 & \cdots & 0 & \cdots & a_{M-2,M-L-1} & \cdots & \cdots & a_{M-2,M-1} & 0 \\ 0 & \cdots & \cdots & 0 & 0 & a_{M-1,M-L} & \cdots & \cdots & a_{M-1,M} \\ 0 & \cdots & \cdots & 0 & 0 & 0 & a_{M,M-L+1} & \cdots & a_{MM} \end{bmatrix}, \tag{6.18}$$

where $a_{ij} = 0$ if $(i-j) \geq L$ or $(i-j) < -1$, and the set of all other entries is generic. We show that any matrix of this form is full rank with high probability for any positive integer value of the connectivity parameter L. The statement holds trivially for the case where $M = 1$, since $a_{11} \neq 0$ with high probability. Hence, in the rest of the proof we only consider the case where $M > 1$.

For a matrix of the form in (6.18), assume that there exists a linear combination of the rows that equals zero and has coefficients $\alpha_1, \alpha_2, \ldots, \alpha_M$, where α_i is the coefficient of the ith row and not all the coefficients equal zero. It follows that $\alpha_{M-1}a_{M-1,M} + \alpha_M a_{MM} = 0$. For the case where $\alpha_{M-1} = 0$, since $a_{MM} \neq 0$ with high probability, it almost surely follows that $\alpha_M = 0$. Also, if $a_{M-2,M-1} \neq 0$ then $\alpha_{M-2} = 0$. We conclude that if $\alpha_{M-1} = 0$, then for almost all realizations of the elements it must be the case that $\alpha_i = 0, \forall i \in [M]$. Hence, we only consider the case where $\alpha_{M-1} \neq 0$. Consider the following new matrix obtained by replacing the last two rows by one row that is a linear combination of them in the direction that nulls the last entry. More precisely, the new matrix has the following form:

$$\begin{bmatrix} a^{(1)}_{11} & a^{(1)}_{12} & 0 & 0 & 0 & \cdots & \cdots & 0 \\ a^{(1)}_{21} & a^{(1)}_{22} & a^{(1)}_{23} & 0 & 0 & \cdots & \cdots & 0 \\ \vdots & & & & & & & \vdots \\ 0 & \cdots & 0 & a^{(1)}_{M-2,M-L-1} & \cdots & \cdots & a^{(1)}_{M-2,M-1} & 0 \\ 0 & \cdots & 0 & 0 & a^{(1)}_{M-1,M-L} & \cdots & a^{(1)}_{M-1,M-1} & 0 \end{bmatrix}, \tag{6.19}$$

where $\forall j \in [M]$, $a_{ij}^{(1)} = a_{ij}$ $\forall i \in [M-2]$, and $a_{M-1,j}^{(1)} = a_{M-1,j} + \frac{\alpha_M}{\alpha_{M-1}} a_{Mj} = a_{M-1,j} - \frac{a_{M-1,M}}{a_{MM}} a_{Mj}$. Note that $\alpha_1, \alpha_2, \ldots, \alpha_{M-1}$ are the coefficients for a linear combination of the rows of the new matrix that equals zero. In particular, it follows that the following $(M-1) \times (M-1)$ matrix is rank deficient:

$$\begin{bmatrix} a_{11}^{(1)} & a_{12}^{(1)} & 0 & 0 & 0 & \cdots & \cdots \\ a_{21}^{(1)} & a_{22}^{(1)} & a_{23}^{(1)} & 0 & 0 & \cdots & \cdots \\ \vdots & & & & & & \vdots \\ 0 & \cdots & 0 & a_{M-2,M-L-1}^{(1)} & \cdots & \cdots & a_{M-2,M-1}^{(1)} \\ 0 & \cdots & 0 & 0 & a_{M-1,M-L}^{(1)} & \cdots & a_{M-1,M-1}^{(1)} \end{bmatrix}. \tag{6.20}$$

Note that $a_{ij}^{(1)} = 0$ if $(i-j) \geq L$ or $(i-j) < -1$, and the set of all other entries is generic; hence, the form in (6.20) is the same as the form in (6.18) with M replaced by $M-1$. By repeated application of the above argument, we find that a matrix of the following form is rank deficient:

$$\begin{bmatrix} a_{11}^{(M-2)} & a_{12}^{(M-2)} \\ a_{21}^{(M-2)} & a_{22}^{(M-2)} \end{bmatrix}, \tag{6.21}$$

where $\forall k \in \{2,3,\ldots,M-2\}$, $\forall j \in [M]$, $a_{ij}^{(k)} = a_{ij}^{(k-1)}$ $\forall i \in [M-k-1]$, and $a_{M-k,j}^{(k)} = a_{M-k,j}^{(k-1)} - \frac{a_{M-k,M-k+1}^{(k-1)}}{a_{M-k+1,M-k+1}^{(k-1)}} a_{M-k+1,j}^{(k-1)}$. We note that in each step of the argument, the set of non-zero entries remains generic, and hence the set of elements in the matrix of the form in (6.21) is generic. It follows that a matrix of the form in (6.21) is full rank with high probability, thereby reaching a contradiction to the assumption of rank deficiency of the matrix of the form in (6.18) for almost all realizations of its elements.

6.8.2 Proof of Non-Singularity of Any $MN \times MN$ Submatrix of $\boldsymbol{M}_N$

The matrix $\boldsymbol{M}_N$ has the form

$$\begin{bmatrix} a_{11} & \cdots & a_{1N} & 0 & \cdots & \cdots & \cdots & \cdots & \cdots & 0 \\ a_{21} & \cdots & \cdots & \cdots & a_{2,2N} & 0 & \cdots & \cdots & \cdots & 0 \\ \vdots & & & & & & & & & \vdots \\ a_{M-1,1} & \cdots & \cdots & \cdots & \cdots & \cdots & a_{M-1,(M-1)N} & 0 & \cdots & 0 \\ a_{M1} & \cdots & \cdots & \cdots & \cdots & \cdots & \cdots & \cdots & \cdots & a_{M,MN} \\ \vdots & & & & & & & & & \vdots \\ a_{M(N-1)+2,1} & \cdots & \cdots & \cdots & \cdots & \cdots & \cdots & \cdots & \cdots & a_{M(N-1)+2,MN} \\ 0 & \cdots & 0 & a_{M(N-1)+3,N+1} & \cdots & \cdots & \cdots & \cdots & \cdots & a_{M(N-1)+3,MN} \\ \vdots & & & & & & & & & \vdots \\ 0 & \cdots & \cdots & \cdots & \cdots & \cdots & 0 & a_{MN+1,(M-1)N+1} & \cdots & a_{MN+1,MN} \end{bmatrix}, \tag{6.22}$$

where $N > 1$, and the set of all the elements that are not identically zeros in (6.22) is *generic*. The proof follows similar steps to those followed in the proof above of the non-singularity of $\boldsymbol{M}_1$. For the case where $M = 1$, the set of all elements of any $N \times N$ submatrix of any matrix of the form in (6.22) is generic, and hence the statement holds for this case. We only consider the case where $M > 1$ in the rest of the proof.

For any matrix that has the form in (6.22), consider a submatrix that is obtained by removing one of the rows, and assume that there exists a linear combination of the remaining rows that equals zero. This implies that there is a linear combination of the $MN + 1$ rows of the original matrix that equals zero and has coefficients $\alpha_1, \alpha_2, \ldots, \alpha_{MN+1}$, where α_i is the coefficient of the ith row, and that there exists $i^* \in [MN+1]$ such that $\alpha_{i^*} = 0$.

Let β_N be the number of rows where the last N entries are not identically zero, not including the row indexed i^* whose corresponding coefficient $\alpha_{i^*} = 0$. If $\beta_N \leq N$ then all the corresponding coefficients are zeros almost surely, i.e., $\alpha_i = 0, \forall i \in \{M, M+1, \ldots, MN+1\}$. It follows that

$$\begin{bmatrix} \alpha_1 & \alpha_2 & \cdots & \alpha_{M-1} \end{bmatrix} \begin{bmatrix} a_{11} & \cdots & a_{1N} & 0 & \cdots & \cdots & \cdots & \cdots & \cdots & 0 \\ a_{21} & \cdots & \cdots & \cdots & a_{2,2N} & 0 & \cdots & \cdots & \cdots & 0 \\ \vdots & & & & & & & & & \vdots \\ a_{M-1,1} & \cdots & \cdots & \cdots & \cdots & \cdots & a_{M-1,(M-1)N} & 0 & \cdots & 0 \end{bmatrix} = \begin{bmatrix} 0 & 0 & \cdots & 0 \end{bmatrix}, \tag{6.23}$$

and it follows from (6.23) that $\alpha_i = 0\ \forall i \in [M-1]$ almost surely, thereby contradicting the assumption that not all the coefficients are zeros. Therefore, we only consider the case where $\beta_N > N$. Consider the following matrix obtained by replacing the rows in (6.22) with a generic set of elements in the last N entries, by N fewer rows that form a basis for a subspace whose vector elements have zeros in the last N entries:

$$\begin{bmatrix} a_{11}^{(1)} & \cdots & a_{1,MN}^{(1)} \\ \vdots & & \vdots \\ a_{x1}^{(1)} & \cdots & a_{x,MN}^{(1)} \end{bmatrix}, \tag{6.24}$$

where

$$x = \begin{cases} (M-1)N & \text{if } i^* \in \{(M-1)N+1, \ldots, MN+1\}, \\ (M-1)N+1 & \text{otherwise,} \end{cases} \tag{6.25}$$

and $a_{ij}^{(1)} = a_{ij}$, $\forall i \in [M-1], j \in [MN]$. In order to describe the remaining elements $a_{ij}^{(1)}, i \in \{M, M+1, \ldots, x\}, j \in [MN]$, we first define the matrices $\boldsymbol{A}$, $\boldsymbol{B}$, $\boldsymbol{C}$, and $\boldsymbol{D}$, as follows:

$$\boldsymbol{A} = \begin{bmatrix} a_{M,(M-1)N+1} & \cdots & a_{M,MN} \\ \vdots & & \vdots \\ a_{x,(M-1)N+1} & \cdots & a_{x,MN} \end{bmatrix}. \tag{6.26}$$

The $N \times N$ matrix $\boldsymbol{B}$ is defined as follows for the case where $i^* \in \{(M-1)N+1,\dots,MN+1\}$:

$$\boldsymbol{B} = \begin{bmatrix} a_{(M-1)N+1,(M-1)N+1} & \cdots & a_{(M-1)N+1,MN} \\ \vdots & & \vdots \\ a_{i^*-1,(M-1)N+1} & \cdots & a_{i^*-1,MN} \\ a_{i^*+1,(M-1)N+1} & \cdots & a_{i^*+1,MN} \\ \vdots & & \vdots \\ a_{MN+1,(M-1)N+1} & \cdots & a_{MN+1,MN} \end{bmatrix}, \tag{6.27}$$

and for the case where $i^* \in [(M-1)N]$:

$$\boldsymbol{B} = \begin{bmatrix} a_{(M-1)N+2,(M-1)N+1} & \cdots & a_{(M-1)N+2,MN} \\ \vdots & & \vdots \\ a_{MN+1,(M-1)N+1} & \cdots & a_{MN+1,MN} \end{bmatrix}. \tag{6.28}$$

Since we consider the case where the number of rows with a generic set of elements in the last N entries is greater than N, it follows that $\boldsymbol{B}$ is full rank almost surely. Therefore, the following definition of the matrix $\boldsymbol{C}$ is valid:

$$\boldsymbol{C} = -\boldsymbol{A}\boldsymbol{B}^{-1}. \tag{6.29}$$

For the case where $i^* \in \{(M-1)N+1,\dots,MN+1\}$, the $N \times MN$ matrix $\boldsymbol{D}$ is defined as follows:

$$\boldsymbol{D} = \begin{bmatrix} a_{(M-1)N+1,1} & \cdots & a_{(M-1)N+1,MN} \\ \vdots & & \vdots \\ a_{i^*-1,1} & \cdots & a_{i^*-1,MN} \\ a_{i^*+1,1} & \cdots & a_{i^*+1,MN} \\ \vdots & & \vdots \\ a_{MN+1,1} & \cdots & a_{MN+1,MN} \end{bmatrix}, \tag{6.30}$$

and for the case where $i^* \in [(M-1)N]$:

$$\boldsymbol{D} = \begin{bmatrix} a_{(M-1)N+2,1} & \cdots & a_{(M-1)N+2,MN} \\ \vdots & & \vdots \\ a_{MN+1,1} & \cdots & a_{MN+1,MN} \end{bmatrix}. \tag{6.31}$$

The elements $a_{ij}^{(1)}, i \in \{M, M+1,\dots,x\}, j \in [MN]$ are obtained as follows:

$$\begin{bmatrix} a_{M1}^{(1)} & \cdots & a_{M,MN}^{(1)} \\ \vdots & & \vdots \\ a_{x1}^{(1)} & \cdots & a_{x,MN}^{(1)} \end{bmatrix} = \begin{bmatrix} a_{M1} & \cdots & a_{M,MN} \\ \vdots & & \vdots \\ a_{x1} & \cdots & a_{x,MN} \end{bmatrix} + \boldsymbol{C}\boldsymbol{D}. \tag{6.32}$$

We next show that the new matrix in (6.24) has the following form:

$$\begin{bmatrix} a_{11}^{(1)} & \cdots & a_{1N}^{(1)} & 0 & \cdots & \cdots & & \cdots & & \cdots & \cdots & 0 \\ a_{21}^{(1)} & \cdots & \cdots & \cdots & a_{2,2N}^{(1)} & 0 & & \cdots & & \cdots & \cdots & 0 \\ \vdots & & & & & & & & & & & \vdots \\ a_{M-1,1}^{(1)} & \cdots & \cdots & \cdots & \cdots & \cdots & a_{M-1,(M-1)N}^{(1)} & & 0 & \cdots & & 0 \\ a_{M1}^{(1)} & \cdots & \cdots & \cdots & \cdots & \cdots & a_{M,(M-1)N}^{(1)} & & 0 & \cdots & & 0 \\ \vdots & & & & & & & & & & & \vdots \\ a_{x1}^{(1)} & \cdots & \cdots & \cdots & \cdots & \cdots & a_{x,(M-1)N}^{(1)} & & 0 & \cdots & & 0 \end{bmatrix}, \tag{6.33}$$

where the set of all the elements that are not marked with zeros is generic. Moreover,

$$\begin{bmatrix} \alpha_1 & \alpha_2 & \cdots & \alpha_x \end{bmatrix} \begin{bmatrix} a_{11}^{(1)} & \cdots & a_{1,MN}^{(1)} \\ \vdots & & \vdots \\ a_{x1}^{(1)} & \cdots & a_{x,MN}^{(1)} \end{bmatrix} = \begin{bmatrix} 0 & 0 & \cdots & 0 \end{bmatrix}. \tag{6.34}$$

Since the first $M-1$ rows in (6.22) have zero entries in the last N positions, it follows that

$$\begin{bmatrix} \alpha_M & \alpha_{M+1} & \cdots & \alpha_{MN+1} \end{bmatrix} \begin{bmatrix} a_{M,(M-1)N+1} & \cdots & a_{M,MN} \\ \vdots & & \vdots \\ a_{MN+1,(M-1)N+1} & \cdots & a_{MN+1,MN} \end{bmatrix} = \begin{bmatrix} 0 & 0 & \cdots & 0 \end{bmatrix}. \tag{6.35}$$

Let $\boldsymbol{E}$ be the $1 \times N$ vector defined as

$$\boldsymbol{E} = \begin{cases} \begin{bmatrix} \alpha_{(M-1)N+1} & \cdots & \alpha_{i^*-1} & \alpha_{i^*+1} & \alpha_{MN+1} \end{bmatrix} & \text{if } i^* \in \{(M-1)N+1, \ldots, MN+1\}, \\ \begin{bmatrix} \alpha_{(M-1)N+2} & \cdots & \alpha_{MN+1} \end{bmatrix} & \text{otherwise.} \end{cases} \tag{6.36}$$

The equality in (6.35) implies that

$$\begin{bmatrix} \alpha_M & \cdots & \alpha_x \end{bmatrix} \boldsymbol{A} = -\boldsymbol{E}\boldsymbol{B}, \tag{6.37}$$

and consequently

$$\begin{bmatrix} \alpha_M & \cdots & \alpha_x \end{bmatrix} \boldsymbol{C} = \boldsymbol{E}. \tag{6.38}$$

It follows that

$$\begin{aligned}
&\begin{bmatrix}\alpha_M & \cdots & \alpha_x\end{bmatrix}\begin{bmatrix}a^{(1)}_{M1} & \cdots & a^{(1)}_{M,MN}\\ \vdots & & \vdots\\ a^{(1)}_{x1} & \cdots & a^{(1)}_{x,MN}\end{bmatrix}\\
&= \begin{bmatrix}\alpha_M & \cdots & \alpha_x\end{bmatrix}\begin{bmatrix}a_{M1} & \cdots & a_{M,MN}\\ \vdots & & \vdots\\ a_{x1} & \cdots & a_{x,MN}\end{bmatrix} + \begin{bmatrix}\alpha_M & \cdots & \alpha_x\end{bmatrix}\boldsymbol{CD}\\
&= \begin{bmatrix}\alpha_M & \cdots & \alpha_x\end{bmatrix}\begin{bmatrix}a_{M1} & \cdots & a_{M,MN}\\ \vdots & & \vdots\\ a_{x1} & \cdots & a_{x,MN}\end{bmatrix} + \boldsymbol{ED}\\
&= \begin{bmatrix}\alpha_M & \cdots & \alpha_{MN+1}\end{bmatrix}\begin{bmatrix}a_{M1} & \cdots & a_{M,MN}\\ \vdots & & \vdots\\ a_{MN+1,1} & \cdots & a_{MN+1,MN}\end{bmatrix}.
\end{aligned} \tag{6.39}$$

We can see that (6.34) follows from (6.39) and the fact that $a^{(1)}_{ij} = a_{ij}\ \forall i \in [M-1], j \in [MN]$.

To prove (6.33), we first validate the positions of the zero entries, and then prove that the set of all remaining elements is generic. The positions of the zero entries in the first $M-1$ rows follows from (6.22) and the fact that $a^{(1)}_{ij} = a_{ij}\ \forall i \in [M-1], j \in [MN]$. To show that all the remaining rows in (6.33) have zeros in the last N positions, consider the following equality that follows from (6.32):

$$\begin{aligned}
\begin{bmatrix}a^{(1)}_{M,(M-1)N+1} & \cdots & a^{(1)}_{M,MN}\\ \vdots & & \vdots\\ a^{(1)}_{x,(M-1)N+1} & \cdots & a^{(1)}_{x,MN}\end{bmatrix} &\overset{(a)}{=} \begin{bmatrix}a_{M,(M-1)N+1} & \cdots & a_{M,MN}\\ \vdots & & \vdots\\ a_{x,(M-1)N+1} & \cdots & a_{x,MN}\end{bmatrix} + \boldsymbol{CB}\\
&= \boldsymbol{A} + \boldsymbol{CB}\\
&= \boldsymbol{A} - \boldsymbol{AB}^{-1}\boldsymbol{B}\\
&= \begin{bmatrix}0 & \cdots & \cdots & 0\\ \vdots & & & \vdots\\ 0 & \cdots & \cdots & 0\end{bmatrix},
\end{aligned} \tag{6.40}$$

where (a) follows as the matrix $\boldsymbol{B}$ is formed by taking the last N columns of the matrix $\boldsymbol{D}$. In order to prove that the set of all elements in (6.33) that are not identically zeros is generic, we first note that

$$a^{(1)}_{ij} = a_{ij} + f(\{a_{kq} : x < k \leq MN+1 \text{ or } M(N-1) < q \leq MN\}),$$
$$\forall i \in [x], j \in [M(N-1)]. \tag{6.41}$$

In particular, for all $i \in [x], j \in [M(N-1)]$, the element a_{ij} contributes only to $a^{(1)}_{ij}$ among the set $\mathcal{S} = \{a^{(1)}_{ij} : i \in [x], j \in [M(N-1)]\}$. Since the set of elements that are not identical to zeros in (6.33) is a subset of $\mathcal{S}$, it follows from (6.41) that the former set is generic as a result of the fact that the set $\{a_{ij} : i \in [MN+1], j \in [MN]\}$ is generic. The proof of the statement in (6.33) and (6.34) is complete. Moreover, the same conclusions hold for the submatrix obtained by removing the last N columns in (6.33), i.e., for the matrix

$$\begin{bmatrix} a_{11}^{(1)} & \cdots & a_{1N}^{(1)} & 0 & \cdots & \cdots & \cdots & 0 \\ a_{21}^{(1)} & \cdots & \cdots & \cdots & a_{2,2N}^{(1)} & 0 & \cdots & 0 \\ \vdots & & & & & & & \vdots \\ a_{M-1,1}^{(1)} & \cdots & \cdots & \cdots & \cdots & \cdots & \cdots & a_{M-1,(M-1)N}^{(1)} \\ a_{M1}^{(1)} & \cdots & \cdots & \cdots & \cdots & \cdots & \cdots & a_{M,(M-1)N}^{(1)} \\ \vdots & & & & & & & \vdots \\ a_{x1}^{(1)} & \cdots & \cdots & \cdots & \cdots & \cdots & \cdots & a_{x,(M-1)N}^{(1)} \end{bmatrix}, \tag{6.42}$$

the set of all elements that are not identically zero is generic, and

$$\begin{bmatrix} \alpha_1 & \alpha_2 & \cdots & \alpha_x \end{bmatrix} \begin{bmatrix} a_{11}^{(1)} & \cdots & a_{1,(M-1)N}^{(1)} \\ \vdots & & \vdots \\ a_{x1}^{(1)} & \cdots & a_{x,(M-1)N}^{(1)} \end{bmatrix} = \begin{bmatrix} 0 & 0 & \cdots & 0 \end{bmatrix}. \tag{6.43}$$

We next show that $\alpha_1, \ldots, \alpha_x$ are not all zeros. If we assume otherwise, it follows that

$$\begin{bmatrix} \alpha_{x+1} & \cdots & \alpha_{MN+1} \end{bmatrix} \begin{bmatrix} a_{x+1,1} & \cdots & a_{x+1,MN} \\ \vdots & & \vdots \\ a_{MN+1,1} & \cdots & a_{MN+1,MN} \end{bmatrix} = \begin{bmatrix} 0 & 0 & \cdots & 0 \end{bmatrix}. \tag{6.44}$$

Since $\alpha_{i^*} = 0$, then it follows that

$$\boldsymbol{ED} = \begin{bmatrix} 0 & 0 & \cdots & 0 \end{bmatrix}. \tag{6.45}$$

In particular, since $\boldsymbol{B}$ is formed by taking the last N columns of $\boldsymbol{D}$, $\boldsymbol{EB}$ is an all-zero vector, which almost surely implies that $\boldsymbol{E}$ is an all-zero vector, since $\boldsymbol{B}$ is full rank. We conclude from this argument that $\alpha_1, \ldots, \alpha_x$ cannot be all zeros as otherwise $\alpha_i = 0\ \forall i \in [MN+1]$, which contradicts the assumption. We have proved so far that in the case that any $MN \times MN$ submatrix of any matrix of the form in (6.22) is rank deficient, it follows that a matrix of the form in (6.42) is rank deficient. Note also that $\{(i,j) : a_{ij}^{(1)} = 0\} \subseteq \{(i,j) : a_{ij} = 0\}$. We can then repeat the above argument by replacing the rows in (6.42) whose last N entries are not identically zero by N fewer rows whose last N entries are identically zero and then removing the last N columns, to finally show that the matrix in (6.46) is rank deficient,

$$\begin{bmatrix} a_{11}^{(M-1)} & \cdots & a_{1N}^{(M-1)} \\ \vdots & & \vdots \\ a_{N1}^{(M-1)} & \cdots & a_{NN}^{(M-1)} \end{bmatrix}, \tag{6.46}$$

where the set of all elements in (6.46) is generic, which implies that any matrix in the form (6.46) is full rank with high probability. Therefore, the assumption that there exists a rank-deficient $MN \times MN$ submatrix of $\boldsymbol{M}_N$ is not true with high probability.

6.9 Notes and Further Reading

In [79], a special case of the locally connected network with $L = 1$ (Wyner's asymmetric network) was considered, where the message assignment strategy was fixed to the spiral strategy discussed in Section 5.4, i.e., each message is assigned to the transmitter with the same index as well as $M - 1$ following transmitters. The asymptotic per-user DoF was then characterized as $\frac{M}{M+1}$. This shows for our problem that

$$\tau_1(M) \geq \frac{M}{M+1}. \tag{6.47}$$

In [80, Remark 2], a message assignment strategy was described to enable the achievability of an asymptotic per-user DoF as high as $\frac{2M-1}{2M}$. It can be easily verified that this is indeed true, and hence we can conclude from this work that

$$\tau_1(M) \geq \frac{2M-1}{2M}. \tag{6.48}$$

The main difference in the strategy described in [80, Remark 2] from the spiral message assignment strategy considered in [79] is that, unlike the spiral strategy, messages are assigned to transmitters in an asymmetric fashion, where we say that a message assignment is symmetric if and only if for all $j,i \in [K], j > i$, the transmit set $\mathcal{T}_j$ is obtained by shifting forward the indices of the elements of the transmit set $\mathcal{T}_i$ by $(j-i)$. We showed in Section 6.3.2 that both the message assignment strategy analyzed in [79] and the one suggested in [80] are suboptimal for $L = 1$, and the value of $\tau_1(M)$ is in fact strictly larger than the bounds in (6.47) and (6.48). The key idea that enabled the result presented in this chapter is that each message need not be available at the transmitter carrying its own index.

In Section 6.7, we discussed the problem of CoMP transmission in networks with multiple-antenna transmitters. Communication scenarios with cooperating multiple-antenna transmitters have been considered in [81] and [82] under the umbrella of the X-channel. However, in the X-channel, mutually exclusive parts of each message are given to different transmitters. This was extended in [83] to allow each part of each message to be available at more than one transmitter. In [84], the MIMO X-channel was studied in the setting where transmitters share further side information.

Finally, it is worth noting that we implicitly assumed in our discussion in this chapter that the transmit beams at all the transmitters are designed jointly. This kind of coordination is also referred to in the literature as *transmitter cooperation* or CoMP, even without the sharing of messages (see, e.g., [85]).

7 Backhaul Load Constraint

In the previous two chapters, we have considered cooperative transmission schemes that are constrained by the number of transmitters to which each message can be assigned. While meeting a backhaul capacity limit by a *per-message* constraint can have its analytical advantages, such as reducing a difficult information-theoretic problem to a simple combinatorial one, this approach suffers from a few important drawbacks.

For one, the maximum transmit set size constraint may not reflect most practical scenarios for two reasons. First, because fractional reuse across different resource (e.g., frequency or time) slots can be used to achieve an equal load on the backhaul per message, it may not make sense to impose a maximum transmit set size constraint for each channel use. Secondly, the maximum transmit set size constraint may not reflect the nature of the backhaul link, as we will discuss further below.

Also, as we have discussed in the previous chapter, the optimal solutions for the maximum transmit set size constraint may not fully utilize the constraint; i.e., for the optimal solution, some messages need not be assigned to the allowed maximum number of transmitters. We have seen in Chapter 6 that this is the case for the optimal schemes for linear interference networks, as well as for general locally connected networks where we impose the restriction of using only zero-forcing coding schemes.

The appropriate constraint to consider for the assignment of messages to transmitters should depend on the nature of the backhaul link used in practice (see, e.g., [69, 86]). For example, in the context of heterogeneous networks, the backhaul can be a wireless network, and an overall backhaul load constraint would be a more appropriate choice. On the other hand, for the case of wireline or optical fiber backhaul links, the maximum transmit set size constraint can be useful. However, even with wireline backhaul links, an average transmit set size constraint can allow for flexible solutions that interleave the use of the backhaul links over multiple communication sessions. In general, a constraint that bounds the average transmit set size is more relevant to practice than imposing a maximum constraint on each transmit set size. We show in this chapter how the solutions for the CoMP transmission problem provided under the maximum transmit set size constraint can be used to find solutions under an average transmit set size or backhaul load constraint.

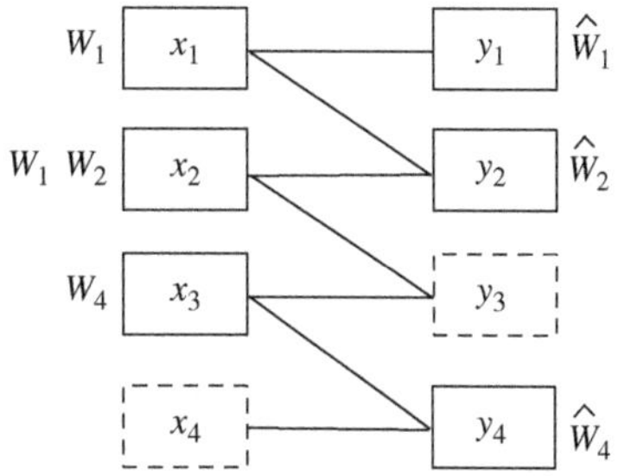

Fig. 7.1 Achieving $\frac{3}{4}$ per-user DoF with no extra backhaul load. Only signals corresponding to the first subnetwork in a general K-user network are shown. The signals in the dashed boxes are deactivated.

7.1 Example: CoMP with No Extra Backhaul Load

Consider the following assignment of messages to transmitters in a Wyner's linear network. The first message W_1 is assigned to the first two transmitters, i.e., the transmit set $\mathcal{T}_1 = \{1,2\}$. The second message is only assigned to the second transmitter, $\mathcal{T}_2 = \{2\}$. The third message is inactive, and the fourth message is assigned to the third transmitter, $\mathcal{T}_3 = \emptyset$ and $\mathcal{T}_4 = \{3\}$. The fourth transmitter is inactive, and hence we can repeat the pattern of message assignments for every successive four users in the network, and claim that the achieved per-user DoF for the first four users is a lower bound on the asymptotic per-user DoF. Further, because the sum of the transmit set sizes equals the number of users, $\sum_{i=1}^{4} |\mathcal{T}_i| = 4$, we notice that the required backhaul load is the same as that required to assign each message to exactly one transmitter.

The first message is transmitted from the first transmitter to the first receiver with no interference. Also, because W_1 is available at the second transmitter, and the channel state information is assumed to be available at the transmitters, the interference caused by W_1 at the second receiver can be canceled. W_2 can be then transmitted from the second transmitter to the second receiver with no interference. Finally, message W_4 can be transmitted from the third transmitter to the fourth receiver with no interference. The achieved DoF for the first four users is then three, $\sum_{i=1}^{4} d_i \geq 3$. It then follows that the asymptotic per-user DoF $\tau \geq \frac{3}{4}$. We illustrate the scheme in Figure 7.1.

We can see the contrast of this example to the case where each message can only be available at one transmitter. In the latter case, there is also no extra load on the backhaul, but the asymptotic per-user DoF is $\tau(M = 1) = \frac{2}{3}$, as discussed in the previous chapter. What this example suggests, then, is that by allowing for a more *flexible* backhaul constraint that allows some messages to be available at more than one transmitter, at the cost of not transmitting other messages in the considered channel use, we can use CoMP transmission to achieve significant degrees of freedom gains that scale with the size of the network without incurring additional overall load on the backhaul.

Consider the following alternative assignment to achieve the same asymptotic per-user DoF of $\frac{3}{4}$, with no additional load on the backhaul. In the previous chapter, we characterized all optimal message assignments for linear networks, under a maximum transmit set size constraint of M. For the case where $M = 1$, we could achieve an asymptotic per-user DoF of $\frac{2}{3}$, while requiring a backhaul load of $\frac{2}{3}$, by splitting the network into subnetworks of three users each, and using the transmit sets $\mathcal{T}_1 = \{1\}$, $\mathcal{T}_2 = \emptyset$, $\mathcal{T}_3 = \{2\}$ for the first subnetwork, and a similar assignment for each successive subnetwork. For the case where $M = 2$, we could achieve an asymptotic per-user DoF of $\frac{4}{5}$, with the following choice of transmit sets: $\mathcal{T}_1 = \{1,2\}, \mathcal{T}_2 = \{2\}, \mathcal{T}_3 = \emptyset, \mathcal{T}_4 = \{3\}, \mathcal{T}_5 = \{3,4\}$, and using a similar pattern of message assignment for each successive set of five users (note that this is possible, because the fifth transmitter is inactive). The required backhaul load for this assignment is then $\frac{\sum_{i=1}^{5}|\mathcal{T}_i|}{5} = \frac{6}{5}$. By using the assignment of $M = 1$ for a fraction of $\frac{3}{8}$ of the network and the assignment of $M = 2$ for the remaining part of the network, we could achieve the asymptotic per-user DoF of $\frac{3}{8} \cdot \frac{2}{3} + \frac{5}{8} \cdot \frac{4}{5} = \frac{3}{4}$ with a backhaul load of $\frac{3}{8} \cdot \frac{2}{3} + \frac{5}{8} \cdot \frac{6}{5} = 1$, i.e., no extra backhaul load. We notice from this result that schemes that were identified under a maximum transmit set size constraint can guide us to characterize optimal schemes under the average transmit set size (backhaul load) constraint.

In Section 7.2, we extend the insights obtained in the above example to the case where a general overall backhaul load constraint is imposed for linear networks. We then show optimality of the presented scheme with respect to the backhaul load constraint. In Section 7.4, we show that these insights can be useful for analyzing practical models of hexagonal infrastructural cellular networks, essentially by viewing the cellular network as a set of linear subnetworks, and eliminating interference between the subnetworks.

7.2 Flexible Backhaul Design and the Optimal DoF Scheme

We consider a backhaul load constraint that is defined through the average transmit set size as

$$\frac{\sum_{i=1}^{K}|\mathcal{T}_i|}{K} \leq B. \tag{7.1}$$

We use the notation $\tau^{\text{avg}}(B)$ to denote the asymptotic per-user DoF of Wyner's linear network under the backhaul load constraint. From the example in Section 7.1, we know that $\tau^{\text{avg}}(B=1) \geq \frac{3}{4}$. We start this section by showing how this result can be generalized for arbitrary positive values of B. In particular, we show that this lower bound follows:

$$\tau^{\text{avg}}(B) \geq \frac{4B-1}{4B}, \ \forall B \in \mathbf{Z}^{+}. \tag{7.2}$$

Comparing this result to the $\frac{2M}{2M+1}$ asymptotic per-user DoF under a maximum transmit set size constraint of M, we see that a gain is achieved when using the more relaxed backhaul load constraint.

Similar to the coding scheme explained in Chapter 6, we split the network into subnetworks, each consisting of a set of successive transmitter–receiver pairs, and the last transmitter in each subnetwork is deactivated to eliminate interference between subnetworks. If the subnetworks have the same size, and the messages intended for

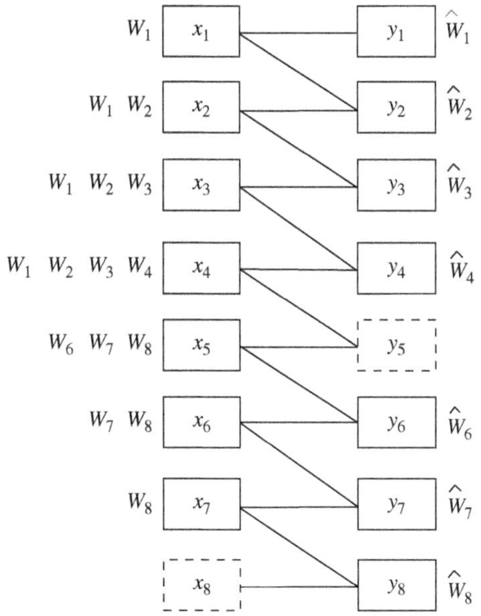

Fig. 7.2 Achieving $\frac{7}{8}$ per-user DoF with a backhaul load constraint $B = 2$. Only signals corresponding to the first subnetwork in a general K-user network are shown. The signals in the dashed boxes are deactivated. ©[2017] IEEE. Reprinted, with permission, from [87].

all receivers in a subnetwork are confined to be available only at transmitters within the same subnetwork, then the per-user DoF achieved in the first subnetwork is a lower bound on the asymptotic per-user DoF. The size of the subnetwork is the same as the denominator of the target per-user DoF, and the number of active messages (the ones whose transmit set is not the empty set) in each subnetwork is the same as the numerator of the target per-user DoF. For the maximum transmit set size constraint of M considered in Chapter 6, the target per-user DoF was $\frac{2M}{2M+1}$, and hence each subnetwork had $2M+1$ transmitter–receiver pairs, and $2M$ messages were active in each. Here, the target per-user DoF is $\frac{4B-1}{4B}$, and hence we consider subnetworks whose size is $4B$ and activate $4B-1$ messages in each subnetwork.

We explain the coding scheme for the first subnetwork, consisting of the first $4B$ users, and the scheme used for the remainder of the network will be clear by analogy. For each message $W_j, j \in \{1, 2, \ldots, 2B\}$, the transmit set $\mathcal{T}_j = \{j, j+1, \ldots, 2B\}$. In other words, message W_j is assigned to the transmitter having the same index, as well as all following transmitters up to the transmitter with index $2B$. Using this assignment, the transmit beams for W_j can be designed such that W_j is delivered from the jth transmitter to the jth receiver, and the interference caused by W_j at all receivers with indices in the range $\{j+1, j+2, \ldots, 2B\}$ is canceled. In Figure 7.2, we illustrate the message assignment for the special case when the value of the backhaul load constraint $B = 2$. In this case, $2B = 4$, and we notice how each of the first four messages is assigned according to the above explanation, leading to interference-free signals observed at each of the first four receivers.

Note that each subnetwork has an even number of indices. The message whose index follows the middle is inactive. For the first subnetwork, message W_{2B+1} is inactive. For each message $W_j, j \in \{2B+2, 2B+3, \ldots, 4B\}$, the transmit set $\mathcal{T}_j = \{j-1, j-2, \ldots, 2B+1\}$, i.e., message W_j is assigned to the transmitter having the previous index, as well as all preceding transmitters down to the transmitter with index $2B+1$. We now observe that the part of the subnetwork consisting of transmitters $\{4B-1, 4B-2, \ldots, 2B+1\}$ and receivers $\{4B, 4B-1, \ldots, 2B+2\}$ is identical to the part consisting of the first $2B-1$ transmitter–receiver pairs. To clarify this point, consider the example of Figure 7.2: W_8 is transmitted through x_7 to y_8 without interference, as W_1 is transmitted through x_1 to y_1 without interference. The interference caused by W_8 at y_7 and y_6 is canceled through x_6 and x_5, with respect to order, as the interference caused by W_1 at y_2 and y_3 is canceled through x_2 and x_3, respectively. We can then see that interference-free communication is achieved for messages $W_j, j \in \{2B+2, 2B+3, \ldots, 4B\}$, in a similar fashion to how it is achieved for messages $W_j, j \in \{1, 2, \ldots, 2B-1\}$. It becomes clear that with this message assignment, interference-free communication is achieved for $4B-1$ messages in each subnetwork, and hence the achieved per-user DoF of $\frac{4B-1}{4B}$.

7.2.1 Using Schemes Developed under Per-Message Backhaul Constraint

We have shown in Section 7.1 that the optimal per-user DoF under the backhaul load constraint could be achieved by using a convex combination of the schemes that were shown to be optimal in Chapter 6 under a per-message backhaul constraint that limits the maximum transmit set size by a bound M. The schemes that are optimal for the cases when $M = 1$ and $M = 2$ were used to achieve the per-user DoF under a backhaul load of unity, $B = 1$. We show in this section that this argument can be generalized by using a convex combination of the schemes that are optimal for the cases when $M = 2B-1$ and $M = 2B$, to achieve the value of $\tau^{\text{avg}}(B)$.

If we take a closer look at the schemes found to be optimal under the maximum transmit set size constraint, we will find that these schemes do not fully exploit the constraint; in other words, their backhaul load is strictly below the maximum value M. In fact, the optimal scheme under a per-message constraint of M requires a backhaul load of $\frac{M(M+1)}{2M+1}$. We can derive this value by considering the following argument. The network is split into subnetworks, each of size $2M+1$. Each of the first M messages is assigned to the transmitter carrying the same index, and all following transmitters up to the Mth transmitter in the subnetwork. This requires a total backhaul load of $\sum_{i=1}^{M} i = \frac{M(M+1)}{2}$. The middle message in each subnetwork is inactive, and the remaining part of the subnetwork is a mirrored version of the first part. For example, in the first subnetwork, messages $W_j, j \in \{2M+1, 2M, \ldots, M+2\}$ require the same backhaul load as messages $W_j, j \in \{1, 2, \ldots, M\}$, with respect to order. Hence, the total backhaul load required for all $2M+1$ messages in a subnetwork is $M(M+1)$. The average backhaul load is then $\frac{M(M+1)}{2M+1}$.

We now know that the scheme that is optimal under a maximum transmit set size constraint of $M = 2B-1$ requires a backhaul load of $\frac{2B(2B-1)}{4B-1}$, and the scheme for $M = 2B$ requires a backhaul load of $\frac{2B(2B+1)}{4B+1}$. By using the scheme of $M = 2B-1$ for a fraction of $\frac{4B-1}{8B}$ of the network, and the scheme of $M = 2B+1$ for the remainder, the total backhaul load becomes $\frac{4B-1}{8B} \cdot \frac{2B(2B-1)}{4B-1} + \frac{4B+1}{8B} \cdot \frac{2B(2B+1)}{4B+1} = B$. Since the per-user DoF under the maximum transmit set size constraint $\tau(M) = \frac{2M}{2M+1}$, the achieved per-user DoF using the aforementioned convex combination is then $\frac{4B-1}{8B} \cdot \frac{4B-2}{4B-1} + \frac{4B+1}{8B} \cdot \frac{4B}{4B+1} = \frac{4B-1}{4B}$, which is the same as the one we showed earlier in Section 7.2.

7.2.2 Upper Bound Proof

We show in this section why the lower bound of (7.2) is also an upper bound, and hence we have an exact characterization of the asymptotic per-user DoF for linear networks under the backhaul load constraint:

$$\tau^{\text{avg}}(B) = \frac{4B-1}{4B}, \ \forall B \in \mathbf{Z}^+. \tag{7.3}$$

The proof of the upper bound consists of two steps. First, we generalize the argument from Chapter 6 to show the upper bound under a maximum transmit set size constraint, and show here that we can make a stronger statement. If there is only a subset of messages, whose transmit set sizes are bounded by a maximum value of M, then we can draw an information-theoretic upper bound on the per-user DoF in terms of M and the size of that subset of messages. The second step is to show that an average transmit set size constraint of B implies that there will always be a subset of messages $\mathcal{S}$ with transmit set size bounded by a maximum value M, and the values of $|\mathcal{S}|$ and M give the desired upper bound.

LEMMA 7.1 (**[88, Lemma** 3]) *For any K-user linear interference channel with DoF η, if there exists a subset of messages $\mathcal{S} \subseteq [K]$ such that each message in $\mathcal{S}$ is available at a maximum of M transmitters, i.e., $|\mathcal{T}_i| \leq M, \forall i \in \mathcal{S}$, then the DoF is bounded by*

$$\eta \leq K - \frac{|\mathcal{S}|}{2M+1} + C_K, \tag{7.4}$$

where $\lim_{K\to\infty} \frac{C_K}{K} = 0$.

What Lemma 7.1 implies for linear networks is that if we consider a sequence of K-user networks, with the value of K increasing to infinity, and for each network we have a set $\mathcal{S}_K$ of messages whose transmit set sizes are at most M, and the limiting value of the fraction $\lim_{K\to\infty} \frac{\mathcal{S}_K}{K} = s$, then the asymptotic per-user DoF is upper-bounded by $1 - \frac{s}{2M+1}$. Note that for the special case when $\mathcal{S}_K = \{1,2,\ldots,K\}$, where we impose the maximum transmit set size constraint of M on all messages, we have $s = 1$ and $\tau(M) \leq 1 - \frac{1}{2M+1}$.

Following the same footsteps as the argument for the upper bound on $\tau(M)$ in Section 6.5, we use Lemma 5.2, which we explain here for convenience. What Lemma 5.2 implies is that if we are allowed to ignore the effect of the Gaussian noise, and then be able to reconstruct all the transmit signals from only the received signals $y_{\mathcal{A}}$, then the degrees of freedom is upper-bounded by the cardinality $|\mathcal{A}|$.

To prove the upper bound on $\tau(M)$ in Section 6.5, we used Lemma 5.2 with the set $\mathcal{A} = \{i \in [K] : i \bmod (2M+1) \neq M+1\}$. The idea was that if the network is split into subnetworks, and each subnetwork has $2M+1$ successive users, then the message with the middle index in each subnetwork (the $(M+1)$th message) can only be available at transmitters within the subnetwork. Further, this message cannot be available at the transmitter with the last index in the subnetwork. We know that this is true for any *useful* message assignment from the result of Corollary 6.1. The proof then proceeded by reconstructing the last transmit signal in each subnetwork, since these signals rely only on the messages $W_{\mathcal{A}}$. Having the transmit signals $\{x_i : i \bmod (2M+1) = 0\}$ and the received signals $\{y_i : i \in \mathcal{A}\}$, and ignoring the Gaussian noise, all the transmit signals can be reconstructed by simple linear operations.

Lemma 7.1 can be proved by extending the argument used to prove the $\frac{2M}{2M+1}$ upper bound on $\tau(M)$. When using Lemma 5.2, we use a similar set to the set $\{i \in [K] : i \bmod (2M+1) \neq M+1\}$, but instead of considering all K indices of the network, we restrict our attention to the set $\mathcal{S}$, where messages have a maximum transmit set size constraint. We make the following definition: Let $g_{\mathcal{S}} : \mathcal{S} \to \{1, 2, \ldots, |\mathcal{S}|\}$ be a function that returns the ascending order of any element in the set $\mathcal{S}$, e.g., $g_{\mathcal{S}}(\min\{i : i \in \mathcal{S}\}) = 1$ and $g_{\mathcal{S}}(\max\{i : i \in \mathcal{S}\}) = |\mathcal{S}|$; then we define the set $\mathcal{A}$ such that the complement set $\bar{\mathcal{A}} = \{i : i \in \mathcal{S}, g_{\mathcal{S}}(i) = (2M+1)(j-1) + M + 1, j \in \mathbf{Z}^+\}$. So, instead of splitting the network into subnetworks each consisting of successive $2M+1$ indices, we split the network here into subnetworks each consisting of successive $2M+1$ indices in $\mathcal{S}$ and all other indices in between them. As before, in each subnetwork there is only one element in $\bar{\mathcal{A}}$. Instead of having that element as the middle or $(M+1)$th index among consecutive $2M+1$ indices, we have it here as the middle index among all $2M+1$ indices in the subnetwork that are members of $\mathcal{S}$. The same argument then carries over to show that the DoF is upper-bounded by the cardinality of the chosen set $\mathcal{A}$,

$$\begin{aligned} \eta &\leq |\mathcal{A}| \\ &= \left(1 - \frac{|\mathcal{S}|}{2M+1}\right) K + o(K). \end{aligned} \tag{7.5}$$

Corollary 6.1 implies that there is no decrease in the sum-rate if we restrict the choice of message assignments to those that satisfy the following conditions. Any message in $\bar{\mathcal{A}}$ cannot be available at the last transmitter of any subnetwork. Hence, we can reconstruct these transmit signals. Then, simple linear operations suffice to reconstruct all remaining transmit signals, using Gaussian noise signals and the received signals $y_{\mathcal{A}}$. The bound in (7.5) then follows from the conclusion of Lemma 5.2.

The second step of the proof is of a combinatorial nature. The goal is to show that if a message assignment satisfies a backhaul load constraint of B, then there has to be a *large* subset of messages $\mathcal{S}$ whose transmit set sizes are below a *small* value M. The values of $|\mathcal{S}|$ and M should enable us to use Lemma 7.1 to imply that $\tau(B) \leq \frac{4B-1}{4B}$. The following lemma captures this combinatorial fact needed to complete the proof.

LEMMA 7.2 *[88, Lemma 4] For any message assignment for a K-user channel with an average transmit set size constraint B, there exists an integer $M \in \{0, 1, \ldots, K\}$, and a subset $\mathcal{S} \subseteq [K]$ whose size $|\mathcal{S}| \geq \frac{2M+1}{4B} K$, such that each message in $\mathcal{S}$ is available at a maximum of M transmitters, i.e., $|\mathcal{T}_i| \leq M, \forall i \in \mathcal{S}$.*

Direct application of Lemmas 7.1 and 7.2 leads to the upper bound $\tau(B) \leq \frac{4B-1}{4B}$. We now discuss why the latter is true. Consider the simplest case when $B = 1$. We argue that Lemma 7.2 holds in this case with a value $M = 1$. In other words, for any message assignment that does not require extra backhaul load, at least $\frac{3}{4}$ of the messages in the network are not assigned to more than one transmitter. Assume that this is not true; then there would be a fraction of more than $\frac{1}{4}$ of the messages that are available at two or more transmitters. In this case, the only way that the average transmit set size constraint $\sum_{i=1}^{K} |\mathcal{T}_i| \leq BK = K$ is satisfied is that there is at least an equal fraction of more than $\frac{1}{4}$ of the messages that are inactive whose transmit sets are empty. But if we have a fraction of more than $\frac{1}{4}$ of the messages that are inactive, then the statement of Lemma 7.1 applies for $M = 0$ and we cannot achieve a per-user DoF that is greater than $\frac{3}{4}$. Now that we see how this simple combinatorial fact is true when $B = 1$, we explain in the following why it is true for any positive integer value of B.

For every integer $j \in \{0, 1, \ldots, K\}$, we define R_j as the fraction of messages that are available at exactly j transmitters. More precisely,

$$R_j = \frac{|\{i : i \in [K], |\mathcal{T}_i| = j\}|}{K}. \tag{7.6}$$

If $\sum_{j=2B}^{K} R_j \leq \frac{1}{4B}$, then more than $\frac{4B-1}{4B}K$ users have a transmit set whose size is at most $2B - 1$, and Lemma 7.2 follows with $M = 2B - 1$. We can assume then that $\sum_{j=2B}^{K} R_j > \frac{1}{4B}$. In what follows, we show that this assumption implies that $\sum_{j=0}^{M} R_j > \frac{2M+1}{4B}$ for some integer $M \in \{0, \ldots, 2B-2\}$, and hence Lemma 7.2 would follow with that value of M.

If we look carefully at the message assignment defined earlier in Section 7.2 and used to achieve the lower bound $\tau(B) \geq \frac{4B-1}{4B}$, and if we define R_j^* as the fraction of messages that are available at exactly j transmitters for that assignment, then we find that $R_0^* = R_{2B}^* = \frac{1}{4B}$, $R_j^* = \frac{1}{2B}$ $\forall j \in \{1, \ldots, 2B-1\}$, and no message is assigned to more than $2B$ transmitters. Further, this assignment meets the backhaul load constraint tightly. In other words, $\sum_{j=1}^{K} jR_j^* = B$. Recall that we are considering message assignments that satisfy the condition

$$\sum_{j=2B}^{K} R_j > \frac{1}{4B}$$

$$= R_{2B}^* = \sum_{j=2B}^{K} R_j^*. \tag{7.7}$$

Since we know by definition that $\sum_{j=1}^{K} R_j^* = \sum_{j=1}^{K} R_j$, (7.7) implies that there exists an integer $M \in \{0, 1, \ldots, 2B-1\}$ such that $R_M > R_M^*$; let m be the smallest such integer. If $\sum_{j=0}^{m} R_j > \sum_{j=0}^{m} R_j^* = \frac{2m+1}{4B}$, then Lemma 7.2 holds with that value of m, and hence we can assume for the rest of the proof that $\sum_{j=0}^{m} R_j \leq \sum_{j=0}^{m} R_j^*$. Now, since we have the aforementioned assumption, and also from the choice of m we know that $\forall j \in \{0, 1, \ldots, m-1\}, R_j \leq R_j^*$, we can construct another message assignment by removing elements from some transmit sets whose size is m, such that the new assignment has a smaller average transmit set size, and has transmit sets $\mathcal{T}_i^*$ where $\forall j \in \{0,1,\ldots,m\}, |\{i : i \in [K], |\mathcal{T}_i^*| = j\}| \leq R_j^*$. By successive application of the above argument, we can construct a message assignment that satisfies the backhaul load constraint of B, and has transmit sets $\mathcal{T}_i^*$ where $\forall j \in \{0,1,\ldots,2B-1\}, |\{i : i \in [K], |\mathcal{T}_i^*| = j\}| \leq R_j^*$ and $|\{i : i \in [K], |\mathcal{T}_i^*| \geq 2B\}| \geq R_{2B}^*$. But then the new assignment must have a larger average transmit set size than that of the assignment achieving the $\frac{4B-1}{4B}$ lower bound. More precisely, the following has to be true:

$$\begin{aligned} \frac{\sum_{i=1}^{K} |\mathcal{T}_i^*|}{K} &= \frac{1}{K}\sum_{j=1}^{K} j|\{i : i \in [K], |\mathcal{T}_i^*| = j\}| \\ &> \sum_{j=1}^{K} jR_j^* \\ &= B. \end{aligned} \tag{7.8}$$

It then follows that the newly constructed message assignment violates the backhaul load constraint, and hence we reach a contradiction that proves Lemma 7.2.

7.3 General Locally Connected Networks

We saw in Section 7.2.1 that the use of a convex combination of the schemes that are optimal under the maximum transmit set size constraint can lead to optimal schemes under the backhaul load constraint. We now show that this approach can also provide good achievable schemes for the more general locally connected channel model, where each transmitter is connected to the receiver having the same index as well as L following receivers, where $L > 1$.

Consider the schemes illustrated in Chapter 6 to achieve the lower bound of $\tau_L(M) \geq \frac{2M}{2M+L}$. The message assignments used by these schemes do not meet the maximum transmit set size constraint tightly. More precisely, $|\mathcal{T}_i| < M$ for most users. Only two users in each subnetwork have a transmit set of size M, which constitutes a fraction of $\frac{2}{2M+L}$ of the users in the network. For all remaining users, the transmit set size is strictly less than M. A more careful look at these assignments would reveal that the network is split into subnetworks, each of size $2M+L$, and each subnetwork has L inactive messages (with empty transmit sets). The $2M$ messages that are active in each subnetwork are split into two equal-sized groups. The sum of the transmit set sizes in each of the two groups is the same, so it suffices to calculate it for one of them. Consider for example, the first group of M users in the first subnetwork. This group consists of users $\{1,2,\ldots,M\}$, and the transmit set $\mathcal{T}_i = \{i, i+1, \ldots, M\}$ for each user i in this group. The sum of transmit set sizes is hence given by

$$\sum_{i=1}^{M} |\mathcal{T}_i| = \frac{M(M+1)}{2}. \tag{7.9}$$

Table 7.1 Achievable per-user DoF values for locally connected channels with a backhaul constraint $\sum_{i=1}^{K} |\mathcal{T}_i| \leq K$. For each L, the schemes used and the convex combination in which they are used are shown.

	Scheme 1	Scheme 2	Ratio	$\tau_L(B=1) \geq$
$L=2$	$\tau_L(M=2) \geq \frac{2}{3}$, $\tau_L^{\text{avg}}(B=1) \geq \frac{2}{3}$	—	—	$\frac{2}{3}$
$L=3$	$\tau_L(M=2) \geq \frac{4}{7}$, $\tau_L^{\text{avg}}\left(B=\frac{6}{7}\right) \geq \frac{4}{7}$	$\tau_L(M=3) \geq \frac{2}{3}$, $\tau_L^{\text{avg}}\left(B=\frac{4}{3}\right) \geq \frac{2}{3}$	$7:3$	$\frac{3}{5}$
$L=4$	$\tau_L(M=2) \geq \frac{1}{2}$, $\tau_L^{\text{avg}}\left(B=\frac{3}{4}\right) \geq \frac{1}{2}$	$\tau_L(M=3) \geq \frac{3}{5}$, $\tau_L^{\text{avg}}\left(B=\frac{6}{5}\right) \geq \frac{3}{5}$	$4:9$	$\frac{5}{9}$
$L=5$	$\tau_L(M=2) \geq \frac{4}{9}$, $\tau_L^{\text{avg}}\left(B=\frac{2}{3}\right) \geq \frac{4}{9}$	$\tau_L(M=3) \geq \frac{6}{11}$, $\tau_L^{\text{avg}}\left(B=\frac{12}{11}\right) \geq \frac{6}{11}$	$3:11$	$\frac{11}{21}$
$L=6$	$\tau_L(M=3) \geq \frac{1}{2}$, $\tau_L^{\text{avg}}(B=1) \geq \frac{1}{2}$	—	—	$\frac{1}{2}$

This further implies that the backhaul load is given by

$$\begin{aligned} B &= \frac{\sum_{i=1} K|\mathcal{T}_i|}{K} \\ &= \frac{\sum_{i=1} 2M+L|\mathcal{T}_i|}{2M+L} \\ &= \frac{M(M+1)}{2M+L}. \end{aligned} \tag{7.10}$$

We can then use these schemes to draw conclusions about what we can achieve for each value of the local connectivity parameter L, under any backhaul load constraint B. More specifically, by using convex combinations of the schemes that are optimal under a maximum transmit set size constraint M, we can identify achievable values for the asymptotic per-user DoF, with no additional overall load on the wireless backhaul ($B=1$). In Table 7.1, we characterize the schemes employed to achieve an asymptotic per-user DoF that is greater than or equal to half for each value of $L \leq 6$, and state the required ratio for the convex combination to achieve the target value. For example, for the case when $L=2$, we use the scheme that is optimal for a maximum transmit set size constraint of $M=2$ for 70% of the network, and the scheme that is optimal for a maximum transmit set size constraint of $M=3$ for 30% of the network. The achieved value for the asymptotic per-user DoF will then be $\frac{2}{3}$ with a backhaul load of unity.

We note that the lower bounds stated in Table 7.1 are achieved through the use of only zero-forcing transmit beamforming. In other words, there is no need for the symbol extensions that are used in the asymptotic interference alignment scheme of [46]. In [71, Theorem 8], it was shown that for $L \geq 2$, by allowing each message to be available at one transmitter, the asymptotic per-user DoF is $\frac{1}{2}$; it was also shown in [71, Theorem 6] that the $\frac{1}{2}$ per-user DoF value cannot be achieved through zero-forcing transmit forming for $L \geq 3$. In contrast, it can be seen in Table 7.1 that for $L \leq 6$, the $\frac{1}{2}$ per-user DoF value can be achieved through zero-forcing transmit beamforming and a flexible design of the backhaul links, without incurring additional overall load on the backhaul ($B = 1$).

7.4 Cellular Network Models

For linear networks with $L = 1,2,3,4,5,6$, we have shown that compared to the case of no cooperation ($M = 1$), a greater per-user DoF can be achieved with delay-free one-shot schemes under an average backhaul load $B = 1$, i.e., without incurring additional backhaul load. We now investigate whether these results hold for denser networks that may be more relevant to practice.

7.4.1 Example: Two-Dimensional Wyner Network

Through this example, we illustrate that the insights for linear interference networks may apply in denser networks by treating the denser network as a set of interfering linear networks. We consider the two-dimensional network depicted in Figure 7.3(a), where each transmitter is connected to four cell edge receivers. The precise channel model for a K-user channel is as follows:

$$h_{ij} \text{ is not identically 0 if and only if}$$
$$i \in \left\{ j, j+1, j+\left\lfloor \sqrt{K} \right\rfloor, j+\left\lfloor \sqrt{K} \right\rfloor + 1 \right\}. \tag{7.11}$$

Under the backhaul load constraint $\frac{\sum_{i=1}^{K} |\mathcal{T}_i|}{K} \leq 1$, a per-user DoF of $\frac{5}{9}$ can be achieved using only zero-forcing transmit beamforming. This can be done by deactivating every third row of transmitters, and splitting the rest of the network into noninterfering linear subnetworks – see Figure 7.3(b). In each subnetwork, a backhaul load constraint of $\frac{3}{2}$ is imposed. For example, the following constraint is imposed on the first row of users: $\frac{\sum_{i=1}^{\lfloor \sqrt{K} \rfloor} |\mathcal{T}_i|}{\lfloor \sqrt{K} \rfloor} \leq \frac{3}{2}$. A convex combination of the schemes that are characterized as optimal with maximum transmit set size constraints $M = 2$ and $M = 3$ is then used to achieve a per-user DoF of $\frac{5}{6}$ in each active subnetwork while satisfying a backhaul load constraint of $\frac{3}{2}$. The scheme for $M = 2$ for a linear network has a backhaul load of $B = \frac{6}{5}$ and a per-user DoF of $\frac{4}{5}$, and the scheme for $M = 3$ has a backhaul load of $B = \frac{12}{7}$ and a per-user DoF of $\frac{6}{7}$. We can then use the scheme for $M = 2$ for a fraction of $\frac{5}{12}$ of the network, and the scheme for $M = 3$ for the remaining part of the network, and achieve the target per-user DoF of $\frac{5}{6}$ with a backhaul load of $\frac{3}{2}$. Since $\frac{2}{3}$ of the subnetworks are active, a per-user DoF of $\frac{5}{9}$ is achieved while satisfying a backhaul load constraint of unity.

We observe that a per-user DoF greater than $\frac{1}{2}$ is achieved by using simple zero-forcing schemes with an average backhaul load of one without the need for

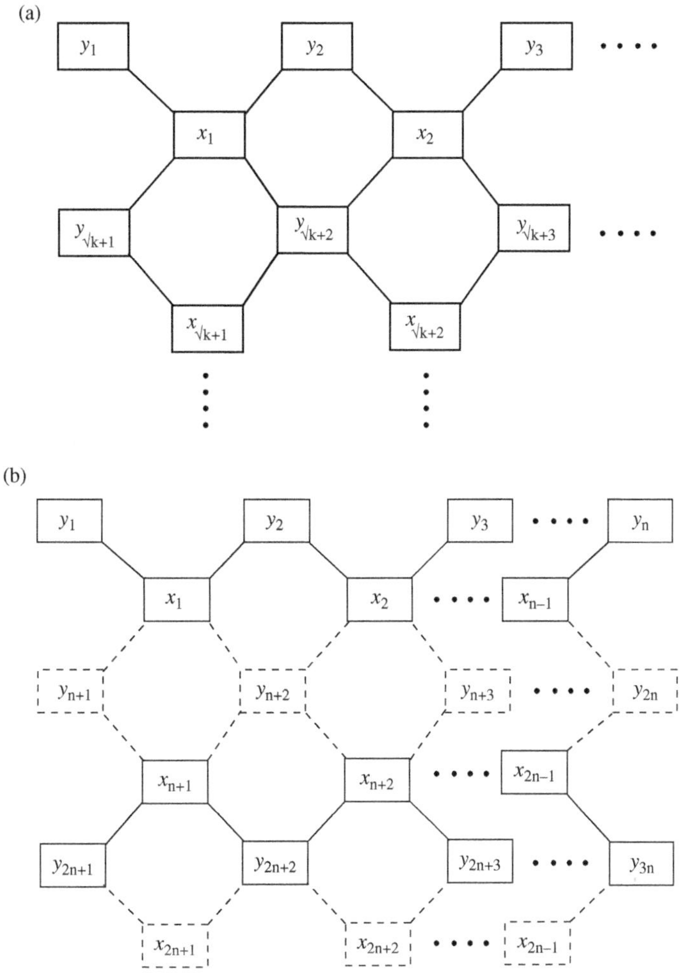

Fig. 7.3 Two-dimensional interference network. (a) The channel model. (b) An example coding scheme where dashed boxes and lines represent inactive nodes and edges; the network is split into noninterfering linear subnetworks. ©[2017] IEEE. Reprinted, with permission, from [87].

interference alignment. We could reach this conclusion by leveraging the schemes that were discovered for linear networks in more complex network topologies, by splitting the network into linear subnetworks and deactivating a subset of *edge transmitters* to eliminate inter-subnetwork interference. In the next section, we explore whether we can extend this idea to the practically relevant hexagonal cellular network model.

7.4.2 Hexagonal Cellular Networks with No Cooperation

Information-Theoretic Upper Bound

In order to evaluate the gains of the added flexibility offered by combining CoMP transmission and the backhaul load constraint, we first study the information-theoretic

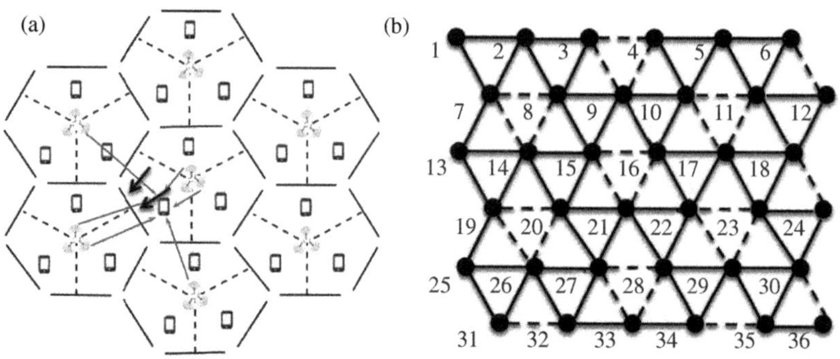

Fig. 7.4 (a) Cellular network. (b) Connectivity graph. The dotted lines in (b) represent the interference between sectors belonging to the same cell. ©[2017] IEEE. Reprinted, with permission, from [89].

limit when each message could be available only at one transmitter. This is the path that we followed for linear and locally connected networks, and we follow it here for hexagonal cellular network models. We first discuss the model of network connectivity. This is a sectored K-user cellular network with three sectors per cell, as shown in Figure 7.4(a). We assume a local interference model, where the interference at each receiver is only due to the basestations in the neighboring sectors. It is assumed that the sectors belonging to the same cell do not interfere with each other, the justification being that the interference power due to sectors in the same cell is usually far lower than the interference from out-of-cell users located in the sector's line of sight.

The cellular model is represented by an undirected connectivity graph $G(V,E)$ as shown in Figure 7.4(b), where each vertex $u \in V$ corresponds to a transmitter–receiver pair. For any node a, the transmitter, receiver, and intended message (word) corresponding to the node are denoted by T_a, R_a, and W_a, respectively. An edge $e \in E$ between two vertices $u, v \in V$ corresponds to a channel existing between T_u and R_v, as well as a channel existing between T_v and R_u. The dotted lines denote interference between sectors that belong to the same cell, and is ignored in our model. To simplify the presentation, without much loss of generality, we consider only $K-$user networks where $\sqrt{K}$ is an integer, and nodes are numbered as in Figure 7.4(b). (In the figure, $\sqrt{K} = 6$.) Since we are studying the performance in the asymptotic limit of the number of users, the assumption is not restrictive.

Define $\tau_c(M)$ as the asymptotic per-user DoF for the considered cellular network, with a maximum transmit set size constraint of M. In this section, we show that $\tau_c(M = 1) \leq \frac{1}{2}$. In other words, one can only achieve $\frac{1}{2}$ per-user DoF in a large cellular network if each message can only be available at a single transmitter. We present the following lemma from [71] for the case when each message can be available at only one transmitter ($M = 1$). The lemma gives a relation between the DoF of the message being delivered by a transmitter and the DoF corresponding to the messages intended for the receivers connected to that transmitter. Here, $\mathcal{R}_j$ denotes the set of receivers that are connected to transmitter T_j.

LEMMA 7.3 (**[71, Lemma 5]**) *If* $\mathcal{T}_i = \{X_j\}$, *then* $d_i + d_k \leq 1$, $\forall k \in \mathcal{R}_j$.

What Lemma 7.3 captures is the simple fact that if a message is being delivered to its destination through a single transmitter, then this message has a conflict with all the messages intended for receivers connected to that transmitter. Since any of the aforementioned receivers can observe its own message and the message carried by the designated transmitter, then their sum DoF is unity. We now make few definitions to simplify presentation. Each transmitter–receiver pair in the network is referred to as a node.

DEFINITION 7.1 *For a set of nodes* $a_1, a_2, \ldots, a_n$, *we define* $a_1 \circ a_2 \circ \cdots \circ a_n$ *to denote that the per-user DoF of messages* $W_{a_1}, W_{a_2}, \ldots, W_{a_n}$ *is at most half, i.e.,*

$$\sum_{i=1}^{n} d_i \leq \frac{n}{2}. \tag{7.12}$$

DEFINITION 7.2 *If a and b are two nodes such that they are connected in the connectivity graph, and the transmitter of node a has the message for node b, i.e.,* $a \in \mathcal{T}_b$, *we say that it is a useful message assignment from a to b and this is denoted by* $a \to b$.

Consider any chain a, b, c such that $a \to b \to c$. We have $\mathcal{T}_b = \{a\}$ and $a, b \in \mathcal{R}_a$, and hence, from Lemma 7.3, we have $a \circ b$. Similarly, we have $\mathcal{T}_c = \{b\}$ and $a, b, c \in \mathcal{R}_b$, and, from Lemma 7.3, we have $a \circ c$ and $b \circ c$. Thus, we have $a \circ b \circ c$ for any $a \to b \to c$.

THEOREM 7.4 *For the considered hexagonal cellular network model, the following bound holds for the case of no cooperation:*

$$\tau_c(M = 1) \leq \frac{1}{2}.$$

We explain the proof of Theorem 7.4 in the rest of this section. Consider the division of the network into triangles as shown in Figure 7.5. We say that a triangle is in state S_i if exactly i of the messages of the triangle are assigned to transmitters within the triangle,

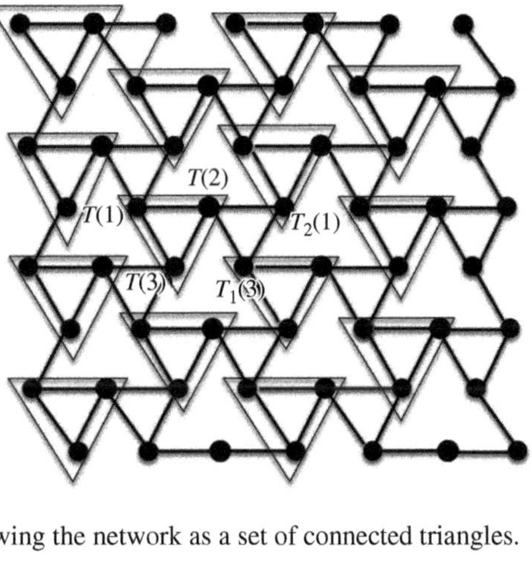

Fig. 7.5 Converse argument: viewing the network as a set of connected triangles.

$0 \le i \le 3$. Let $\mathcal{S}_i$ denote the set of all triangles in state S_i. We highlight the following types of nodes:

Self-serving nodes: We refer to a node as a self-serving node if the message to the receiver corresponding to the node is assigned to its own transmitter. We define $\mathcal{SS}_1$ as the set of all self-serving nodes belonging to triangles in state S_1.
Outsider nodes: We refer to a node as an outsider node if no message within its triangle is assigned to its transmitter, and also its message is not assigned within its triangle. Let $\mathcal{O}$ denote the set of all outsider nodes. Note that every triangle in state S_0 consists of three outsider nodes, every triangle in state S_1 has at least one outsider node, and a triangle in state S_2 may contain an outsider node.

In any triangle T, $T(1), T(2), T(3)$ denote the top left node, the top right node, and the bottom node, respectively. We define a middle triangle which is formed by the connected nodes of three different neighboring triangles. For example, $T(2), T_1(3), T_2(1)$ in Figure 7.5 are members of a middle triangle. For any node b, let M_b denote the middle triangle containing node b.

In Algorithm 7.1, we define a strategy for including nodes in a set $\mathcal{S}$ such that at any stage, the per-user DoF of the nodes already included in $\mathcal{S}$ is upper-bounded by $\frac{1}{2}$. Note that at the end of the algorithm, all nodes are included in $\mathcal{S}$. To facilitate the understanding of Algorithm 7.1, we observe the following:

- If $T(i) \in \mathcal{SS}_1$, then $T(i)$ is a self-serving node and since the triangle T associated with it is in state S_1, the other nodes in the triangle $T((i+1) \bmod 3), T((i+2) \bmod 3)$ are outsider nodes. Without loss of generality, we have $T(i) \circ T((i+1) \bmod 3)$, and hence we can include nodes $T(i), T((i+1) \bmod 3)$ in the set $\mathcal{S}$, as in line 4.
- For any middle triangle with nodes $T(i), T_1((i+1) \bmod 3), T_2((i+2) \bmod 3)$ containing at least two outsider nodes, we have $T(i) \circ T_1((i+1) \bmod 3)$, $T_1((i+1) \bmod 3) \circ T_2((i+2) \bmod 3)$, $T(i) \circ T_2((i+2) \bmod 3)$. The aforementioned three inequalities imply the inequality $T(i) \circ T_1((i+1) \bmod 3) \circ T_2((i+2) \bmod 3)$. Note that the three inequalities hold because the message of an outsider node can only be available at a transmitter in the middle triangle containing the node; we discuss this argument in detail below. Let the two outsider nodes be $T(i)$ and $T_1((i+1) \bmod 3)$. Since $T_1((i+1) \bmod 3)$ is an outsider node, we have the following possibilities: Suppose the message to node $T_1((i+1) \bmod 3)$ is not available at either $T_2((i+2) \bmod 3)$ or $T(i)$ and hence cannot be transmitted. Then we have $d_{T_1((i+1) \bmod 3)} = 0$ and so $T(i) \circ T_1((i+1) \bmod 3)$ and $T_2((i+2) \bmod 3) \circ T_1((i+1) \bmod 3)$ hold. Suppose the message for $T_1((i+1) \bmod 3)$ is transmitted by any of the other nodes in the middle triangle. Then, from Lemma 7.3, we have $T(i) \circ T_1((i+1) \bmod 3)$ and $T_2((i+2) \bmod 3) \circ T_1((i+1) \bmod 3)$. Similarly, for the other outsider node $T(i)$, we have $T(i) \circ T_2((i+2) \bmod 3)$. Hence we have $T(i) \circ T_1((i+1) \bmod 3) \circ T_2((i+2) \bmod 3)$.

  If the triangle contains three outsider nodes, we include nodes $T(i), T_1((i+1) \bmod 3), T_2((i+2) \bmod 3)$ in the set $\mathcal{S}$, as in line 10. If it contains only two outsider nodes

Algorithm 7.1 Strategy for including all nodes in $\mathcal{S}$ such that the half per-user DoF upper bound proof is simplified.

1: Initialize $\mathcal{S} \leftarrow \emptyset$
2: **while** $\mathcal{SS}_1 \backslash \mathcal{S} \neq \emptyset$ **do**
3:     **for** $b \in \mathcal{SS}_1$ where $b = T(i)$ for some triangle T **do**
4:         $\mathcal{S} \leftarrow \mathcal{S} \cup \{T(i), T((i+1) \text{ mod } 3)\}$
5:     **end for**
6: **end while**
7: **while** $\mathcal{O} \backslash \mathcal{S} \neq \emptyset$ **do**
8:     **for** $b \in \mathcal{O} \backslash \mathcal{S}$ where $b = T(i)$ for some triangle T and M_b contains nodes $T_1((i+1) \text{ mod } 3)$ and $T_2((i+2) \text{ mod } 3)$ **do**
9:         **if** $M_{T(i)} \backslash \mathcal{S}$ contains 3 outsider nodes **then**
10:             $\mathcal{S} \leftarrow \mathcal{S} \cup \{T(i), T_1((i+1) \text{ mod } 3), T_2((i+2) \text{ mod } 3)\}$
11:         **else if** $M_{T(i)} \backslash \mathcal{S}$ contains 2 outsider nodes $T(i)$ and j where $j \in \{T_1((i+1) \text{ mod } 3), T_2((i+2) \text{ mod } 3)\}$ **then**
12:             $\mathcal{S} \leftarrow \mathcal{S} \cup \{T(i), j\}$
13:         **else if** $M_{T(i)} \backslash \mathcal{S}$ contains $T(i)$ as the only outsider node and message for $T(i)$ is assigned within $M_{T(i)} \backslash \mathcal{S}$ at $j \rightarrow T(i)$ where $j \in \{T_1((i+1) \text{ mod } 3), T_2((i+2) \text{ mod } 3)\}$ **then**
14:             $\mathcal{S} \leftarrow \mathcal{S} \cup \{T(i), j\}$
15:         **else if** $M_{T(i)} \backslash \mathcal{S}$ contains $T(i)$ as the only outsider node and message for $T(i)$ is not assigned within $M_{T(i)} \backslash \mathcal{S}$ **then**
16:             $\mathcal{S} \leftarrow \mathcal{S} \cup \{T(i)\}$
17:         **end if**
18:     **end for**
19: **end while**
20: **while** $\mathcal{S}_1 \cup \mathcal{S}_2 \cup \mathcal{S}_3 \backslash \mathcal{S} \neq \emptyset$ **do**
21:     **for** triangle $T \in \mathcal{S}_1 \cup \mathcal{S}_2 \cup \mathcal{S}_3$ **do**
22:         $\mathcal{S} \leftarrow \mathcal{S} \cup T \backslash \mathcal{S}$
23:     **end for**
24: **end while**

$T(i)$,j, where $j \in \{T_1((i+1) \text{ mod } 3), T_2((i+2) \text{ mod } 3)\}$, we include them in the set $\mathcal{S}$, as in line 12.

- Let $T(i)$ be the only outsider node among $T(i), T_1((i+1) \text{ mod } 3), T_2((i+2) \text{ mod } 3)$. If its message is available at neighboring node j where $j \in \{T_1((i+1) \text{ mod } 3), T_2((i+2) \text{ mod } 3)\}$, i.e., $j \rightarrow T(i)$ as defined in Definition 7.2, then we have $j \circ T(i)$ and include nodes $T(i)$,j in the set $\mathcal{S}$, as in line 14.
- If the message of outsider node $T(i)$ is not assigned within nodes $T_1((i+1) \text{ mod } 3), T_2((i+2) \text{ mod } 3)$, we have $d_{T(i)} = 0$ and we include $T(i)$ in the set $\mathcal{S}$, as in line 16.

We now consider the case where the message of $T(i)$ is assigned to a node in the set $M_{T(i)} \cap \mathcal{S}$, and show that the per-user DoF of the set $\mathcal{S}$ is still upper-bounded by $\frac{1}{2}$ when we add only the node $T(i)$ to the set $\mathcal{S}$. Suppose $j \rightarrow T(i)$ where $j \in$

$\{T_1((i+1) \bmod 3), T_2((i+2) \bmod 3)\}$ but $j \in \mathcal{S}$. This is possible only when j was included in $\mathcal{S}$ according to line 4 in the algorithm. Let j be the self-serving node and $j+1$ be the outsider node included in line 4. We have $j \circ T(i)$, $T(i) \circ j+1$, and we have $j \circ j+1$ from before. We also have $T(i) \circ j+1$ from Lemma 7.3 since $\mathcal{T}_{T(i)} = \{j\}$ and $j+1 \in \mathcal{R}_j$. Hence, $T(i)$ can be included without any increase in the per-user DoF. The same argument holds even if j was the outsider node and $j+1$ the self-serving node included in line 4. Note that the only other possibility through which nodes $j, j+1$ were previously included in the set $\mathcal{S}$ is when both transmitters j and $j+1$ are carrying messages for the only remaining outsider nodes $T(i)$ and $T'(k)$ in their respective middle triangles. In that case we see that $T(i) \circ j$, $T'(k) \circ (j+1)$, and the per-user DoF is still upper-bounded by $\frac{1}{2}$.

Consider all triangles in $\mathcal{S}_1 \cup \mathcal{S}_2 \cup \mathcal{S}_3$. If T denotes such a triangle with nodes $T(1), T(2), T(3)$, let t denote the set of nodes in T but not included in $\mathcal{S}$ by line 19. Note that for triangles in $\mathcal{S}_2 \cup \mathcal{S}_3$ with nodes $T(1), T(2), T(3)$, we have $T(1) \circ T(2), T(2) \circ T(3), T(1) \circ T(3)$, and hence $T(1) \circ T(2) \circ T(3)$ from Lemma 7.3; we will use this fact below. Consider the following cases for any triangle T that has one or more nodes in the set $t = T \backslash \mathcal{S}$:

- The set t contains only one node $T(i)$. We first find two nodes $T(k), j$ where $T(k) \rightarrow j$ that were previously added to $\mathcal{S}$ according to line 14, and show that $T(i) \circ T(k) \circ j$ holds. We then show that nodes $T(k)$ and j do not appear in any other such combination, and hence adding $T(i)$ to $\mathcal{S}$ does not change the per-user DoF upper bound of $\frac{1}{2}$ for the set $\mathcal{S}$.

  Note that, by definition, a triangle in state S_2 or S_3 has at least two messages assigned within the triangle and thus has at least two non-outsider nodes. Hence, if $T \in \mathcal{S}_2 \cup \mathcal{S}_3$, then there exists at least one node $k \in \{(i+1) \bmod 3, (i+2) \bmod 3\}$ such that $T(k)$ is not an outsider node and we have $T(i) \circ T(k)$. If $T \in \mathcal{S}_1$, since all the self-serving nodes and outsider nodes have already been included in $\mathcal{S}$, we have $T(i) \rightarrow T(k)$ or $T(k) \rightarrow T(i)$ for some $k \in \{(i+1) \bmod 3, (i+2) \bmod 3\}$.

  Since $T(k)$ was a non-outsider node that was previously considered, it must have been added according to line 14. Hence, there is an assignment $T(k) \rightarrow j$ where j is an outsider node in the middle triangle $M_{T(k)}$, $j \in \{T_1((k+1) \bmod 3), T_2((k+2) \bmod 3)\}$, and $T(k) \circ j$ was considered. We have $T(k) \circ j$ and $T(i) \circ j$ from Lemma 7.3, since $\mathcal{T}_j = \{T(k)\}$ and $T(k), T(i) \in \mathcal{R}_{T(k)}$. Hence, we have $j \circ T(i) \circ T(k)$.

  Note that neither j nor $T(k)$ is part of any other such combination. This is true for $T(k)$ because all the nodes in its triangle have already been considered. Since $T(k) \rightarrow j$ and j has been added to the set $\mathcal{S}$ according to line 14, outsider node j cannot be part of any such combination that does not involve $T(k)$. Thus, we include $t = \{T(i)\}$ in the set $\mathcal{S}$ as in line 22, while the per-user DoF of the set $\mathcal{S}$ is upper-bounded by $\frac{1}{2}$.
- The set t contains two nodes $T(i), T((i+1) \bmod 3)$ for some $i \in \{1,2,3\}$. If $T \in \mathcal{S}_1$, then one of the nodes is carrying a message for the other and we have $T(i) \circ T((i+1) \bmod 3)$. If $T \in \mathcal{S}_2 \cup \mathcal{S}_3$, we have $T(i) \circ T((i+1) \bmod 3)$ and we include $t = \{T(i), T((i+1) \bmod 3)\}$ in the set S, as in line 22.

- The set t contains three nodes $T(1), T(2), T(3)$. This can happen only when $T \in \mathcal{S}_2 \cup \mathcal{S}_3$. In this case, we have $T(1) \circ T(2) \circ T(3)$ and we include $t = \{T(1), T(2), T(3)\}$ in the set $\mathcal{S}$, as in line 22.

This shows that, using a traditional approach for interference management, the maximum asymptotic per-user DoF for the considered hexagonal cellular network model is at most $\frac{1}{2}$. Further, the only known way this DoF value can be approached is in the limit as the length of symbol extension goes to infinity, as in the asymptotic interference alignment scheme of [46]. We show next that one-shot zero-forcing schemes that do not require symbol extension can only be used to achieve an asymptotic per-user DoF of at most $\frac{3}{7}$. However, if cooperative transmission is enabled with flexible message assignments that do not overload the backhaul, an asymptotic per-user DoF of $\frac{1}{2}$ can be achieved using these simple zero-forcing schemes.

Optimal Zero-Forcing Schemes

We restrict our attention here to the class of zero-forcing schemes, and characterize lower and upper bounds for the maximum achievable per-user DoF.

THEOREM 7.5 *The following bounds hold under restriction to zero-forcing schemes for the asymptotic per-user DoF of hexagonal cellular networks with no cooperation:*

$$\frac{1}{3} \leq \tau_c^{zf}(M = 1) \leq \frac{3}{7}. \tag{7.13}$$

We spend the rest of this section proving Theorem 7.5.

Notice that the network can be divided into disjoint triangles as shown in Figure 7.6. In each triangle, by deactivating two nodes (1 and 2 in Figure 7.6), the network decomposes into $K/3$ isolated nodes that each achieves a DoF of one, thus achieving a per-user DoF of $\frac{1}{3}$ in the network.

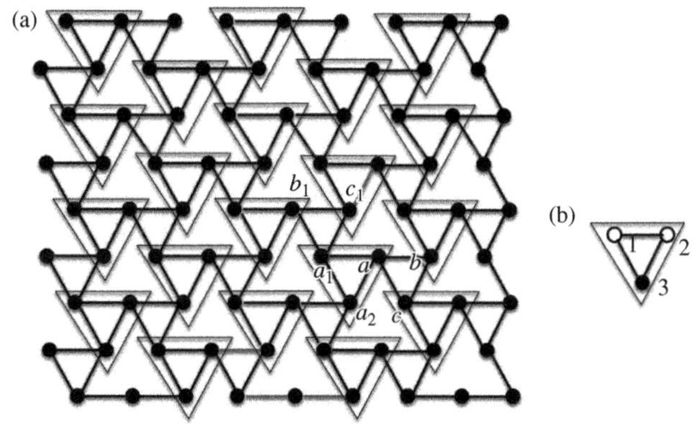

Fig. 7.6 (a) Division of network into triangular subnetworks. (b) By deactivating nodes 1 and 2, a per-user DoF of $\frac{1}{3}$ is achieved. ©[2017] IEEE. Reprinted, with permission, from [89].

Zero-Forcing Upper Bound

Consider the decomposition of the network into disjoint triangles as in Figure 7.6. For any zero-forcing coding scheme, we note that any triangle in the network is in one of the following states:

State 0 (inactive triangle): All transmitters and receivers in the triangle are inactive.
State 1 (self-serving triangle): Exactly one transmitter in the triangle sends a message to exactly one receiver within the triangle. None of the other transmitters or receivers can be active in this triangle.
State 2 (serving triangle): At least one transmitter in the triangle is activated to serve a receiver in another triangle, and there are no active receivers in the considered triangle.
State 3 (served triangle): At least one receiver in the triangle is activated as it is being served by a transmitter in another triangle, and there are no active transmitters within the considered triangle.

For triangles in states 1 and 0, the number of active receivers is bounded by the number of triangles, i.e., a fraction of $\frac{1}{3}$ of the number of users.

For every transmitter c that is active in a triangle, say T_1, that is in state 2, there exists a neighboring receiver b in a different triangle, say T_2, in state 3 that is being served by it and a neighboring node a in a third triangle T_3, whose transmitter and receiver are both inactive. We now consider the following cases for the state of triangle T_3:

Case 1: T_3 is in state 2 or 3. The remaining neighbors of a,b,c are the nodes in their own triangles. We now know that $d_a+d_b+d_c \leq 1$, because the receivers with indices a and c are inactive and hence the per-user DoF is at most $\frac{1}{3}$. Further, because none of the nodes a, b, and c have other neighbors except in their own triangles, there is no overcounting when we repeat this procedure to obtain DoF bounds on other similar sets of users.
Case 2: T_3 is in state 1. Suppose that in T_3 there is a node a_2 that serves itself. Then there is another inactive node in T_3 that may form a group similar to a,b,c with its neighbors from different triangles, say b_1,c_1. We note that these two groups are disjoint. So, among the seven nodes $(S_a \cup \{b,c,b_1,c_1\})$, there are at most three active receivers. This is illustrated in Figure 7.7. Suppose T_3 does not contain a self-serving node. Then a is the only node with inactive transmitter and receiver in T_3, and among the five nodes $(T_3 \cup \{b,c\})$ we attain a sum DoF of at most two.
Case 3: T_3 is in state 0. Then, in the set of the five nodes $(T_3 \cup \{b,c\})$, we attain a sum DoF of at most one.

For any scheme, the network can be rearranged into a combination of disjoint groups of three, five, and seven users, and the per-user DoF for each group is at most $\frac{3}{7}$. It follows that $\tau \leq \frac{3}{7}$ holds asymptotically for any choice of cell associations and interference avoidance schemes.

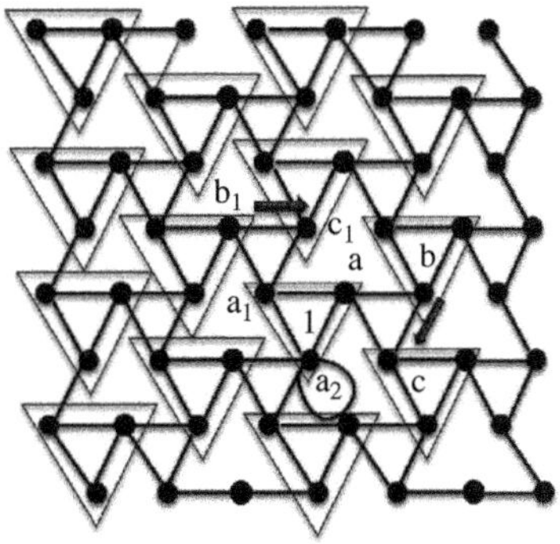

Fig. 7.7 Illustration of the $\frac{3}{7}$ upper bound on the per-user DoF achievable by zero-forcing schemes with no cooperation. The number 1 inside a triangle indicates that it is in state S_1.

7.4.3 Flexible Message Assignment with Cooperation

We now show, through the result in Theorem 7.6, how a smart choice for assigning messages to transmitters, aided by cooperative transmission, can achieve scalable DoF gains through an interference avoidance coding scheme. We achieve this by treating the hexagonal model as interfering locally connected linear networks with $L = 2$. In particular, we show a lower bound on the achievable per-user DoF that is greater than the $\frac{3}{7}$ upper bound of the case without cooperation that was proved in Theorem 7.5.

THEOREM 7.6 *Under the average backhaul constraint $B = 1$, the following lower bound holds for the asymptotic per-user DoF, and is achievable using a zero-forcing scheme:*

$$\tau_{\mathrm{c}}^{\mathrm{avg,zf}}(B=1) \geq \frac{1}{2}. \tag{7.14}$$

Proof Consider a division of the network formed by deactivating a third of the nodes, as shown in Figure 7.8(a). We note that the remaining network consists of noninterfering locally connected subnetworks with connectivity parameter $L = 2$. In each subnetwork, we use the scheme in [71] for $M = 3$ that considers a division of the subnetwork into noninterfering blocks of eight nodes. The message assignment is shown in Figure 7.8(b). This scheme achieves a per-user DoF of $\frac{3}{4}$ with $B = \frac{3}{2}$ in the locally connected linear subnetwork. Since the linear subnetworks only account for $\frac{2}{3}$ of the network, we obtain a per-user DoF of $\frac{1}{2}$ with $B = 1$ in the entire network.

7.5 Summary

In this chapter, we have studied the potential gains offered by cooperative transmission in the downlink of cellular networks, under an average backhaul load constraint. We first characterized the asymptotic per-user DoF in the linear interference network, and showed that the optimal coding schemes rely only on zero-forcing transmit beamforming. Furthermore, the optimal schemes utilize an asymmetric assignment

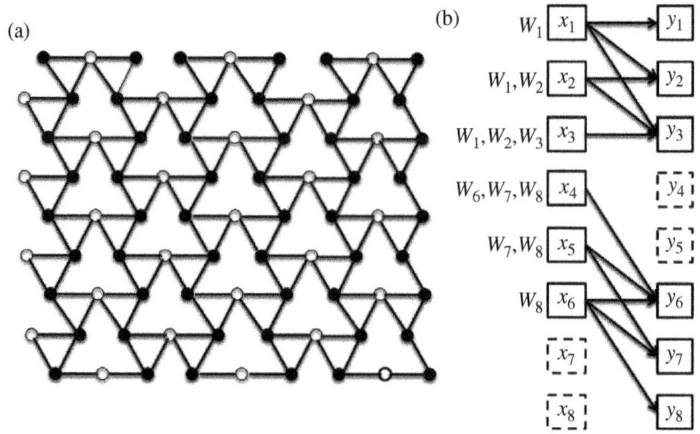

Fig. 7.8 (a) Division of cellular network into subnetworks. (b) The message assignment in each subnetwork. Nodes represented with open circles and signals represented with dashed boxes are inactive.

of messages, such that the backhaul constraint is satisfied, where some messages are assigned to more than B transmitters at the cost of assigning others to fewer than B transmitters, and some messages are not assigned at all. Thus, the average backhaul constraint allows for higher degree of freedom gains compared to the maximum transmit set size constraint, and we have $\tau^{\text{avg}}(B) > \tau(M)$. We then extended these results to more general and practically relevant networks, such as linear interference networks with higher connectivity and hexagonal-sectored cellular networks. We showed that in locally connected linear networks $\tau_L^{\text{avg,zf}}(B = 1) > \tau_L(M = 1)$ for $L \leq 5$, and in hexagonal cellular networks $\tau_{\text{c}}^{\text{avg,zf}}(B = 1) > \tau_{\text{c}}(M = 1)$. The proposed schemes are simple zero-forcing schemes with flexible message assignments that achieve the information-theoretic upper bound of the per-user DoF for the case of no cooperation with an average backhaul load of one message per transmitter, i.e., with no extra backhaul load, and without the need for interference alignment.

We end with the following insights that we gained from our study of CoMP transmission with a backhaul load constraint:

- When we allow for flexible assignment of backhaul resources, by enabling asymmetric allocation of the resources to different users in a single channel use, we obtain significant gains with CoMP transmission, even with the condition that the backhaul load remains the same as that required to assign each message to only one transmitter.
- The above point is particularly relevant in practice, by noting that these gains can be achieved with simple zero-forcing schemes that do not suffer from the coding delay incurred by symbol extensions. However, it is also important to note that we still assume the availability of perfect channel state information at the transmitters, as well as perfect synchronization between the transmitters, to enable complete interference cancellation.
- The above two observations hold not only for simple Wyner-type locally connected networks but also for more complex models of cellular networks that are more

relevant to practice. The key idea enabling this extension is that the more complex hexagonal network can be viewed as a set of interconnected locally connected subnetworks, each with sparse connectivity. By designing a scheduling scheme that eliminates interference between different subnetworks, we can exploit the analysis of the locally connected networks to design schemes that use CoMP transmission with flexible backhaul resource allocation to achieve significant rate gains with minimal delay requirements.

8 Cellular Uplink

Our treatment of the CoMP concept of communication has been restricted so far to transmission schemes and their applications in the cellular downlink. In this chapter, we investigate whether the ideas studied for enabling rate gains and minimizing delay requirements through CoMP can be applied to reception schemes and their applications in the cellular uplink. Throughout our discussion of CoMP transmission, we assumed that messages are distributed through a backhaul network to transmitters. More specifically, we assumed that sharing of information between transmitters occurs through sharing of digital messages, instead of assuming that quantized analog signals are being shared. This is a natural assumption when considering the cellular downlink for the following reasons. First, sharing of analog signals is prone to quantization errors, which complicate the interference management problem, and make it harder to obtain clear insights through information-theoretic analyses. Second, since our purpose is to model fairly general scenarios for cellular networks, and in particular, those anticipated in next generation wireless networks, it is natural to assume that a message is being delivered from a central controller (or basestation controller) to each basestation transmitter. This is the case even if we are not considering cooperative transmission, and hence it is also natural to extend this assumption with a more powerful backhaul to allow for delivering the digital message to more than one basestation transmitter. For the considered setting of cellular uplink, the first reason mentioned above will still hold. However, the second aspect is different. Here, we can think of two different ways for cooperative reception. The first is when sharing of analog received signals is permitted; we call such schemes CoMP reception schemes and we discuss information-theoretic results for this case in Section 8.1. The second is when only sharing of decoded messages is permitted; we call this the message passing model and we discuss information-theoretic results for this case in Section 8.2.

There are noticeable similarities between the message passing model and the CoMP transmission schemes that are based on zero-forcing. By passing one message from one basestation to another through the backhaul, the interference caused by this message at exactly one receiver could be eliminated. However, unlike the downlink case, when the interference caused by a message is canceled at one receiver through cooperation, it does not propagate to other receivers. For example, if message W_1 in a linear network is delivered to its own destination through the first transmitter, and also given to the second transmitter to cancel its interference at the second receiver, the fact that it is now transmitted by the second transmitter will make its interference propagate to the third receiver. In the uplink, that does not happen. If the decoded message W_1 is passed from the first receiver to the second receiver, so that its interference at the second receiver can be canceled, that does not make this message appear at the third receiver as a consequence. This difference is why the answer to the following question is non-trivial: Given limited backhaul resources, and the constraint that we have to fix a limited number of associations between mobile terminals and basestations in a cellular network, how do we choose the associations that maximize the average rate across both downlink and uplink sessions?

Here, we assume in the downlink that a basestation transmitter can have a message for a mobile terminal receiver only if the mobile terminal is associated with the basestation. Also, for uplink, a basestation receiver cannot have the decoded message of a mobile terminal transmitter unless the mobile terminal is associated with the basestation. This constraint of a limited number of associations can make sense in practical settings due to limited backhaul resources, as well as the overhead incurred by transmitting or decoding a given message. We study this problem of joint uplink/downlink design in Section 8.3.

8.1 CoMP Reception

The problem of analyzing the degrees of freedom of CoMP reception schemes was studied in [70] for fully connected interference networks. It was assumed that each message could be decoded using M_r received signals. It was also assumed that each message is being transmitted jointly by M_t transmitters. The following result of [70] characterizes a necessary and sufficient condition on the number of cooperating transmitters and receivers per message in order to achieve full degrees of freedom.

THEOREM 8.1 *Full DoF can be achieved in a K-user fully connected interference network using CoMP transmission and reception if $M_t + M_r \geq K + 1$.*

We spend the rest of this section proving Theorem 8.1, and then discussing the insights it implies. Note that if either $M_t = 1$ or $M_r = 1$, then the statement of the theorem just implies that full DoF is achievable with only transmitter or receiver cooperation if and only if the cooperation is *full*. In other words, if only receivers are allowed to cooperate then each message has to be decoded using all received signals. This case is already well known; the nontrivial case would be when both transmitter and receiver cooperation are employed ($M_t, M_r > 1$). The achieving scheme relies on linear transmit and receive beamforming to cancel interference at undesired receivers.

We now formalize the beamforming problem. Let $\boldsymbol{V}$ and $\boldsymbol{U}$ be the $K \times K$ matrices representing the transmit and receive beams, respectively. The kth column of $\boldsymbol{V}$ represents the beam along which the message W_k is transmitted. Similarly, the kth column of $\boldsymbol{U}$ represents the direction along which W_k is received. Note that the cooperation constraints of M_t and M_r restrict each of the matrices $\boldsymbol{V}$ and $\boldsymbol{U}$ to have only M_t and M_r nonzero entries in each row. We choose these matrices as follows:

$$v_{ik} \neq 0 \Rightarrow i \in \{k, k+1, \ldots, k+M_t-1\}, \tag{8.1}$$

$$u_{ik} \neq 0 \Rightarrow i \in \{k, k+1, \ldots, k+M_r-1\}, \tag{8.2}$$

where the $\Rightarrow$ sign denotes that the left-hand side *implies* the right-hand side. Let $\boldsymbol{H}$ denote the $K \times K$ matrix of channel coefficients. Recall, that we are assuming that elements of $\boldsymbol{H}$ are drawn from a joint continuous distribution. If M_t and M_r satisfy $M_t + M_r \geq K + 1$, then we prove the existence of transmit beamforming matrix $\boldsymbol{V}$ and receive beamforming matrix $\boldsymbol{U}$ satisfying (8.1), and the following holds for almost all realizations of $\boldsymbol{H}$:

$$\boldsymbol{U}^\top \boldsymbol{H} \boldsymbol{V} = \boldsymbol{I}, \tag{8.3}$$

where $\boldsymbol{I}$ is the $K \times K$ identity matrix with unity diagonal entries and zero off-diagonal entries. Observe that the above choice for beamforming matrices $\boldsymbol{V}$ and $\boldsymbol{U}$ achieves K DoF since they create K interference-free AWGN channels, one per message, with each channel having a nonzero SNR. The condition in (8.3) is equivalent to finding the following decomposition for the inverse of the channel matrix:

$$\boldsymbol{H}^{-1} = \boldsymbol{V} \boldsymbol{U}^\top. \tag{8.4}$$

The problem then reduces to that of knowing whether the inverse of the channel matrix admits the decomposition in (8.4). Note that the matrix $\boldsymbol{H}^{-1}$ consists of random elements, and hence we are interested in making probabilistic statements on the existence of the decomposition in (8.4). We discuss this decomposition problem in a more general context in the next section.

8.1.1 Structural Matrix Decomposition

We describe here a general problem of matrix decomposition, where we study when a matrix admits a factorization into two matrices, based on the locations of the nonzero entries in these matrices, or their *structure*. We begin by formally defining this concept of matrix structure.

DEFINITION 8.1 **(S-matrix)** Given a matrix $\boldsymbol{V}$ and a $(0,1)$-matrix $\bar{\boldsymbol{V}}$ of the same size, we say that $\bar{\boldsymbol{V}}$ is a structural matrix (or S-matrix) of $\boldsymbol{V}$ if $\bar{v}_{ij} = 1$ for all i,j such that $v_{ij} \neq 0$.

We now define what we mean by a structural matrix decomposition of a matrix.

DEFINITION 8.2 **(SMD)** Let $\boldsymbol{A}$ be a square matrix, and $\bar{\boldsymbol{V}}, \bar{\boldsymbol{U}}$ be $(0,1)$-matrices of the same size. We say that the matrix $\boldsymbol{A}$ admits a structural matrix decomposition (SMD) with respect to $\bar{\boldsymbol{V}}$ and $\bar{\boldsymbol{U}}$ if $\boldsymbol{A}$ can be factorized as

$$\boldsymbol{A} = \boldsymbol{V} \boldsymbol{U}^\top, \tag{8.5}$$

with $\bar{\boldsymbol{V}}$ and $\bar{\boldsymbol{U}}$ being S-matrices of $\boldsymbol{V}$ and $\boldsymbol{U}$, respectively.

Note that the well-known LU (lower upper) decomposition is a special case of SMD where $\bar{\boldsymbol{V}}$ and $\bar{\boldsymbol{U}}$ are the structures of lower triangular matrices. For example, when all considered matrices are 3×3 matrices, the S-matrices are as follows:

$$\bar{U} = \bar{V} = \begin{bmatrix} 1 & 0 & 0 \\ 1 & 1 & 0 \\ 1 & 1 & 1 \end{bmatrix}. \tag{8.6}$$

If the random matrix $\boldsymbol{A}$ is *generic*, then we know it admits an LU decomposition. We refer the reader to Definition B.1 in Appendix B for what it means when we say that a property holds for a generic value of $\boldsymbol{A}$. We are going to show in the rest of this section that it is also true that if $\boldsymbol{A}$ is generic, then it admits a decomposition with S-matrices $\bar{\boldsymbol{V}}$ and $\bar{\boldsymbol{U}}$ that satisfy the conditions in (8.1).

THEOREM 8.2 *Suppose the square structural matrices $\bar{\boldsymbol{V}}$ and $\bar{\boldsymbol{U}}$ satisfy the conditions*

- *The diagonal entries of $\bar{\boldsymbol{V}}$ and $\bar{\boldsymbol{U}}$ are nonzero.*
- *The matrix $\bar{\boldsymbol{V}} + \bar{\boldsymbol{U}}^\top$ is a full matrix, i.e., all of its entries are nonzero.*

Then, a generic matrix $\boldsymbol{A}$ with the same dimension admits the SMD $\boldsymbol{A} = \boldsymbol{V}\boldsymbol{U}^\top$ with respect to the S-matrices $\bar{\boldsymbol{V}}$ and $\bar{\boldsymbol{U}}$.

We are interested in proving Theorem 8.2 when all considered square matrices have dimension $K \times K$, so that we can apply it by setting $\boldsymbol{A}$ as the matrix of channel coefficients $\boldsymbol{H}$. The statement can be proved by viewing the equation $\boldsymbol{A} = \boldsymbol{V}\boldsymbol{U}^\top$ as a set of polynomial equations,

$$a_{ij} = f_{ij}(\boldsymbol{t}), \forall i,j \in \{1,2,\ldots,K\}, \tag{8.7}$$

where $\boldsymbol{t}$ consists of the following set of variables:

$$\{v_{ij} : \bar{v}_{ij} = 1\} \cup \{u_{ij} : i \neq j \text{ and } \bar{u}_{ij} = 1\}. \tag{8.8}$$

In other words, $\boldsymbol{t}$ consists of the entries of $\boldsymbol{V}$ and $\boldsymbol{U}$ that could take arbitrary values. Note that we excluded the diagonal entries of $\boldsymbol{U}$, $\{u_{kk}, k \in \{1,2,\ldots,K\}\}$, from $\boldsymbol{t}$ because if $\boldsymbol{A} = \boldsymbol{V}\boldsymbol{U}^\top$ then it is also true that $\boldsymbol{A} = (\boldsymbol{V}\boldsymbol{\Lambda})(\boldsymbol{U}\boldsymbol{\Lambda}^{-1})^\top$ for any full diagonal matrix $\boldsymbol{\Lambda}$, and hence we fix the diagonal entries of $\boldsymbol{U}$ to be all unity so that the decomposition is unique if it exists. We know from Lemma B.2 in Appendix B that the system of polynomial equations in (8.7) admits a solution for a generic $\boldsymbol{A}$ if and only if the Jacobian matrix $\boldsymbol{J}_f$ of the polynomial map

$$\boldsymbol{f} : \mathbb{C}^{N_\mathrm{v}} \to \mathbb{C}^{K \times K} \tag{8.9}$$

has full row rank of K^2 at some point $\boldsymbol{t}^*$, where N_v is the number of variables in $\boldsymbol{t}$. We note that $\boldsymbol{J}_f$ may have a row rank of K^2 because it follows from Theorem 8.2 that $N_\mathrm{v} \geq K^2$, since for each position for a matrix entry indexed by the pair $(i,j) \in \{1,2,\ldots,K\} \times \{1,2,\ldots,K\}$, it is the case that either $\bar{u}_{ij} = 1$ or $\bar{v}_{ij} = 1$. We now set the point $\boldsymbol{t}^*$ such that $\boldsymbol{U}^* = \boldsymbol{V}^* = \boldsymbol{I}$, the identity matrix.

We proceed with the proof by finding a full rank $K^2 \times K^2$ submatrix of $\boldsymbol{J}_f$ at the point $\boldsymbol{t}^*$. The submatrix we choose is the one corresponding to the K^2 variables $\{t_{ij} : t_{ij} =$

v_{ij} or $t_{ij} = u_{ji}, i,j \in \{1,2,\ldots,K\}\}$. The entries of $\boldsymbol{J}_f$ in that submatrix are given by the partial derivatives,

$$\begin{aligned}\frac{\partial a_{pq}}{\partial t_{ij}} &= \frac{\partial f_{pq}(\boldsymbol{t})}{\partial t_{ij}} \\ &= \frac{\partial \sum_{\ell=1}^{K} v_{p\ell} u_{q\ell}}{\partial t_{ij}} \\ &= \sum_{\ell=1}^{K} \frac{\partial (v_{p\ell} u_{q\ell})}{\partial t_{ij}}. \end{aligned} \tag{8.10}$$

If $t_{ij} = v_{ij}$, then the following holds:

$$\begin{aligned}\frac{\partial a_{pq}}{\partial t_{ij}} &= \sum_{\ell=1}^{K} \frac{\partial (v_{p\ell} u_{q\ell})}{\partial t_{ij}} \\ &= \delta_{pi} u_{qj}^{*} \\ &= \delta_{pi} \delta_{qj}, \end{aligned} \tag{8.11}$$

where δ_{ij} is the Kronecker delta function. Also, if $t_{ij} = u_{ji}$, then the following holds:

$$\begin{aligned}\frac{\partial a_{pq}}{\partial t_{ij}} &= \sum_{\ell=1}^{K} \frac{\partial (v_{p\ell} u_{q\ell})}{\partial t_{ij}} \\ &= v_{pi}^{*} \delta_{qj} \\ &= \delta_{pi} \delta_{qj}. \end{aligned} \tag{8.12}$$

We hence have

$$\frac{\partial a_{pq}}{\partial t_{ij}} = \begin{cases} 1 & \text{if } (p,q) = (i,j), \\ 0 & \text{otherwise.} \end{cases} \tag{8.13}$$

It then follows that the $K^2 \times K^2$ considered submatrix of $\boldsymbol{J}_f$ is the identity matrix, at the considered point $\boldsymbol{t}^*$, and hence $\boldsymbol{J}_f$ has full row rank at this point, and the system of polynomial equations in (8.7) admits a solution for a generic $\boldsymbol{A}$.

We have thus proved Theorem 8.2, and it remains to establish that the choice of transmit and receive sets in (8.1) lead to structural matrices that satisfy the conditions in Theorem 8.2. Then, Theorem 8.1 is proved. It is worth noting that in the case where we are restricted to the choice of transmit and receive sets in (8.1), then it was shown in [70] that the sufficient condition of Theorem 8.1 is also necessary.

8.2 Message Passing Model

A theoretical framework for cooperative reception in cellular uplink was introduced in [90] and [91], based on passing decoded messages from one basestation receiver to another. The passed message could be used to eliminate interference caused by this message at an undesired receiver. This model has the obvious advantage of sharing digital messages, unlike the model discussed in Section 8.1 where multiple analog received signals are jointly used to decode one message. However, this message passing model can suffer from propagation delays due to its successive decoding nature. For example, in a linear network where each transmitter is connected to the receiver having the same index as well as one following receiver, passing each message from its destined receiver to the following receiver suffices for achieving full DoF interference-free communication. However, the propagation delay in that example is proportional to the number of users K, as each message can only be decoded after all messages with lower indices have been decoded.

There are key differences between message passing cooperation for the uplink and the cooperative transmission scheme we discussed for the downlink in Chapter 6. First, as in the linear network example we discussed above, having each message available at a number of basestations equal to the number of interfering signals at each receiver suffices to achieve full DoF in the uplink model. We know that this is not true for the downlink. For example, for a locally connected network with connectivity parameter L, where each mobile terminal is connected to the basestation with the same index as well as L following basestations, having each message available at L extra basestation transmitters can only lead us to achieve a per-user DoF of $\frac{2}{3}$ using zero-forcing schemes in the downlink. The reason for the higher uplink rate is that interference does not propagate through message passing decoding, unlike sharing of messages in the downlink, where each new origin of a message results in an added set of receivers, at which the message causes interference. On the other hand, the uplink rate can also be lower than the downlink for other network topologies. For example, if we have a fully connected interference channel, and each mobile terminal can be associated with all K basestations, then in the downlink we can obtain full degrees of freedom. But message passing decoding cannot resolve interference in a fully connected setting, and no cooperation gain will be achieved in the uplink. In general, if we form a directed graph with messages as vertices, and an arrow exists from node a to node b if the transmitter carrying message W_a is connected to the destination of node b, then message passing decoding cannot resolve all the interference in any cycle.

Let $\mathcal{R}_i$ be the set of receivers where the decoded message W_i can be available. If each message can be passed from its original basestation receiver to $M-1$ extra receivers ($|\mathcal{R}_i| \leq M, \ \forall i \in [K]$) for a locally connected network with connectivity parameter L, then we know that the following lower bounds hold for the asymptotic per-user DoF $\tau_{\mathrm{U}}(L,M)$ (we add the subscript U for uplink) [92] when we are restricted to zero-forcing schemes that use message passing decoding to eliminate interference, and deactivate any receiver where interference is not eliminated:

$$\tau_{\mathrm{U}}^{\mathrm{zf}}(L,M) \geq \begin{cases} 1 & L+1 \leq M, \\ \frac{M+1}{L+2} & \frac{L}{2} \leq M \leq L, \\ \frac{2M}{2M+L} & 1 \leq M \leq \frac{L}{2}-1. \end{cases} \tag{8.14}$$

The cell association that is used to achieve the above is as follows. When $M \geq L+1$, each mobile terminal is associated with the $L+1$ basestations connected to it. The last basestation in the network, with index K, decodes the last message and then passes it on to the L other basestations connected to the Kth mobile terminal, eliminating all interference caused by that mobile terminal. Each basestation then decodes its message and passes it on to the other basestations, eliminating the interference caused by the message. Thus, one degree of freedom is achieved for each user.

In the second range, $\frac{L}{2} \leq M \leq L$, the cell association that is used to achieve an asymptotic per-user DoF value of $\frac{M+1}{L+2}$ is as follows. The network is split into subnetworks, each with consecutive $L+2$ transmitter–receiver pairs. In each subnetwork, we decode the last $M+1$ words. For each $i \in \{L+2, L+1, \ldots, L+2-M+1\}$, message W_i is associated with $\{i, i-1, \ldots, L+2-M+1\} \subseteq \mathcal{R}_i$. Thus, the last M words are decoded. The basestations with indices in the set $\{2, 3, \ldots, L+2-M\}$ are inactive as there is interference from the last transmitter in the subnetwork that cannot be eliminated. The first basestation decodes W_{L+2-M}. To eliminate the interference caused by the transmitters in the set $\mathcal{S} = \{L+2-M+1, L+2-M+2, \ldots, L+1\}$ at the first basestation of the subnetwork, we add the first basestation to each $\mathcal{R}_i, \forall i \in \mathcal{S}$. Now, for messages with indices in the set $\mathcal{S}$, we have used up $\alpha_i = 2+i-(L+2-M+1)$ associations; the factor of two comes from the basestation resolving W_i and the first basestation of the subnetwork. But each transmitter with indices in the set $\mathcal{S} \backslash \{L+1\}$ also interferes with the subnetwork directly preceding this subnetwork. For all $i \in \mathcal{S} \backslash \{L+1\}$, the message W_i interferes with the bottom $L+1-i$ basestations of the preceding subnetwork, which is precisely the number of associations left for the respective message, i.e., $M - \alpha_i = L+1-i$, thus inter-subnetwork interference can be eliminated at those basestations.

In the third range, $1 \leq M \leq \frac{L}{2} - 1$, the cell association that achieves the lower bound of $\frac{2M}{2M+L}$ is similar to the one described in Section 6.3 for the downlink. The network is split into disjoint subnetworks, each with consecutive $2M+L$ transmitter–receiver pairs. For the uplink, we consider two sets of indices for transmitters $\mathcal{A}_{\mathrm{T}} = \{1, 2, \ldots, M\}$ and $\mathcal{B}_{\mathrm{T}} = \{M+L+1, M+L+2 \ldots, 2M+L\}$, and the corresponding sets of receivers $\mathcal{A}_{\mathrm{R}} = \{1, 2, \ldots, M\}$ and $\mathcal{B}_{\mathrm{R}} = \{M+1, M+L+2 \ldots, 2M\}$. For each $i \in \mathcal{A}_{\mathrm{T}}$, the message W_i is associated with the receivers receiving it in $\mathcal{A}_{\mathrm{R}}$. Receiver i decodes W_i, and the other associations in $\mathcal{R}_i$ exist for eliminating interference. Similarly, for each $j \in \mathcal{B}_{\mathrm{T}}$, the message W_j is associated with the receivers receiving it in $\mathcal{B}_{\mathrm{R}}$, but now receiver $j-L$ decodes W_j, and the other associations in $\mathcal{R}_j$ are for eliminating interference.

We note that (8.14) implies that the per-user DoF achieved through message passing cooperation in the uplink is at least as good as the $\frac{2M}{2M+L}$ per-user DoF achieved through zero-forcing in the downlink, for any values of M and L. Further, for any value of $M > \frac{L}{2}$, we are guaranteed to achieve a higher per-user DoF for the uplink. The reasons for this are the key differences that we discussed earlier in this section. The achieved rate in the uplink is higher because interference does not propagate through message passing decoding. Further, the structure of locally connected networks, as well as the optimal downlink message assignment, do not give rise to cycles in the directed message conflict graph.

By examining (8.14), we note that $\tau_{\mathrm{U}}(L,M)$ is characterized for $M \geq L+1$. Also, for $M < \frac{L}{2}$, while we can only prove that $\tau_{\mathrm{U}}^{\mathrm{zf}} \geq \frac{2M}{2M+L}$, we know that $\tau_{\mathrm{U}}(L,M) \geq \frac{1}{2}$ using the asymptotic interference alignment scheme of [46], without exploiting cooperation. We provide a converse proof for a special case of the second range of (8.14). When $M = L$, the optimal zero-forcing per-user DoF for the uplink can be characterized as

$$\tau_{\mathrm{U}}^{\mathrm{zf}}(L,L) = \frac{L+1}{L+2}. \tag{8.15}$$

We begin by dividing the network into subnetworks of $L+2$ consecutive transmitter–receiver pairs. We observe that in any subnetwork, if we have $M+1 = L+1$ consecutive active receivers (basestations), then the transmitter connected to all these receivers must be inactive, because a message's interference cannot be canceled at M or more receivers. Let Γ_{BS} be the set of subnetworks where all $M+2$ receivers are active, and Φ_{BS} be the set of subnetworks with at most M active receivers. Similarly, let Γ_{MT} and Φ_{MT} be the subnetworks with $M+2$ active transmitters and at most M active transmitters, with respect to order. To be able to achieve a higher per-user DoF than (8.15), it must be true that both conditions hold: $|\Gamma_{\mathrm{BS}}| > |\Phi_{\mathrm{BS}}|$ and $|\Gamma_{\mathrm{MT}}| > |\Phi_{\mathrm{MT}}|$. Now note that for any subnetwork that belongs to Γ_{BS}, at most M transmitters will be active, because the interference caused by any message cannot be canceled at M or more receivers. Hence, $\Gamma_{\mathrm{BS}} \subseteq \Phi_{\mathrm{MT}}$. Further, the same logic applies to conclude that for any subnetwork with $M+1$ active receivers, the number of active transmitters is at most $M+1$, and hence $\Gamma_{\mathrm{MT}} \subseteq \Phi_{\mathrm{BS}}$. It follows that if $|\Gamma_{\mathrm{BS}}| > |\Phi_{\mathrm{BS}}|$, then $|\Gamma_{\mathrm{MT}}| < |\Phi_{\mathrm{MT}}|$, and hence the statement is proved.

8.3 Joint Downlink–Uplink Backhaul Design

The key difference in this section from previous discussions is that we identify the optimal cell associations that enable the achievability of the best average rate over both uplink and downlink. In other words, we do not allow the cell association to change between the uplink and downlink sessions. This could reflect a practical scenario, where cell associations are fixed over both uplink and downlink sessions either to save the overhead or to set up or allocate backhaul links that are required to associate a mobile terminal with multiple basestations. The locally connected channel model here has K basestations and K mobile terminals. The ith basestation is connected to mobile terminals with indices in the set $\{i, i+1, \ldots, i+L\}$. For each $i \in [K]$, let $\mathcal{C}_i \subseteq [K]$ be the set of basestations with which mobile terminal i is associated, i.e., those basestations that carry the terminal's message in the downlink and will have its decoded message for the uplink. The transmitters in $\mathcal{C}_i$ cooperatively transmit the message W_i to mobile terminal i in the downlink. In the uplink, one of the basestation receivers in $\mathcal{C}_i$ will decode W_i and pass it to the remaining receivers in the set. The results we discuss in this section are based on the backhaul constraint $|\mathcal{C}_i| \leq M, \forall i \in [K]$. We let $\tau_{\mathrm{UD}}(M,L)$ denote the average per-user DoF over both uplink and downlink.

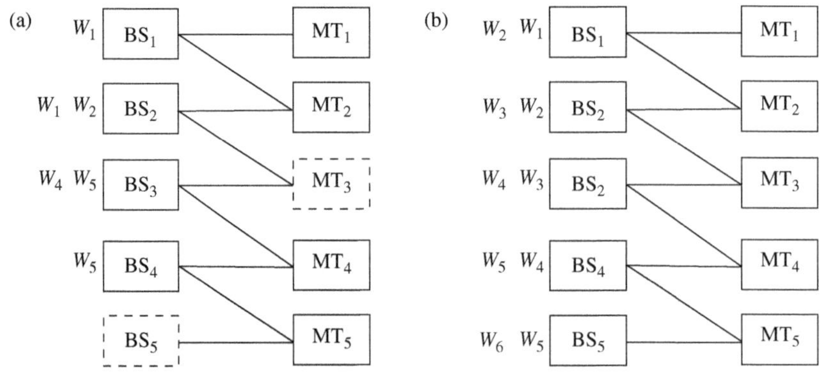

Fig. 8.1 Optimal cell associations for the case of $M = 2$: (a) downlink; (b) uplink. ©[2017] IEEE. Reprinted, with permission, from [93].

8.3.1 Example: Wyner's Network with Two Cell Associations

We explain the answer to the considered problem for $L = 1$ through the simple scenario when each mobile terminal can be associated with two basestations. The optimal downlink cell association for this case is illustrated in Figure 6.3. For the uplink, the optimal cell association in this case is simply associating each mobile terminal with the two basestations connected to it, and this guarantees achieving one degree of freedom per user as follows: The last basestation decodes the last message and then passes it to basestation $K - 1$. Starting with basestation $K - 1$ and moving in the direction of decrementing the basestation index, each basestation will decode the message with the same index and then pass it to the basestation with a previous index. The optimal cell association for the uplink is illustrated in Figure 8.1(b).

Now consider the average per-user DoF achieved over both the uplink and downlink sessions. The cell association of Figure 8.1(a) enables achieving a downlink per-user DoF of $\frac{4}{5}$, but the maximum per-user DoF that can be achieved using this same cell association is also $\frac{4}{5}$, yielding an average per-user DoF of $\frac{4}{5}$. The cell association depicted in Figure 8.1(b) enables achieving an uplink per-user DoF of unity, while the maximum achievable downlink per-user DoF is $\frac{2}{3}$, yielding a higher average of $\frac{5}{6}$ per-user DoF.

We observe through the example of Figure 8.1(b) that using the cell association that is optimal for the uplink, an average per-user DoF of $\frac{5}{6}$ can be achieved for the case when $M = 2$. In general, this value can be generalized to $\frac{4M-3}{4M-2}$, and we can prove that this is the best achievable average per-user DoF for any choice of cell associations that satisfies the backhaul constraint.

8.3.2 Average Per-User DoF for Wyner's Network

We characterize in this section the average per-user DoF over both uplink and downlink sessions for Wyner's linear networks ($L = 1$).

THEOREM 8.3 *For the linear interference network with M cell associations allowed for each mobile terminal, the average per-user DoF is given by*

$$\tau_{\mathrm{UD}}(M,L=1) = \begin{cases} \frac{2}{3} & M=1, \\ \frac{4M-3}{4M-2} & M \geq 2. \end{cases} \tag{8.16}$$

We observe from (8.16) that for every value of $M \geq 2$, the average per-user DoF is higher than the $\frac{2M}{2M+1}$ per-user DoF value obtained when optimizing for the downlink only, due to the higher gains that are possible through uplink cooperation for locally connected networks. Through message passing between basestation receivers in the uplink, interference can be eliminated if all the basestations connected to a mobile terminal are associated with it, and this is possible when $M \geq 2$; hence, one degree of freedom per user can be achieved. Further, as we will explain in detail below, the average per-user DoF when $M \geq 2$ is an average of a unity uplink per-user DoF and the downlink per-user DoF with reduced backhaul capacity. More precisely, the following holds for all $M \geq 2$:

$$\tau_{\mathrm{UD}}(M,L=1) = \frac{1+\tau(M-1,L=1)}{2} \tag{8.17}$$

$$= \frac{1+\frac{2(M-1)}{2(M-1)+1}}{2} \tag{8.18}$$

$$= \frac{4M-3}{4M-2}. \tag{8.19}$$

We now provide the proof that $\tau_{\mathrm{UD}} \geq \frac{4M-3}{4M-2}$. For the case where $M=1$, we divide the network into subnetworks, each consisting of three successive indices. In each subnetwork, the mobile terminal with the middle index is deactivated and each other mobile terminal enjoys interference-free communication in both the uplink and downlink sessions. The cell association enabling interference-free communication is

$$\mathcal{C}_i = \begin{cases} \{i\} & i \bmod 3 = 1, \\ \{i-1\} & i \bmod 3 = 0. \end{cases} \tag{8.20}$$

It follows that $\frac{2}{3}$ per-user DoF is achieved in each of the downlink and uplink sessions. For the case where $M=2$, we associate each mobile terminal with the two basestations connected to it. More precisely,

$$\mathcal{C}_i = \begin{cases} \{1\} & i=1, \\ \{i,i-1\} & i \geq 2. \end{cases} \tag{8.21}$$

For the uplink, one degree of freedom per user can be achieved since for $1 \leq i \leq K-1$, basestation i receives the decoded message W_{i+1} from basestation $i+1$, and hence, with the availability of the channel state information at the basestations, it can cancel the interference due to message W_{i+1} and decode message W_i without interference.

For the downlink, $\frac{2}{3}$ per-user DoF is achieved by dividing the network into subnetworks, each consisting of three indices, with the last basestation transmitter and mobile terminal receiver in each subnetwork both deactivated. The first mobile terminal in each subnetwork gets interference-free communication because the last basestation transmitter in the preceding subnetwork is inactive. The second basestation transmitter in each subnetwork uses its knowledge of the first message to cancel its interference at the second mobile terminal receiver. Hence, two mobile terminal receivers in each subnetwork have interference-free communication. The average achieved per-user DoF for $M = 2$ is hence $\frac{1+\frac{2}{3}}{2} = \frac{5}{6}$, which is the value stated in (8.16).

For $M > 2$, we first associate each mobile terminal with the two basestations connected to it, and hence one per-user degree of freedom is achieved in the uplink, as in the case of $M = 2$. For the downlink, we now explain how to achieve $\frac{2M-2}{2M-1}$ per-user DoF, yielding an average per-user DoF of $\frac{4M-3}{4M-2}$. Recall that each mobile terminal is already associated with the two basestations connected to it; we assign the remaining $M-2$ associations by using the downlink scheme described in Section 6.3 for the case when each message can be available at $M-1$ basestation transmitters, and note that this is possible since in that downlink scheme, each active mobile terminal is associated with at least one of the basestations connected to it. More specifically, the network is split into subnetworks, each consisting of successive $2M-1$ indices. The middle mobile terminal receiver and the last basestation transmitter in each subnetwork are inactive. We explain the cell associations for the first subnetwork and the rest will follow in a similar fashion. Define the following two subsets of indices in the first subnetwork: $\mathcal{S}_1 = \{1,2,\ldots,M-1\}, \mathcal{S}_2 = \{M+1, M+2,\ldots,2M-1\}$; then the cell associations are determined as follows:

$$\mathcal{C}_i = \{i-1, i\} \cup \begin{cases} \{i, i+1, \ldots, M-1\} & i \in \mathcal{S}_1, \\ \{i-1, i-2, \ldots, M\} & i \in \mathcal{S}_2. \end{cases} \tag{8.22}$$

The proof that the cell associations described in the cases listed in (8.22) enable the achievability of $2M-2$ degrees of freedom in each subnetwork follows from the downlink case discussed in Section 6.3 with a cooperation constraint that allows each message to be available at $M-1$ basestation transmitters.

We now discuss the upper bound proof of Theorem 8.3. For the case where $M = 1$, the upper bound follows from the fact that the maximum per-user DoF for each of the downlink and uplink sessions is $\frac{2}{3}$, even if we are allowed to change the cell association between the uplink and downlink. The proof of the downlink case is provided in Section 6.5. The proof of the uplink case is similar to the downlink case, so we omit it here for brevity and instead focus in the rest of the section on the remaining and more difficult case of $M \geq 2$.

Before making the main argument, we first need the following auxiliary lemmas for finding a converse for the uplink scenario.

LEMMA 8.4 *Given any cell association and any coding scheme for the uplink, the per-user DoF cannot be increased by adding an extra association of mobile terminal i to basestation j, where $j \notin \{i, i-1\}$.*

The lemma states that associating any mobile terminal to a basestation that is not connected to it cannot be *useful* for the uplink case. The key fact validating this lemma is that, unlike the downlink case, the knowledge of a message at a basestation cannot allow for the possibility of propagating the interference caused by this message beyond the two original receivers that are connected to the transmitter responsible for delivering the message. In other words, no matter what cell association we use for mobile terminal i, the message W_i will not cause interference at any basestation except basestations i and $i-1$, and hence having this message at any other basestation cannot help either in decoding the message or in canceling interference.

Lemma 8.4 gives us two possibilities for choosing the cell association of mobile terminal i; either we associate it with both basestations i and $i-1$, or only with one of these basestations. We use the following lemma to upper-bound the degrees of freedom for the latter case.

LEMMA 8.5 *If either mobile terminal i or mobile terminal $i+1$ is not associated with basestation i, then it is either the case that the received signal y_i can be ignored in the uplink without affecting the sum-rate, or it is the case that the uplink sum DoF for messages W_i and W_{i+1} is at most one.*

We now explain why Lemma 8.5 holds. If neither W_i nor W_{i+1} is associated with basestation i, then it is clear that y_i can be ignored in the uplink. Further, if only one of the two messages is associated with basestation i but is not decodable from y_i in the uplink, then we also can ignore this received signal. We now focus on the remaining case when exactly one of W_i and W_{i+1} is associated with basestation i and can be successfully decoded from y_i in the uplink. The proof is similar to the proof of [71, Lemma 5]. First, assume without loss of generality that message W_i is the message associated with basestation i. Assuming a reliable communication scheme, if we are given the received signal y_i then message W_i can be decoded reliably, and hence the transmit signal x_i can be reconstructed. Since the channel state information is available at basestation i, and y_i only depends on x_i, x_{i+1}, and z_i, then the remaining uncertainty in reconstructing the signal x_{i+1} is only due to the Gaussian noise. Since the uncertainty in the Gaussian noise does not reduce the degrees of freedom, we ignore it in our argument. Now, W_{i+1} can be recovered since we know x_{i+1}. Since we could recover both W_i and W_{i+1} using only one received signal, it follows that the sum DoF for messages W_i and W_{i+1} is at most one.

We use the above lemmas and consider the case where each mobile terminal can be associated with two basestations, i.e., $M = 2$. Fix a cell association and divide the indices of the network into sets, each consisting of three consecutive indices. Note that we only care about the ratio of the degrees of freedom to the number of users in large networks, so there is no loss in generality in assuming that K is a multiple of 3. For each subnetwork, if the middle basestation is only associated with at most one of the mobile terminals that are connected to it, then it follows from Lemma 8.5 that the uplink per-user DoF for users in the subnetwork is at most $\frac{2}{3}$. If the middle basestation is associated with more than two mobile terminals connected to it, then we can show for the downlink that, given only the second and third received signals in this subnetwork, one can reconstruct all three transmit signals; this can be proven in this case, and then we can use Lemma 5.2 since, given the second and third received signals, we can reconstruct the middle transmit signal, and then from the connectivity of the linear interference network we can reconstruct the third transmit signal from the third received signal, and the first transmit signal from the second received signal.

We now make the main argument to prove the converse for $M=2$. In order to achieve an average per-user DoF that is greater than $\frac{5}{6}$, the uplink per-user DoF has to exceed $\frac{2}{3}$. Assume that the uplink per-user DoF for a given cell association is x, where $x \geq \frac{2}{3}$, and divide the network into subnetworks, each consisting of three consecutive indices. It follows from the argument in the previous paragraph that for at least a fraction of $(3x-2)$ of the subnetworks, the middle basestation is associated with both the second and third mobile terminals. It hence follows that for a fraction of $(3x-2)$ subnetworks in the downlink, two mobile-terminal-received signals suffice to reconstruct all three basestation transmit signals. We can then reconstruct all transmit signals in the downlink from a number of transmit and received signals that equals a fraction of the number of users given by $\frac{2}{3}(3x-2)+1-(3x-2)=\frac{5}{3}-x$ for large networks. The per-user DoF in the downlink is then at most $\frac{5}{3}-x$, and hence the average per-user DoF is at most $\frac{5}{6}$. The formal information-theoretic argument used to prove the downlink upper bound exploits Lemma 5.2.

We now generalize the above argument to prove the converse for the case where $M \geq 2$. For the average per-user DoF to exceed $\frac{4M-3}{4M-2}$, the uplink per-user DoF has to exceed $\frac{2M-2}{2M-1}$. Fix a cell association for which the uplink per-user DoF equals x, where $x \geq \frac{2M-2}{2M-1}$, and divide the network into subnetworks, each consisting of $2M-1$ successive indices. If the middle basestation in each subnetwork is associated with at most one of the two mobile terminals connected to it, then Lemma 8.5 implies that the uplink per-user DoF in this subnetwork is at most $\frac{2M-2}{2M-1}$. It follows that for at least a fraction of $(2M-1)x-(2M-2)$ of the subnetworks, the middle basestation is associated with the two mobile terminals connected to it. For these subnetworks, we can show that the knowledge of $2M-2$ received signals suffices to reconstruct all $2M-1$ transmit signals, and hence the per-user DoF in the downlink is at most $\frac{4M-3}{2M-1}-x$. It follows that the average per-user DoF is at most $\frac{4M-3}{4M-2}$.

8.3.3 Average Per-User DoF for General Locally Connected Networks

We present zero-forcing schemes, with the goal of optimizing the average rate across both uplink and downlink for arbitrary values of $L \geq 2$. The corresponding per-user DoF inner bounds are given by

$$\tau_{\mathrm{UD}}^{\mathrm{zf}}(M,L) \geq \begin{cases} \frac{1}{2}\left(1+\left(\frac{\lceil \frac{L}{2} \rceil+\delta+M-(L+1)}{M}\right)\right) & L+1 \leq M, \\ \frac{2M}{2M+L} & 1 \leq M \leq L, \end{cases} \tag{8.23}$$

where $\delta=(L+1) \bmod 2$.

The coding scheme that achieves the inner bound for the second range of (8.23) is essentially the union of the scheme described in Section 6.3 and the scheme that achieves the third range of (8.14). The network is split into disjoint subnetworks, each with consecutive $2M+L$ transmitter–receiver pairs. We consider two sets of basestations $\mathcal{A}_{\mathrm{BS}}=\{1,2,\ldots,M\}$ and $\mathcal{B}_{\mathrm{BS}}=\{M+1,M+L+2,\ldots,2M\}$, and two sets of mobile terminals $\mathcal{A}_{\text{MT}} = \{1,2,\ldots,M\}$ and $\mathcal{B}_{\text{MT}} = \{M+L+1, M+L+2,\ldots,2M+L\}$. Now, for each $i \in \mathcal{A}_{\text{MT}}$, $\mathcal{C}_i = \mathcal{A}_{\text{BS}}$. Similarly, for each $j \in \mathcal{B}_{\text{MT}}$, $\mathcal{C}_j = \mathcal{B}_{\text{BS}}$. Thus, for the downlink we can get the optimal per-user DoF described in Section 6.3, and for the uplink we can get the inner bound stated in the third range of (8.14).

For the case where $M \geq L+1$, the coding scheme that achieves the expression described in (8.23) is as follows: First, we associate each mobile terminal with the $L+1$ basestations connected to it. This achieves the per-user DoF value of unity during the uplink in the same way as the scheme that achieves it in Section 8.2. Hence, we know so far that $\mathcal{C}_i \supseteq \{i, i-1, i-2, \ldots, i-L\} \cap [K], \forall i \in [K]$. During the downlink, we divide the network into disjoint subnetworks; each consists of M consecutive transmitter–receiver pairs. This allows us to create in each subnetwork a MISO broadcast channel. Let χ be the number of transmitter–receiver pairs with an inactive node between the last active basestation of one subnetwork and the first active mobile terminal in the following subnetwork. Then we observe that, in order to eliminate inter-subnetwork interference, it has to be the case that $\chi \geq L$. Because the achieved DoF in any subnetwork is bound by the minimum of the number of active transmitters and the number of active receivers in the subnetwork, we set the number of inactive mobile terminals to be the same as the number of inactive basestations. Let that number be ϵ; then $2\epsilon = \chi \geq L$. Since minimizing ϵ will maximize the achieved DoF, we set $\epsilon = \lceil \frac{L}{2} \rceil$. As we are leaving the first ϵ mobile terminals inactive in the subnetwork, the first basestation (call it basestation p) will be transmitting message $W_{p+\epsilon}$ to the the mobile terminal $p+\epsilon$. For this broadcast channel to work, each active basestation must be associated with all active mobile terminals in the subnetwork, so that all interfering signals can be eliminated at each mobile terminal receiver.

When $\delta = 0$, basestation $p+\epsilon$ will be delivering W_{p+L+1}, whose mobile terminal was not associated with basestation p through the uplink assignment, so we will need to add p to $\mathcal{C}_{p+L+1}$. Thus we can only have $M-(L+1)-1$ active basestations among the basestations in the subnetwork that have indices greater than $p+\epsilon$. This is because if basestation j, where $j \geq p+\epsilon+M-(L+1)$, is active then to ensure that the interference caused by it does not propagate, we have to have $p \in \mathcal{C}_{j+\epsilon}$, but then $|\mathcal{C}_{j+\epsilon}| > M$, i.e., mobile terminal $j+\epsilon$ has to be associated with more basestations than the backhaul constraint allows for. So in each subnetwork, we will have a total of $\Delta = \epsilon + M - (L+1)$ words transmitted without interference, out of a total of $\Delta + \epsilon = M$ words in the subnetwork.

When $\delta = 1$, basestation $p+\epsilon+1$ will be delivering W_{p+L+1}, but to ensure that the mobile terminals connected to this basestation other than mobile terminal $p+L+1$ do not suffer from interference, we need to add p to $\mathcal{C}_{p+L+1}$ and $p+\epsilon+1$ to $\mathcal{C}_{p+\epsilon}$. Thus we can only have $M-(L+1)$ active basestations among the basestations whose indices are greater than $p+\epsilon$ in the subnetwork. Otherwise, if basestation j', where $j' \geq p+\epsilon+M-L$, were active then we would need to have $j' \in \mathcal{C}_{p+\epsilon}$, which results in $|\mathcal{C}_{p+\epsilon}| = M+1 > M$. So in each subnetwork we will have a total of $\Delta = \epsilon + \delta + M - (L+1)$ words transmitted without interference, out of a total of $\Delta + \epsilon = M$ words. Figures 8.2 and 8.3 serve as examples for the above scheme. Figure 8.2 uses values of $M = 3$ and $L = 3$; using this scheme, we get a per-user DoF of $\frac{2}{3}$, which is equivalent

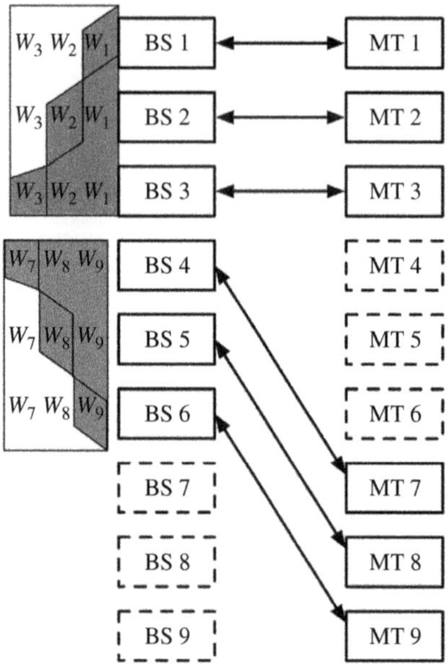

Fig. 8.2 Scheme for average uplink and downlink communication when $M \leq L$. The shading groups message assignments based on their utility for the uplink or downlink.

to $\frac{2M}{2M+L}$. Figure 8.3 uses values of $M = L+2$ for $L = 5$ and $L = 4$ for parts (a) and (b), respectively. The achieved per-user DoF values are $\frac{4}{7}$ or $\frac{4}{6}$, respectively.

8.4 Discussion

In our study of locally connected networks, we have focused on large network analyses, and whether cooperation can lead to degrees of freedom gains that scale with the number of users in the network. Whether cooperation is used for the downlink or the uplink, we observe that using only zero-forcing schemes, cooperation can be used to achieve a per-user DoF of $\frac{1}{2}$, as long as each mobile terminal can be associated with a number of basestations that is at least half the number of interfering signals per receiver, i.e., $M \geq \frac{L}{2}$. The significance of the per-user DoF value of $\frac{1}{2}$ is that this is the value achieved by asymptotic interference alignment without cooperation [46]. Hence, we can conclude that cooperative zero-forcing exceeds the performance of non-cooperative interference alignment when $M > \frac{L}{2}$. This is a clear advantage of using cooperative communication, as the practical obstacles in the way of implementing one-shot zero-forcing and harnessing some of its gains are fewer than those associated with the symbol extension schemes required by asymptotic interference alignment. An intuitive reason for why we cannot show gains for zero-forcing cooperative transmission or message passing decoding when $M < \frac{L}{2}$ is that each mobile terminal is connected to L basestations, and no matter what the pattern of cell association is, interference can be eliminated using zero-forcing in the downlink or message passing in the uplink

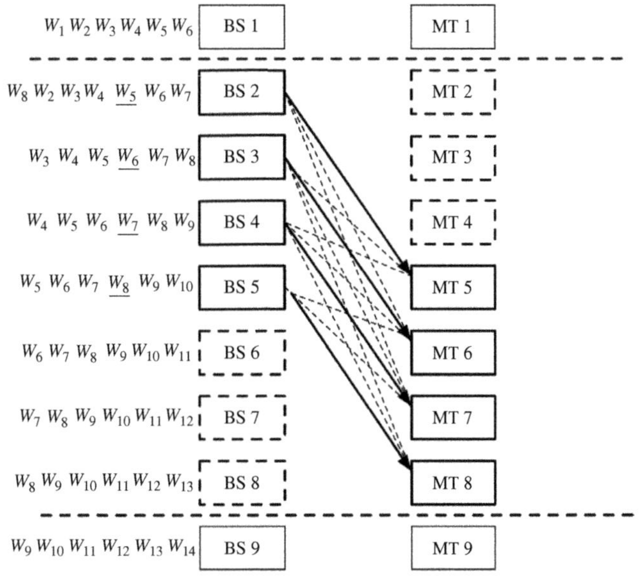

(a) L odd

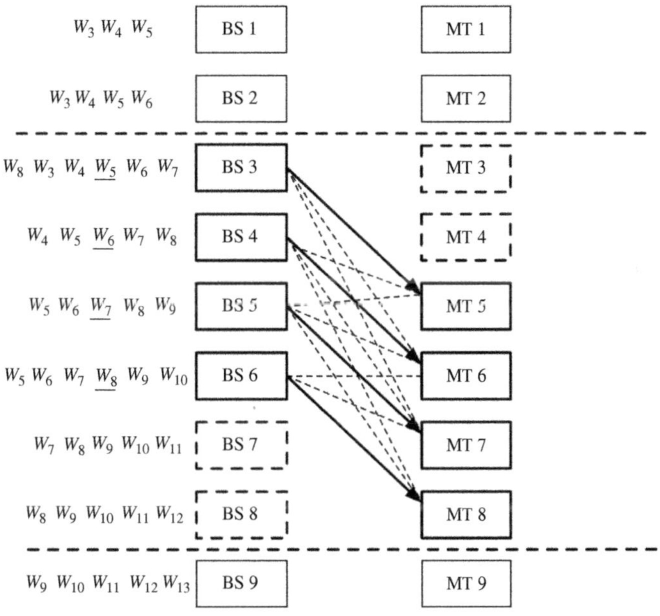

(b) L even

Fig. 8.3 Scheme for downlink, with all the associations needed for optimal uplink, that achieves the lower bound defined in (8.23) when $M \geq L+1$.

from at most M basestations with an index less than the designated basestation and M basestations with an index greater than the designated one. Hence, regardless of the choice of cell associations, if $M < \frac{L}{2}$ then cooperation can *at best* lead us to an equivalent channel where at least two interference links exist per receiver. We know from [71, Theorem 8] that the per-user DoF is $\frac{1}{2}$ if each message can be available at one basestation transmitter in the downlink for $L \geq 2$, and the same conclusion would also hold for the uplink.

We further observe that the $\frac{2M}{2M+L}$ inner bound we presented for the average uplink–downlink per-user DoF in (8.23) for the case when $M \leq L$ matches what is achieved in each of the downlink and uplink sessions. In fact, the cell association would remain the same even if we are optimizing for the downlink only or the uplink only. There is a more general pattern behind this observation. All the zero-forcing schemes presented for the downlink in Chapters 6 and 7 have a dual uplink scheme that is based on message passing decoding. Hence, we also have the conclusions about the backhaul load constraint of Chapter 7 when designing cooperative schemes for the uplink. Specifically, using a backhaul load of unity (no overall load on the backhaul), we could achieve $\frac{1}{2}$ per-user DoF using zero-forcing for values of the connectivity parameter $L \leq 6$, and this holds if we are designing cooperative schemes for the downlink only or the uplink only or the average. The reason for this duality in the presented schemes is that if we look carefully at the message assignment to basestation transmitters in the downlink, and consider a bipartite graph where messages are represented in one partite set and transmitters in the other partite set, and an edge exists between a message node and a transmitter node if the corresponding message is assigned to the transmitter, then there exist no cycles in all of the schemes we presented for the downlink. That implies that there is a valid *decoding order* for the dual message passing uplink scheme that uses the same cell associations.

We have discussed the duality between the cell association decisions needed to optimize the uplink and downlink degrees of freedom in locally connected networks. It is also worth mentioning that because we assume reciprocity of the channel, i.e., the channel coefficient remains the same whether the basestation is transmitting or the mobile terminal is transmitting, it is also true that the channel coefficients that need to be learned by each basestation are the same in the discussed cases, for both zero-forcing cooperative transmission and message passing cooperative reception.

9 Dynamic Interference Management

An important requirement of next-generation (5G) wireless systems is the ability to autonomously adjust to varying environmental conditions. Our focus in this chapter is to analyze information-theoretic models of interference networks that capture the effect of deep fading conditions through introducing random link erasure events in blocks of communication time slots. More specifically, in order to consider the effect of long-term fluctuations (deep fading or shadowing), we assume that communication takes place over blocks of time slots, and independent link erasures occur with probability p in each block.

We can observe through the results presented in Chapters 5–8 that conclusions related to the optimal associations of mobile terminals to basestations and the achievable DoF differ dramatically based on the network topology. For example, under the maximum transmit set size constraint for the downlink, local cooperation cannot lead to a gain in the achieved asymptotic per-user DoF for the fully connected channel. However, local cooperation is optimal for locally connected channels and can lead to achieving scalable DoF gains, and the optimal assignment of messages to transmitters depends on the connectivity parameter L. In practice, the topology may change due to deep fading conditions (see, e.g., [6]) or even intentionally to exploit spectrum opportunities (see, e.g., [94]). In this chapter, we extend our DoF results to dynamic interference networks where a fixed assignment of messages is selected to achieve average DoF optimal performance in networks with changing topology.

In [95], the authors analyzed the average capacity for a point-to-point channel model where slow changes result in varying severity of noise. We apply a similar concept to interference networks by assuming that slowly changing deep fading conditions result in link erasures. We consider the linear interference network ($L = 1$) that was introduced in Chapter 6, and look at two fading effects: long-term fluctuations that result in link erasures over a complete block of time slots, and short-term fluctuations that allow us to assume that any specific joint realization for the non-zero channel coefficients will take place with zero probability. We study the problem of achieving the optimal average degrees of freedom under a maximum transmit set size constraint (5.16). We note that the problem studied in Chapter 6 reduces to the case of no erasures. Here, we extend the schemes in Chapter 6 to consider the occurrence of link erasures, and propose new schemes that lead to achieving better average DoF at high probabilities of erasure.

9.1 Dynamic Interference Channel

We study a dynamic interference channel model that was introduced in [96]. In order to consider the effect of long-term fluctuations (shadowing), we assume that communication takes place over blocks of time slots, and let p be the probability of block erasure. In each block, we assume that for each pair (i,j) of receiver–transmitter indices where the channel is not identically zero, the channel coefficient $h_{ij} = 0$ with probability p. For Wyner's linear network, $h_{ij} = 0$ with probability p for each $(i,j) \in [K] \times [K]$ such that $i \in \{j, j+1\}$. As for the case of no erasures, short-term channel fluctuations allow us to assume that in each time slot, all non-zero channel coefficients are drawn independently from a continuous distribution. We also assume that global channel state information is available at all transmitters and receivers.

We use $\eta_p(K,M)$ to denote the DoF of a K-user channel with block erasure probability p and a maximum transmit set size constraint M, and $\tau_p(M)$ to denote the asymptotic per-user DoF. We call a message assignment strategy *optimal* for a given erasure probability p if there exists a sequence of coding schemes achieving $\tau_p(M)$ using the transmit sets defined by the message assignment strategy. A message assignment strategy is *universally optimal* if it is optimal for all values of p.

Studying dynamic interference networks can be rather difficult because of the complex combinatorial nature of the problem. In particular, we need to consider all possible network realizations and all possible message assignments satisfying the considered cooperation constraint. The justification for our choices of both linear networks and the maximum transmit set size constraint is their role in simplifying the problem, as well as the proven soundness of these assumptions through the studies presented in Chapters 6 and 7 for the case of no erasures. We have observed in Chapter 7 that optimal solutions obtained for linear networks with the maximum transmit set size constraint could lead to optimal solutions for the more difficult and more relevant problem with an average transmit set size constraint. We have also observed in Chapter 7 that coding schemes obtained for linear networks could be used in cellular networks, by perceiving the cellular network as a set of interfering linear subnetworks and then using an interference avoidance scheme with fractional reuse to eliminate inter-subnetwork interference.

We characterize $\tau_p(M = 1)$ in Section 9.2, and show that there is no universally optimal message assignment strategy for the case of $M = 1$. We then extend the insights obtained to derive bounds on $\tau_p(M = 2)$, the simplest case with CoMP transmission.

9.2 Optimal Cell Association

We first consider the case where each receiver can be served by only one transmitter. This corresponds to the problem of associating mobile users with cells in a cellular downlink scenario. We start by discussing orthogonal schemes (TDMA-based) for this problem, and then show that the proposed schemes are optimal. It will be useful in the rest of this section to view each realization of the network where some links are erased as a series of subnetworks that do not interfere with each other. We say that a set of k

users with successive indices $\{i, i+1, \ldots, i+k-1\}$ form a subnetwork if the following two conditions hold: The first condition is that $i=1$ or it is the case that message W_{i-1} does not cause interference at y_i, either because the direct link between the transmitter carrying W_{i-1} and receiver $(i-1)$ is erased, or the transmitter carrying W_{i-1} is not connected to the ith receiver. Second, $i+k-1=K$ or it is the case that message W_{i+k-1} does not cause interference at y_{i+k}, because the carrying transmitter is not connected to one of the receivers $(i+k-1)$ and $(i+k)$.

We say that the subnetwork is *atomic* if the transmitters carrying messages for users in the subnetwork have successive indices and, for any transmitter t carrying a message for a user in the subnetwork and receiver r such that $r \in \{t, t+1\}$ and $r \in \{i, i+1, \ldots, i+k-1\}$, the channel coefficient $h_{rt} \neq 0$.

For $i \in [K]$, let m_i be the number of messages available at the ith transmitter, and let $\boldsymbol{m}^K = (m_1, m_2, \ldots, m_K)$. It is clear that the sequence $\boldsymbol{m}^K$ can be obtained from the transmit sets $\mathcal{T}_i, i \in [K]$; it is also true, as stated in Lemma 9.1, that the converse holds. We use the notion of *useful* message assignments from Section 6.4.1. For $M=1$, a useful message assignment will have each message assigned to one of the two transmitters that can be connected to its designated receiver.

LEMMA 9.1 *For any useful message assignment where each message is assigned to exactly one transmitter, i.e.,* $|\mathcal{T}_i| = 1, \forall i \in [K]$, *the transmit sets* $\mathcal{T}_i$, $i \in [K]$, *are uniquely characterized by the sequence* $\boldsymbol{m}^K$.

Proof Since each message can only be available at one transmitter, then this transmitter has to be connected to the designated receiver. More precisely, $\mathcal{T}_i \subset \{i-1, i\}, \forall i \in \{2, \ldots, K\}$, and $\mathcal{T}_1 = \{1\}$. It follows that each transmitter carries at most two messages, and the first transmitter carries at least the message W_1, i.e., $m_i \in \{0,1,2\}\ \forall i \in \{2, \ldots, K\}$ and $m_1 \in \{1,2\}$. Assume that $m_i = 1, \forall i \in [K]$; then $\mathcal{T}_i = \{i\}, \forall i \in [K]$. For the remaining case, we know that there exists $i \in \{2, \ldots, K\}$ such that $m_i = 0$, since $\sum_{i=1}^{K} m_i = K$; we handle this case in the rest of the proof.

Let x be the smallest index of a transmitter that carries no messages, i.e., $x = \min\{i : m_i = 0\}$. We now show how to reconstruct the transmit sets $\mathcal{T}_i, i \in \{1, \ldots, x\}$ from the sequence $(m_1, m_2, \ldots, m_x)$. We note that $\mathcal{T}_i \subseteq [x], \forall i \in [x]$, and since $m_x = 0$ it follows that $\mathcal{T}_i \nsubseteq [x], \forall i \notin [x]$. It follows that $\sum_{i=1}^{x-1} m_i = x$. Since $\mathcal{T}_i \subset \{i-1, i\}, \forall i \in \{2, \ldots, x\}$, we know that at most one transmitter in the first $x-1$ transmitters carries two messages. Since $\sum_{i=1}^{x-1} m_i = x$, and $m_i \in \{1,2\}, \forall i \in [x-1]$, it follows that there exists an index $y \in [x-1]$ such that $m_y = 2$ and $m_i = 1, \forall i \in [x-1] \backslash \{y\}$. It is now clear that the yth transmitter carries messages W_y and W_{y+1}, each transmitter with an index $j \in \{y+1, \ldots, x-1\}$ is carrying message W_{j+1}, and each transmitter with an index $j \in \{1, \ldots, y\}$ is carrying message W_j. The transmit sets are then determined as follows: $\mathcal{T}_i = \{i\}, \forall i \in [y]$ and $\mathcal{T}_i = \{i-1\}, \forall i \in \{y+1, \ldots, x\}$.

We view the network as a series of subnetworks, where the last transmitter in each subnetwork is either inactive or it is the last transmitter in the network. If the last transmitter in a subnetwork is inactive, then the transmit sets in the subnetwork are determined in a similar fashion to the transmit sets $\mathcal{T}_i, i \in [x]$ in the above scenario. If the last transmitter in the subnetwork is the Kth transmitter, and $m_K = 1$, then each message in this subnetwork is available at the transmitter with the same index.

We use Lemma 9.1 to describe message assignment strategies for large networks through repeating patterns of short ternary strings. Given a ternary string $\boldsymbol{s} = (s_1, \ldots, s_n)$ of fixed length n such that $\sum_{i=1}^{n} s_i = n$, we define $\boldsymbol{m}^K$, $K \geq n$, as follows:

- $m_i = s_{i \bmod n}$ if $i \in \{1, \ldots, n\lfloor \frac{K}{n} \rfloor\}$,
- $m_i = 1$ if $i \in \{n\lfloor \frac{K}{n} \rfloor + 1, \ldots, K\}$.

We now evaluate all possible message assignment strategies satisfying the cell association constraint using ternary strings through the above representation. We restrict our attention to useful message assignments, and note that if there are two transmitters with indices i and j such that $i < j$ and each is carrying two messages, then there is a third transmitter with index k such that $i < k < j$ that carries no messages. It follows that any string defining message assignment strategies that satisfy the cell association constraint has to have one of the following forms:

- $\boldsymbol{s}^{(1)} = (1)$,
- $\boldsymbol{s}^{(2)} = (2, 1, 1, \ldots, 1, 0)$,
- $\boldsymbol{s}^{(3)} = (1, 1, \ldots, 1, 2, 0)$,
- $\boldsymbol{s}^{(4)} = (1, 1, \ldots, 1, 2, 1, 1, \ldots, 1, 0)$.

We now introduce the three candidate message assignment strategies illustrated in Figure 9.1, and we characterize the TDMA per-user DoF achieved through each of them; we will show later that the optimal message assignment strategy at any value of p is given by one of the three strategies introduced. We first consider the message assignment strategy defined by the string having the form $\boldsymbol{s}^{(1)} = (1)$. Here, each message is available at the transmitter having the same index.

LEMMA 9.2 *Under the restriction to the message assignment strategy $\mathcal{T}_{i,K} = \{i\}, \forall K \in \mathbf{Z}^+, i \in [K]$, and orthogonal TDMA schemes, the average per-user DoF is given by*

$$\tau_p^{(1)} = \frac{1}{2}\left(1 - p + (1-p)\left(1 - (1-p)^2\right)^2\right) + \sum_{i=1}^{\infty} \frac{1}{2}\left(1 - (1-p)^2\right)^2 (1-p)^{4i+1}. \tag{9.1}$$

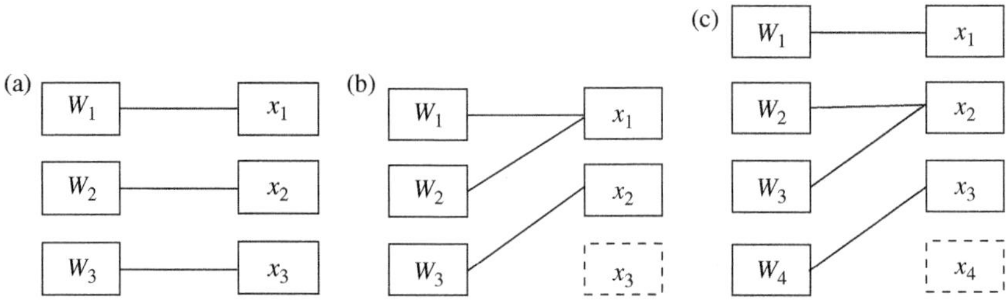

Fig. 9.1 The optimal message assignment strategies for the cell association problem. The dashed boxes represent transmit signals that are inactive in all network realizations. The strategies in (a), (b), and (c) are optimal at high, low, and middle values of the erasure probability p, respectively. ©[2017] IEEE. Reprinted, with permission, from [96].

Proof We will first show how $\frac{1}{2}\left(1-p+(1-p)\left(1-(1-p)^2\right)^2\right)$ DoF can be achieved, and then modify the transmission scheme to show how to achieve $\tau_p^{(1)}$. For each user with an odd index i, message W_i is transmitted whenever the channel coefficient $h_{ii} \neq 0$; the rate achieved by these users contributes to the average per-user DoF by $\frac{1}{2}(1-p)$. For each user with an even index i, message W_i is transmitted whenever the following holds: $h_{ii} \neq 0$, W_{i-1} does not cause interference at y_i, and the transmission of W_i will not disrupt the communication of W_{i+1} to its designated receiver; we note that this happens if and only if $h_{ii} \neq 0$ and $\left(h_{i-1,i-1} = 0 \text{ or } h_{i,i-1} = 0\right)$ and $(h_{i+1,i} = 0 \text{ or } h_{i+1,i+1} = 0)$. It follows that the rate achieved by users with even indices contributes to the average per-user DoF by $\frac{1}{2}(1-p)\left(1-(1-p)^2\right)^2$.

We now discuss a modification of the above scheme to achieve $\tau_p^{(1)}$. As above, users with odd indices have priority, i.e., their messages are delivered whenever their direct links exist, and users with even indices deliver their messages whenever their direct links exist and the channel connectivity allows for avoiding conflict with priority users. However, we make an exception to the priority setting in atomic subnetworks consisting of an odd number of users where the first and last users have even indices; in these subnetworks, one extra DoF is achieved by allowing users with even indices to have priority and deliver their messages. The resulting extra term in the average per-user DoF is calculated as follows: Fixing a user with an even index, the probability that this user is the first user in a subnetwork consisting of an odd number of users in a large network is $\sum_{i=1}^{\infty}\left(1-(1-p)^2\right)^2(1-p)^{4i+1}$; for each of these events, the sum DoF is increased by 1, and hence the term added to the average per-user DoF is equal to half this value, since every other user has an even index.

The optimality of the above scheme within the class of orthogonal TDMA-based schemes follows directly from [97, Theorem 1] for each realization of the network.

We will show later that the above scheme is optimal at high erasure probabilities. In Chapter 6, the optimal message assignment for the case of no erasures was characterized. The per-user DoF was shown to be $\frac{2}{3}$, and is achieved by deactivating every third transmitter and achieving 1 DoF for each transmitted message. We now consider the extension of this message assignment illustrated in Figure 9.1(b), which will be shown later to be optimal for low erasure probabilities.

LEMMA 9.3 *Under the restriction to the message assignment strategy defined by the string* $\mathbf{s} = (2,1,0)$, *and orthogonal TDMA schemes, the average per-user DoF is given by*

$$\tau_p^{(2)} = \frac{2}{3}(1-p) + \frac{1}{3}p(1-p)\left(1-(1-p)^2\right). \tag{9.2}$$

Proof For each user with an index i such that $(i \bmod 3 = 0)$ or $(i \bmod 3 = 1)$, message W_i is transmitted whenever the link between the transmitter carrying W_i and the ith receiver is not erased; these users contribute to the average per-user DoF by a factor of $\frac{2}{3}(1-p)$. For each user with an index i such that $(i \bmod 3 = 2)$, message W_i is transmitted through x_{i-1} whenever the following holds: $h_{i,i-1} \neq 0$, message W_{i-1} is not transmitted because $h_{i-1,i-1}=0$, and the transmission of W_i will not be disrupted by the communication of W_{i+1} through x_i because $(h_{ii}=0)$ or $\left(h_{i+1,i}=0\right)$; these users contribute to the average per-user DoF by a factor of $\frac{1}{3}p(1-p)\left(1-(1-p)^2\right)$. Using the considered message assignment strategy, the TDMA optimality of this scheme follows from [97, Theorem 1] for each network realization.

We now consider the message assignment strategy illustrated in Figure 9.1(c). We will show later that this strategy is optimal for the middle regime of erasure probabilities.

LEMMA 9.4 *Under the restriction to the message assignment strategy defined by the string* $\mathbf{s}=(1,2,1,0)$, *and orthogonal TDMA schemes, the average per-user DoF is given by*

$$\begin{aligned}\tau_p^{(3)} &= \frac{1}{2}(1-p)\\ &+\frac{1}{4}(1-p)\left(1-(1-p)^2\right)\left(1+p+(1-p)^3\right).\end{aligned} \tag{9.3}$$

Proof As in the proof of Lemma 9.2, we first introduce a transmission scheme achieving part of the desired rate, and then modify it to show how the extra term can be achieved. Let each message with an odd index be delivered whenever the link between the transmitter carrying the message and the designated receiver is not erased; these users contribute to the average per-user DoF by a factor of $\frac{1}{2}(1-p)$. For each user with an even index i, if $i \bmod 4=2$, then W_i is transmitted through x_i whenever the following holds: $h_{ii}\neq 0$, message W_{i+1} is not transmitted through x_i because $h_{i+1,i}=0$, and the transmission of W_i will not be disrupted by the communication of W_{i-1} through x_{i-1} because either $h_{i,i-1}=0$ or $h_{i-1,i-1}=0$; these users contribute to the average per-user DoF by a factor of $\frac{1}{4}p(1-p)\left(1-(1-p)^2\right)$. For each user with an even index i such that i is a multiple of 4, W_i is transmitted through x_{i-1} whenever $h_{i,i-1}\neq 0$, and the transmission of W_i will not disrupt the communication of W_{i-1} through x_{i-2} because either $h_{i-1,i-1}=0$ or $h_{i-1,i-2}=0$; these users contribute to the average per-user DoF by a factor of $\frac{1}{4}(1-p)\left(1-(1-p)^2\right)$.

We now modify the above scheme to show how $\tau_p^{(3)}$ can be achieved. Since the ith transmitter is inactive for every i that is a multiple of 4, users $\{i-3,i-2,i-1,i\}$ are separated from the rest of the network for every i that is a multiple of 4, i.e., these users form a subnetwork. We explain the modification for the first four users, and it will be clear how to apply a similar modification for every following set of four users. Consider the event where message W_1 does not cause interference at y_2, because either $h_{11}=0$ or $h_{21}=0$, and it is the case that $h_{22}\neq 0$, $h_{32}\neq 0$, $h_{33}\neq 0$, and $h_{43}\neq 0$; this is the event that users $\{2,3,4\}$ form an atomic subnetwork, and it happens with probability $\left(1-(1-p)^2\right)(1-p)^4$. In this case, we let messages W_2 and W_4 have priority instead of message W_3, and hence the sum DoF for messages $\{W_1,W_2,W_3,W_4\}$ is increased by 1. It follows that an extra term of $\frac{1}{4}\left(1-(1-p)^2\right)(1-p)^4$ is added to the average per-user DoF.

The TDMA optimality of the illustrated scheme follows from [97, Theorem 1] for each network realization.

We now show that under the restriction to TDMA schemes, one of the message assignment strategies illustrated in Lemmas 9.2, 9.3, and 9.4 is optimal at any value of p.

THEOREM 9.5 *For a given erasure probability p, let $\tau_p^{(\mathrm{TDMA})}$ be the average per-user DoF under the restriction to orthogonal TDMA schemes; then, at any value $0 \leq p \leq 1$, the following holds:*

$$\tau_p^{(\mathrm{TDMA})} = \max\left\{\tau_p^{(1)}, \tau_p^{(2)}, \tau_p^{(3)}\right\}, \tag{9.4}$$

where $\tau_p^{(1)}$, $\tau_p^{(2)}$, and $\tau_p^{(3)}$ are given in (9.1), (9.2), *and* (9.3), *respectively.*

Proof The inner bound follows from Lemmas 9.2, 9.3, and 9.4. In order to prove the converse, we need to consider all useful message assignment strategies where each message is assigned to a single transmitter. We know from Lemma 9.2 that the TDMA average per-user DoF achieved through the strategy defined by the string of all ones having the form $\boldsymbol{s}^{(1)} = (1)$ equals $\tau_p^{(1)}$, and hence the upper bound holds in this case.

We now show that the TDMA average per-user DoF achieved through strategies defined by strings of the form $\boldsymbol{s}^{(2)} = (2, 1, \ldots, 1, 0)$ is upper-bounded by a convex combination of $\tau_p^{(1)}$ and $\tau_p^{(2)}$, and hence is upper-bounded by $\max\left\{\tau_p^{(1)}, \tau_p^{(2)}\right\}$. The message assignment strategy considered splits each network into subnetworks consisting of a transmitter carrying two messages followed by a number of transmitters each carrying one message, and the last transmitter in the subnetwork carries no messages. We first consider the case where the number of transmitters carrying single messages is odd. We consider the simple scenario of the message assignment strategy defined by the string $(2, 1, 1, 1, 0)$, and then the proof will be clear for strategies defined by strings of the form $(2, 1, 1, \ldots, 1, 0)$ that have an arbitrary odd number of ones. In this case, it suffices to show that the average per-user DoF in the first subnetwork is upper-bounded by a convex combination of $\tau_p^{(1)}$ and $\tau_p^{(2)}$. The first subnetwork consists of thc first five users: W_1 and W_2 can be transmitted through x_1; W_3, W_4, and W_5 can be transmitted through x_2, x_3, and x_4, respectively; and the transmit signal x_5 is inactive.

We now explain the optimal TDMA scheme for the considered subnetwork. We first explain a simple scheme and then modify it to get the optimal scheme. Each of the messages W_1, W_3, and W_5 is delivered whenever the direct link between its carrying transmitter and its designated receiver is not erased. Message W_2 is delivered whenever message W_1 is not transmitted and message W_3 is not causing interference at y_2. Message W_4 is transmitted whenever W_5 is not causing interference at y_4 and the transmission of W_4 through x_3 will not disrupt the communication of W_3. We now explain the modification. If there is an atomic subnetwork consisting of users $\{2, 3, 4\}$, then we switch the priority setting within this subnetwork, and messages W_2 and W_4 will be delivered instead of message W_3. The TDMA optimality of this scheme for each realization of the network follows from [97, Theorem 1]. Now, we note that the average sum DoF for messages $\{W_1, \ldots, W_5\}$ is equal to their sum DoF in the original scheme plus an extra term due to the modification. The average sum

DoF for messages $\{W_1, W_2, W_5\}$ in the original scheme equals $3\tau_p^{(2)}$, and the sum of the average sum DoF for messages $\{W_3, W_4\}$ and the extra term is upper-bounded by $2\tau_p^{(1)}$. It follows that the average per-user DoF is upper-bounded by $\frac{2}{5}\tau_p^{(1)} + \frac{3}{5}\tau_p^{(2)}$. The proof can be generalized to show that the average TDMA per-user DoF for message assignment strategies defined by strings of the form $s^{(2)}$ with an odd number of ones n is upper-bounded by $\frac{n-1}{n+2}\tau_p^{(1)} + \frac{3}{n+2}\tau_p^{(2)}$.

For message assignment strategies defined by a string of the form $s^{(2)}$ with an even number of ones n, it can be shown in a similar fashion that the TDMA average per-user DoF is upper-bounded by $\frac{n}{n+2}\tau_p^{(1)} + \frac{2}{n+2}\tau_p^{(2)}$. Also, for strategies defined by a string of the form $s^{(3)} = (1,1,\ldots,1,2,0)$ with a number of ones n, the TDMA average per-user DoF is the same as that of a strategy defined by a string of the form $s^{(2)}$ with the same number of ones, and hence is upper-bounded by a convex combination of $\tau_p^{(1)}$ and $\tau_p^{(2)}$. Finally, for strategies defined by a string of the form $s^{(4)} = (1,1,\ldots,1,2,1,1,\ldots,1,0)$ with a number of ones n, it can be shown in a similar fashion that the average per-user DoF is upper-bounded by $\frac{n-2}{n+2}\tau_p^{(1)} + \frac{4}{n+2}\tau_p^{(3)}$.

We now characterize the average per-user DoF for the cell association problem by proving that TDMA schemes are optimal for any candidate message assignment strategy. In order to prove an information-theoretic upper bound on the per-user DoF for each network realization, we use Lemma 5.2 from Chapter 5. Recall that for any set of receiver indices $\mathcal{A} \subseteq [K]$, we use $U_{\mathcal{A}}$ as the set of indices of transmitters that exclusively carry the messages for the receivers in $\mathcal{A}$.

THEOREM 9.6 *The average per-user DoF for the cell association problem is given by*

$$\tau_p(M=1) = \tau_p^{(\mathrm{TDMA})} = \max\left\{\tau_p^{(1)}, \tau_p^{(2)}, \tau_p^{(3)}\right\}, \tag{9.5}$$

where $\tau_p^{(1)}$, $\tau_p^{(2)}$, and $\tau_p^{(3)}$ are given in (9.1), (9.2), *and* (9.3), *respectively.*

Proof In order to prove the statement, we need to show that $\tau_p(M=1) \leq \tau_p^{(\mathrm{TDMA})}$; we do so by using Lemma 5.2 to show that for any useful message assignment strategy satisfying the cell association constraint, and any network realization, the asymptotic per-user DoF is given by that achieved through the optimal TDMA scheme.

Consider message assignment strategies defined by strings having one of the forms $s^{(1)} = (1)$, $s^{(2)} = (2,1,1,\ldots,1,0)$, or $s^{(3)} = (1,1,\ldots,1,2,0)$. We view each network realization as a series of atomic subnetworks, and show that for each atomic subnetwork, the sum DoF is achieved by the optimal TDMA scheme. For an atomic subnetwork consisting of a number of users n, we note that $\left\lfloor \frac{n+1}{2} \right\rfloor$ users are active in the optimal TDMA scheme; we now show that in this case, using Lemma 5.2, the sum DoF for users in the subnetwork is bounded by $\left\lfloor \frac{n+1}{2} \right\rfloor$. Let the users in the atomic subnetwork have the indices $\{i, i+1, \ldots, i+n-1\}$; then we use Lemma 5.2 with the set $\mathcal{A} = \left\{i+2j : j \in \left\{0,1,2,\ldots,\left\lfloor \frac{n-1}{2} \right\rfloor\right\}\right\}$, except for the cases of message assignment strategies defined by strings having one of the forms $s^{(1)} = (1)$ or $s^{(3)} = (1,1,\ldots,1,2,0)$ with an even number of ones, where we use the set $\mathcal{A} = \left\{i+1+2j : j \in \left\{0,1,2,\ldots,\frac{n-2}{2}\right\}\right\}$. We now note that each transmitter that carries a message for a user in the atomic subnetwork and has an index in $\bar{U}_{\mathcal{A}}$ is connected to a receiver in $\mathcal{A}$, and this receiver is connected to one more transmitter with an index in $U_{\mathcal{A}}$; hence, the missing transmit signals $x_{\bar{U}_{\mathcal{A}}}$ can be recovered from $y_{\mathcal{A}} - z_{\mathcal{A}}$ and $x_{U_{\mathcal{A}}}$. The condition in the statement of Lemma 5.2 is then satisfied, allowing us to prove that the sum DoF for users in the atomic subnetwork is upper-bounded by $|\mathcal{A}| = \left\lfloor \frac{n+1}{2} \right\rfloor$.

The proof is similar for message assignment strategies defined by strings that have the form $s^{(4)} = \{1, 1, \ldots, 1, 2, 1, 1, \ldots, 1, 0\}$. However, there is a difference in selecting the set $\mathcal{A}$ for atomic subnetworks consisting of users with indices $\{i, i+1, \ldots, i+x, i+x+1, \ldots, i+n-1\}$, where $1 \leq x \leq n-2$, and messages W_{i+x} and W_{i+x+1} are both available at transmitter $i+x$. In this case, we apply Lemma 5.2 with the set $\mathcal{A}$ defined as above, but including indices $\{i+x, i+x+1\}$ and excluding indices $\{i+x-1, i+x+2\}$. It can be seen that the condition in Lemma 5.2 will be satisfied in this case, and the proven upper bound on the sum DoF for each atomic subnetwork is achievable through TDMA.

In Figure 9.2, we plot $\tau_p(M = 1)$ at each value of p. The result of Theorem 9.6 implies that the message assignment strategies considered in Lemmas 9.2, 9.3, and 9.4 are optimal at high, low, and middle values of the erasure probability p, respectively. We note that in densely connected networks at a low probability of erasure, the *interference-aware* message assignment strategy in Figure 9.1(b) is optimal; through this assignment, the maximum number of interference-free communication links can be created for the case of no erasures. On the other hand, the linear nature of the channel connectivity does not affect the choice of optimal message assignment at high probability of erasure. As the effect of interference diminishes at high probability of erasure, assigning each message to a unique transmitter, as in the strategy in Figure 9.1(a), becomes the only criterion of optimality. At middle values of p, the message assignment strategy in Figure 9.1(c) is optimal; in this assignment, the network is split into four-user subnetworks. In the first subnetwork, the assignment is optimal as the maximum number of interference-free communication links can be created for the two events where there is an atomic subnetwork consisting of users $\{1, 2, 3\}$ or users $\{2, 3, 4\}$.

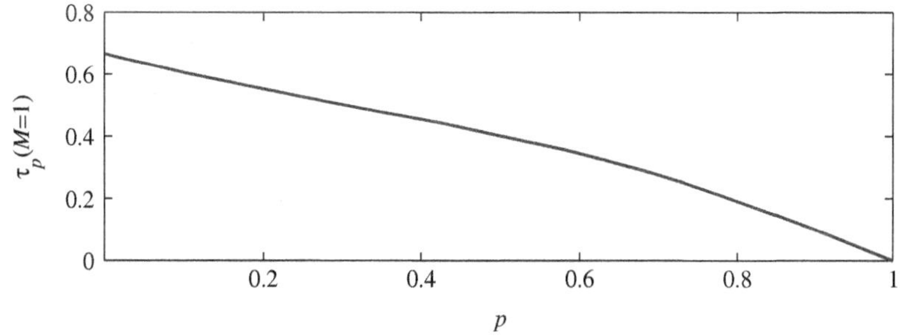

Fig. 9.2 The average per-user DoF for the cell association problem.

9.3 CoMP Transmission

In this section, we are interested in understanding the role of transmitter cooperation in dynamic linear interference networks. In particular, if each message can be assigned to more than one transmitter, without any constraint on their identity, what would be the optimal assignment of messages to transmitters and the corresponding transmission scheme to maximize the average rate over all possible realizations of the network?

We have shown in the previous section that there is no message assignment strategy for the cell association problem where each message is available at one transmitter that is optimal for all values of p. We show in this section that this statement is true even for the case where each message can be available at more than one transmitter ($M > 1$). Recall that for a given value of M, we say that a message assignment strategy is universally optimal if it can be used to achieve $\tau_p(M)$ for all values of p.

THEOREM 9.7 *For any value of the cooperation constraint $M \in \mathbf{Z}^+$, there does not exist a universally optimal message assignment strategy.*

Proof The proof follows from Theorem 9.6 for the case where $M = 1$. We show that for any value of $M > 1$, any message assignment strategy that enables the achievability of $\tau_p(M)$ at high probabilities of erasure is not optimal for the case of no erasures, i.e., cannot be used to achieve $\tau_p(M)$ as $p \to 0$. For any message assignment strategy, consider the value of $\lim_{p\to 1} \frac{\tau_p(M)}{1-p}$ and note that this value equals the average number of transmitters in a transmit set that can be connected to the designated receiver. More precisely,

$$\lim_{p\to 1} \frac{\tau_p(M)}{1-p} = \frac{\sum_{i=1}^{K} |\mathcal{T}_i \cap \{i-1, i\}|}{K}, \tag{9.6}$$

where $\mathcal{T}_i$ in (9.6) corresponds to an optimal message assignment strategy at high probabilities of erasure. It follows that there exists a value $0 < \bar{p} < 1$ such that for any message assignment strategy that enables the achievability of $\tau_p(M)$ for $p \geq \bar{p}$, almost all messages are assigned to the two transmitters that can be connected to the designated receiver, i.e., if we let $\mathcal{S}_K = \left\{i : \mathcal{T}_{i,K} = \{i-1, i\}\right\}$, then $\lim_{K\to\infty} \frac{|\mathcal{S}_K|}{K} = 1$.

We recall from Chapter 6 that for the case of no erasures, the average per-user DoF equals $\frac{2M}{2M+1}$. We also note that, following the same footsteps as in the converse argument for the case of no erasures in Section 6.5, we can show that for any message assignment strategy such that $\lim_{K\to\infty} \frac{|\mathcal{S}_K|}{K} = 1$, the per-user DoF for the case of no erasures is upper-bounded by $\frac{2M-2}{2M-1}$; we do so by using Lemma 5.2 in Chapter 5 for each K-user channel with the set $\mathcal{A}$ defined such that the complement set $\bar{\mathcal{A}} = \{i : i \in [K], i = (2M-1)(j-1) + M, j \in \mathbf{Z}^+\}$.

The condition of optimality identified in the proof of Theorem 9.7 for message assignment strategies at high probabilities of erasure suggests a new role for cooperation in dynamic interference networks. The availability of a message at more than one transmitter may not only be used to cancel its interference at other receivers, but to increase the chances of connecting the message to its designated receiver. This new role leads to three effects at high erasure probabilities: First, the achieved DoF in the considered linear interference network becomes larger than that of K parallel channels; in particular, $\lim_{p\to 1} \frac{\tau_p(M>1)}{1-p} = 2$. Secondly, as the effect of interference diminishes at high probabilities of erasures, all messages can simply be assigned to the two transmitters that may be connected to their designated receiver, and a simple interference avoidance scheme can be used in each network realization. It follows that channel state information is no longer needed at transmitters, and only information about the slow changes in the network topology is needed to achieve the optimal average DoF. Finally, unlike the optimal scheme in Chapter 6 for the case of no erasures, where some transmitters are always inactive, achieving the optimal DoF at high probabilities of erasure requires all transmitters to be used in at least one network realization.

We now restrict our attention to the case where $M = 2$, for simplicity. Here, each message can be available at two transmitters, and transmitted jointly by both of them. We start by studying two message assignment strategies that are optimal in the limits of $p \to 0$ and $p \to 1$, and derive inner bounds on the average per-user DoF $\tau_p(M=2)$ based on the strategies considered. In Chapter 6, the message assignment of Figure 9.3(a) was shown to be DoF optimal for the case of no erasures ($p = 0$). The network is split into subnetworks, each with five consecutive users. The last transmitter of each subnetwork is deactivated to eliminate inter-subnetwork interference. In the first subnetwork, message W_3 is not transmitted, and each other message is received without interference at its designated receiver. Note that the transmit beams for messages W_1 and W_5 contributing to the transmit signals x_2 and x_5, respectively, are designed to cancel the interference at receivers y_2 and y_4, respectively. An analogous scheme is used in each following subnetwork. The value of $\tau_p(M = 2)$ is thus $\frac{4}{5}$ for the case where $p = 0$. In order to prove the following result, we extend the message assignment of Figure 9.3(a) to consider the possible presence of block erasures.

Through discussing the details of the achievability proofs for the message assignment strategies that are optimal in the limits as $p \to 0$ and $p \to 1$, we will get a flavor of DoF analysis for dynamic interference networks, and the complexity of the problem of

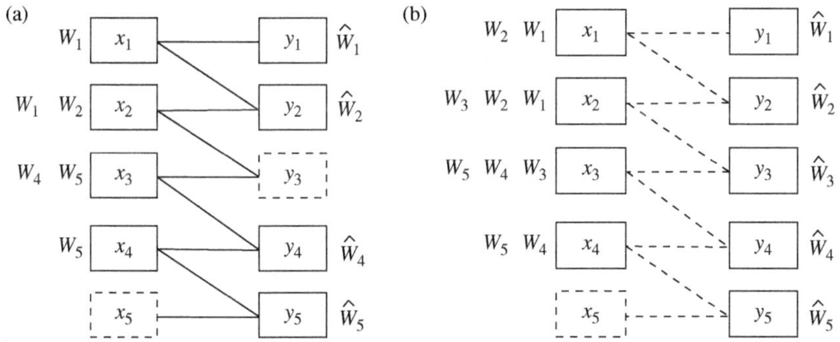

Fig. 9.3 The message assignment in (a) is optimal for a linear network with no erasures ($p = 0$). We extend this message assignment in (b) to consider non-zero erasure probabilities. In both figures, the dashed boxes correspond to inactive signals.

choosing the best message assignments (cell associations) that maximize the average performance. We will then discuss an algorithmic approach to characterize $\tau_p(M=2)$ at all values of $0<p<1$. We start by presenting the assignments that are optimal at extreme values of p in the following two theorems.

THEOREM 9.8 *For $M=2$, the following average per-user DoF is achievable:*

$$\tau_p(M=2) \geq \frac{2}{5}(1-p)(2+Ap), \tag{9.7}$$

where

$$A = p+1-\left((1-p)^2(1-p(1-p))\right)-\frac{1}{2}p(1-p), \tag{9.8}$$

and is asymptotically optimal as $p \to 0$.

Proof We know from Chapter 6 that $\lim_{p\to 0}\tau_p(2)=\frac{4}{5}$, and hence it suffices to show that the inner bound in (9.7) is valid. For each $i \in [K]$, message W_i is assigned as follows:

$$\mathcal{T}_i = \begin{cases} \{i,i+1\} & \text{if } i \equiv 1 \bmod 5, \\ \{i-1,i-2\} & \text{if } i \equiv 0 \bmod 5, \\ \{i-1,i\} & \text{otherwise.} \end{cases}$$

We illustrate this message assignment in Figure 9.3(b). We note that the transmit signals $\{x_i : i \equiv 0 \bmod 5\}$ are inactive, and hence we split the network into five-user subnetworks with no interference between successive subnetworks. We explain the transmission scheme in the first subnetwork and note that a similar scheme applies to each following subnetwork. In the proposed transmission scheme, any receiver is either inactive or receives its desired message without interference, and any transmitter will not transmit more than one message for any network realization. It follows that one DoF is achieved for each message that is transmitted.

Messages W_1, W_2, W_4, and W_5 are transmitted through x_1, x_2, x_3, and x_4, respectively, whenever the coefficients $h_{11}\neq 0$, $h_{22}\neq 0$, $h_{43}\neq 0$, and $h_{54}\neq 0$, respectively. Note that the transmit beam for message W_1 contributing to x_2 can be designed to cancel its interference at y_2. Similarly, the interference caused by W_5 at y_4 can be canceled through x_3. It follows that $(1-p)$ DoF is achieved for each of $\{W_1,W_2,W_4,W_5\}$, and hence $\tau_p(2)\geq\frac{4}{5}(1-p)$. Also, message W_2 is transmitted through x_1 if it cannot be transmitted through x_2 and message W_1 is not transmitted through x_1. More precisely, message W_2 is transmitted through x_1 if $h_{22}=0$ and $h_{21}\neq 0$ and $h_{11}=0$, thereby achieving an extra $p^2(1-p)$ DoF. Similarly, message W_4 can be transmitted through x_4 if $h_{43}=0$ and $h_{44}\neq 0$ and $h_{54}=0$. It follows that

$$\tau_p(2) \geq \frac{4}{5}(1-p)+\frac{2}{5}p^2(1-p). \tag{9.9}$$

Finally, message W_3 will be transmitted through x_3 if message W_4 is not transmitted through x_3 and message W_2 is not causing interference at y_3. Message W_4 is not transmitted through x_3 whenever the coefficient $h_{43}=0$, and message W_2 does not cause interference at y_3 whenever the coefficient $h_{22}=0$ or the coefficient $h_{32}=0$ or

W_2 can be transmitted through x_1. More precisely, message W_3 is transmitted through x_3 if and only if all the following are true:

- $h_{33} \neq 0$ and $h_{43} = 0$.
- $h_{22} = 0$, or $h_{32} = 0$, or it is the case that $h_{11} = 0$ and $h_{21} \neq 0$.

It follows that $f(p)$ DoF is achieved for message W_3, where

$$f(p) = p(1-p)\left(1 - \left((1-p)^2(1-p(1-p))\right)\right). \tag{9.10}$$

Similary, W_3 can be transmitted through x_2 if and only if message W_2 is not transmitted through x_2 and message W_4 is either not transmitted or can be transmitted without causing interference at y_3, i.e., if and only if all the following are true:

- $h_{32} \neq 0$ and $h_{22} = 0$.
- $h_{43} = 0$, or $h_{33} = 0$, or it is the case that $h_{54} = 0$ and $h_{44} \neq 0$.

The above conditions are satisfied with probability $f(p)$. Since we have counted twice the event that $h_{33} \neq 0$ and $h_{43} = 0$ and $h_{32} \neq 0$ and $h_{22} = 0$, it follows that $2f(p) - p^2(1-p)^2$ DoF is achieved for W_3. Summing the DoF achieved for other messages in (9.9), we conclude that

$$\tau_p(2) \geq \frac{4}{5}(1-p) + \frac{2}{5}p^2(1-p) + \frac{1}{5}\left(2f(p) - p^2(1-p)^2\right), \tag{9.11}$$

which is the same inequality as in (9.7).

Although the scheme of Theorem 9.8 is optimal for the case of no erasures ($p = 0$), we know from Theorem 9.7 that better schemes exist at high erasure probabilities. Since in each five-user subnetwork in the scheme of Theorem 9.8 only three users have their messages assigned to the two transmitters that can be connected to their receivers, and two users have only one of these transmitters carrying their messages, we get the asymptotic limit of $\frac{8}{5}$ for the achieved average per-user DoF normalized by $(1-p)$ as $p \to 1$. This leads us to consider an alternative message assignment where the two transmitters carrying each message i are the two transmitters $\{i-1, i\}$ that can be connected to its designated receiver. Such assignment would lead to the ratio $\frac{\tau_p(2)}{1-p} \to 2$ as $p \to 1$. In the following theorem, we analyze a transmission scheme based on this assignment.

THEOREM 9.9 *For $M = 2$, the following average per-user DoF is achievable:*

$$\tau_p(M=2) \geq \frac{1}{3}(1-p)\left(1 + (1-p)^3 + Bp\right), \tag{9.12}$$

where

$$B = 3 + \left(1 + (1-p)^3\right)\left(1 - (1-p)^2 + p(1-p)^3\right) + p\left(1 + (1-p)^2\right), \tag{9.13}$$

and

$$\lim_{p \to 1} \frac{\tau_p(2)}{1-p} = 2. \tag{9.14}$$

Proof For any message assignment, no message can be transmitted if the links from both transmitters carrying the message to its designated receiver are absent, and hence the average DoF achieved for each message is at most $1-p^2$. It follows that $\lim_{p\to 1} \frac{\tau_p(2)}{1-p} \leq \lim_{p\to 1} \frac{(1-p)(1+p)}{1-p} = 2$. We then need only prove that the inner bound in (9.12) is valid. In the achieving scheme, each message is assigned to the two transmitters that may be connected to its designated receiver, i.e., $\mathcal{T}_i = \{i-1, i\}, \forall i \in [K]$. Also, in each network realization, each transmitter will transmit at most one message and any transmitted message will be received at its designated receiver without interference. It follows that one DoF is achieved for any message that is transmitted, and hence the probability of transmission is the same as the average DoF achieved for each message.

Each message W_i such that $i \equiv 0 \bmod 3$ is transmitted through x_{i-1} whenever $h_{i,i-1} \neq 0$, and is transmitted through x_i whenever $h_{i,i-1} = 0$ and $h_{ii} \neq 0$. It follows that n_0 DoF is achieved for each of these messages, where

$$n_0 = (1-p)(1+p). \tag{9.15}$$

We now consider messages W_i such that $i \equiv 1 \bmod 3$. Any such message is transmitted through x_{i-1} whenever $h_{i,i-1} \neq 0$ and $h_{i-1,i-1} = 0$. We note that whenever the channel coefficient $h_{i-1,i-1} \neq 0$, message W_i cannot be transmitted through x_{i-1} as the transmission of W_i through x_{i-1} in this case will prevent W_{i-1} from being transmitted due to either interference at y_{i-1} or sharing the transmitter x_{i-1}. It follows that $n_1^{(1)} = p(1-p)$ DoF is achieved for transmission of W_i through x_{i-1}. Also, message W_i is transmitted through x_i whenever it is not transmitted through x_{i-1} and $h_{ii} \neq 0$ and either $h_{i,i-1} = 0$ or message W_{i-1} is transmitted through x_{i-2}. More precisely, W_i is transmitted through x_i whenever all the following is true: $h_{ii} \neq 0$, and either $h_{i,i-1} = 0$ or it is the case that $h_{i,i-1} \neq 0$ and $h_{i-1,i-1} \neq 0$ and $h_{i-1,i-2} \neq 0$. It follows that $n_1^{(2)} = p(1-p) + (1-p)^4$ is achieved for transmission of W_i through x_i, and hence n_1 DoF is achieved for each message W_i such that $i \equiv 1 \bmod 3$, where

$$n_1 = n_1^{(1)} + n_1^{(2)} = 2p(1-p) + (1-p)^4. \tag{9.16}$$

We now consider messages W_i such that $i \equiv 2 \bmod 3$. Any such message is transmitted through x_{i-1} whenever all the following are true:

- $h_{i,i-1} \neq 0$.
- Either $h_{i-1,i-1} = 0$, or W_{i-1} is not transmitted.
- W_{i+1} is not causing interference at y_i.

The first condition is satisfied with probability $(1-p)$. In order to compute the probability of satisfying the second condition, we note that W_{i-1} is not transmitted for the case when $h_{i-1,i-1} \neq 0$ only if W_{i-2} is transmitted through x_{i-2} and causing interference at y_{i-1}, i.e., only if $h_{i-2,i-3} = 0$ and $h_{i-2,i-2} \neq 0$ and $h_{i-1,i-2} \neq 0$. It follows that the second condition is satisfied with probability $p + p(1-p)^3$. The third condition is not satisfied only if $h_{ii} \neq 0$ and $h_{i+1,i} \neq 0$, and hence will be satisfied with probability at least $1-(1-p)^2$. Moreover, even if $h_{ii} \neq 0$ and $h_{i+1,i} \neq 0$, the third condition can be satisfied if message W_{i+1} can be transmitted through x_{i+1} without causing interference at y_{i+2}, i.e., if $h_{i+1,i+1} \neq 0$ and $h_{i+2,i+1} = 0$. It follows that the third condition will be satisfied with probability $1-(1-p)^2 + p(1-p)^3$, and $n_2^{(1)}$ DoF is achieved by transmission of W_i through x_{i-1}, where

$$n_2^{(1)} = p(1-p)\left(1+(1-p)^3\right)\left(1-(1-p)^2+p(1-p)^3\right). \tag{9.17}$$

Message W_i such that $i \equiv 2 \bmod 3$ is transmitted through x_i whenever $h_{ii} \neq 0$ and $h_{i+1,i} = 0$ and either $h_{i,i-1} = 0$ or W_{i-1} is transmitted through x_{i-2}. It follows that $n_2^{(2)}$ DoF is achieved by transmission of W_i through x_i, where

$$n_2^{(2)} = p(1-p)\left(p + d_1^{(1)}(1-p)\right) \tag{9.18}$$

$$= p^2(1-p)\left(1+(1-p)^2\right), \tag{9.19}$$

and hence $n_2 = n_2^{(1)} + n_2^{(2)}$ DoF is achieved for each message W_i such that $i \equiv 2 \bmod 3$. We finally get

$$\tau_p(2) \geq \frac{n_0 + n_1 + n_2}{3}, \tag{9.20}$$

which is the same inequality as in (9.12).

We plot the inner bounds of (9.7) and (9.12) in Figure 9.4. We note that below a threshold erasure probability $p \approx 0.34$ the scheme of Theorem 9.8 is better, and hence is proposed to be used in this case. For higher probabilities of erasure, the scheme of Theorem 9.9 should be used.

We notice from above that, as $p \to 1$, each message is assigned to the two transmitters connected to its destination to maximize the probability of successful delivery, and, as $p \to 0$, the per-user DoF value goes to $\frac{4}{5}$, and is achieved by splitting the network into subnetworks and avoiding inter-subnetwork interference by deactivating the last transmitter in each subnetwork. And hence, each of the first and last messages in each subnetwork is only assigned to one of the two transmitters connected to its destination, and the other assignment is used at a transmitter not connected to its destination, but connected to another receiver that is prone to interference caused by this message. Further, the middle message in each subnetwork is not transmitted. It was found in [98] that assigning that middle message to only one transmitter connected to its destination, and another transmitter not connected to its destination, leads to better rates than assigning it to the two transmitters connected to its destination at low values of p. This implies that a fraction of $\frac{3}{5}$ of the messages are assigned to only one of the two

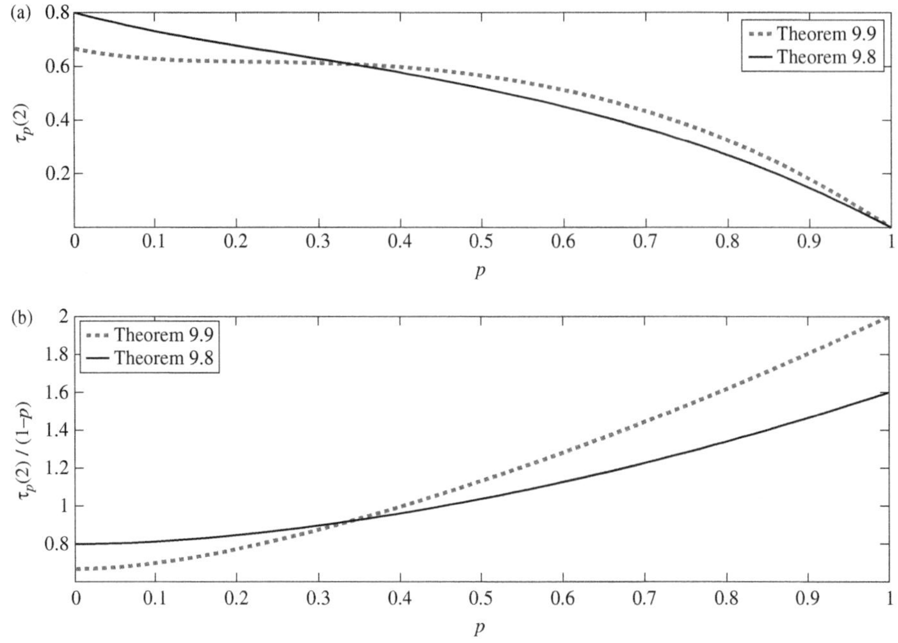

Fig. 9.4 Inner bounds achieved in Theorems 9.8 and 9.9. (a) The achieved per-user DoF. (b) The achieved per-user DoF normalized by $(1-p)$.

transmitters connected to their destination, and the remaining $\frac{2}{5}$ are assigned to the two transmitters connected to their destination. In general, at any value of p from 0 to 1, the assignment achieving the highest per-user DoF using our proposed scheme has a fraction of $f(p)$ of messages that are assigned to only one of the transmitters connected to their destination, and another transmitter used for interference cancellation, and the remaining fraction $1-f(p)$ of messages are assigned to the two transmitters connected to their destination. In [98], it was shown that the value of $f(p)$ decreases monotonically from $\frac{3}{5}$ to 0 as p increases from 0 to 1 for the optimal message assignment strategy under the cooperation constraint $M=2$, which agrees with the intuition about the shifting role of cooperative transmission from canceling interference to increasing the probability of successful delivery as p increases from 0 to 1. We discuss the aforementioned result of [98] in the rest of this section.

An algorithm was presented in [98] that takes as input a given subnetwork of users and the transmit sets for their messages. The output of the algorithm is the transmit signals for transmitters considered in the subnetwork. The chosen transmit signals are based on zero-forcing transmit beamforming that maximizes the average asymptotic per-user DoF value for users within the subnetwork, while guaranteeing that there is no interference caused by the last user of the subnetwork at the first receiver of the following subnetwork. It was shown that this algorithm is optimal, if we are restricted to using zero-forcing schemes.

Table 9.1 Optimal fraction of messages with interference cancellation assignments at every value of p for $M = 2$.

Range of p	Best-performing message assignment
$[0, 0.15]$	$f(p) = \frac{3}{5}$
$[0.16, 0.29]$	$f(p) = \frac{1}{2}$
$(0.29, 0.31)$	$f(p) = \frac{49}{100}$
$[0.31, 0.32]$	$f(p) = \frac{12}{25}$
$[0.33, 0.58]$	$f(p) = \frac{1}{100}$
$[0.59, 1]$	$f(p) = 0$ (as in [96])

The analysis of the algorithm in [98] is similar to the analysis we discussed in the proofs of Theorems 9.8 and 9.9, and hence we skip a discussion of the algorithm and its analysis here. Instead, we discuss simulation results that are based on that algorithm. To compute the average per-user DoF at a certain value of p for a given message assignment, a sufficiently large number n of channel realizations was simulated, where links are erased with probability p, and the algorithm was applied to each realization. The per-user DoF value was then computed as the average number of decoded messages divided by the subnetwork size N.

The simulation was done for a set of message assignments with different fractions of messages that are assigned to one transmitter connected to their desired receiver and another transmitter that can be used to cancel interference, while the remaining fractions of messages are assigned to both transmitters that are connected to their destination. Furthermore, we vary the subnetwork size N.

In Table 9.1, we show the optimal fraction of messages $f(p)$ that should be assigned to only one of the transmitters connected to their destination, and another transmitter used for interference cancellation, according to the simulation results of [98]. Note that in Theorem 9.8 it was shown that an assignment with only $\frac{2}{5}$ of the messages having *interference cancellation* assignments is optimal as $p \rightarrow 0$. Interestingly, in the simulation results of [98], we see that $f(p) = \frac{3}{5}$ for low values of p, as we found another assignment that achieves the same per-user DoF as $p \rightarrow 0$, but performs slightly better on the interval of low-p values $(0, 0.15]$. From the results in Table 9.1, we observe that the optimal fraction $f(p)$ decreases monotonically from $\frac{3}{5}$ to 0 as p goes from 0 to 1.

9.4 Summary

We have explored in this chapter the problem of dynamic interference management in large wireless networks with deep fading conditions. We considered a linear interference network with K transmitter–receiver pairs, where each transmitter can be connected to two neighboring receivers. Long-term fluctuations (shadow fading) in the wireless channel can lead to any link being erased with probability p. The considered rate criterion was the per-user DoF as K goes to infinity. We first studied the case where each message can only be available at a single transmitter, and identified the optimal assignment of messages to transmitters and the corresponding per-user DoF value at each value of the erasure probability p. We then focused on the case where each message can be available at two transmitters. The optimal assignment of messages to transmitters was identified in the two limits $p \to 0$ and $p \to 1$. We also discussed results based on an algorithm in [98] that achieves the optimal zero-forcing average per-user DoF value. A key insight that we can derive from these results is that the role of cooperation shifts from increasing the probability of delivering a message to its intended destination at high values of p, to interference cancellation at low values of p.

10 Recent Advances and Open Problems

Over the years since Claude Shannon [12] introduced the subject, information theory has proven to be a powerful mathematical tool that is also practically relevant in terms of guiding the design of transmitters and receivers for communication systems and networks. From the analysis of the capacity of point-to-point communication links [12] to the more advanced analysis of network information theory [99], there is evidence that information-theoretic tools not only provide performance benchmarks but also inspire new designs for coding schemes and communication protocols. In this book, our focus has been on information-theoretic tools that are geared toward laying a framework for interference management in modern wireless networks. We started with a discussion of the information-theoretic capacity of the two-user interference channel in Chapter 2, which is a long-standing open problem. We do not yet understand how interfering signals should be handled by the decoders at the receivers except in special cases: (i) where the interference power is high and the interfering signal(s) can be decoded, and (ii) and where the interference power is low and the interfering signal(s) can be treated as noise. We then considered the degrees of freedom criterion in the rest of the book, due to its attractive simplicity and utility in providing insights for interference management. We considered general K-user interference channels in Chapter 3, and introduced the concept of asymptotic interference alignment. We then studied in Chapters 5–9 the potential gains of cooperative communication through the lens of DoF analysis in large interference networks. In these studies we assumed a centralized controller, one that has a global view of the network and can make decisions on cell associations and transmission schedules for the whole network, with the goal of maximizing the sum DoF. It is important to note that our study of models with centralized controllers does not imply that we are advocating a centralized approach in next-generation wireless networks over a distributed one that only requires local network knowledge for making decisions. Analyzing fundamental bounds on the performance for the centralized solution can also lead to good design choices for distributed schemes, as well as providing benchmarks for their performance, as we have seen in our study of distributed interference management algorithms in Chapter 4 and Section 5.7.

In this chapter, we describe some recent advances and open problems in interference management that may be of relevance to next-generation wireless networks. We also discuss some of the applications of these networks that have been envisioned.

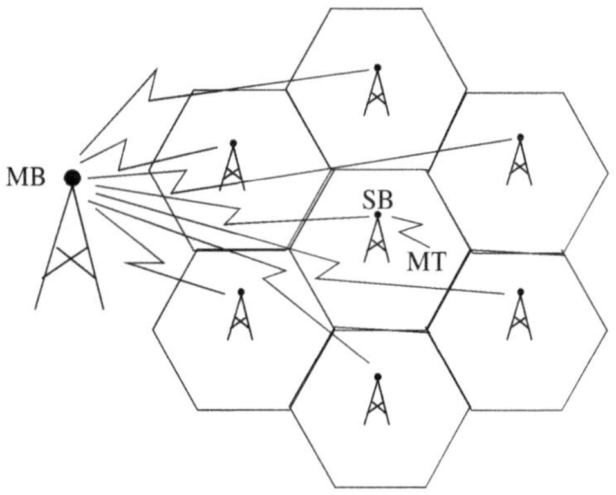

Fig. 10.1 Heterogenous network with a wireless backhaul. Macro-cell basestations (MBs) are assumed to share time–frequency resources with small-cell basestations (SBs), which act as half-duplex analog relays between the MBs and mobile terminals (MTs).

10.1 Heterogeneous Cellular Network with Shared Wireless Backhaul

Heterogeneous networks are envisioned to be a key component of future-generation cellular networks [100]. Heterogeneous networks enable flexible and low-cost deployments and provide a uniform broadband experience to users in the network [101]. Managing interference in such heterogeneous networks is crucial in order to achieve high data rates for the users.[1]

Consider the downlink of a cellular network as a heterogenous network consisting of macro-cell basestations (MBs), small-cell basestations (SBs), and the mobile terminals (MTs), as shown in Figure 10.1. Heterogeneous networks that are built by complementing a macro-cell layer with additional small cells impose new challenges on the backhaul [103]. The best physical location for a small cell often limits the option to use a wired backhaul. In such cases, deploying a wireless backhaul is both faster and more cost effective. One option is to consider a point-to-multipoint wireless backhaul between the MBs and the SBs, where one MB serves several SBs by sharing its antenna resources [104, 105]. Furthermore, in this architecture, we assume that the MBs and the SBs operate on the same frequency band, and that the SBs act as half-duplex analog relays between the MBs and MTs. Therefore, no additional time–frequency resources are consumed by the wireless backhaul.

There are two layers in the network, the wireless *backhaul layer* between MBs and SBs and the *transmission layer* between SBs and MTs. Each MB is associated with a cluster of SBs and is responsible for conveying messages to the MTs connected to the SBs within the cluster. The MBs are equipped with a sufficiently large number of antennas to be able to: (i) beamform separate signals to each of the active SBs within the cluster, and (ii) zero-force interference caused at SBs outside the cluster.

[1] A study of interference management in multi-hop wireless networks is relevant to the wireless backhaul problem. See [102] for an overview.

Coordinated multi-point transmission (see Chapters 6 and 7) can be implemented in the transmission layer from the SBs to the MTs. It is important to note that in this heterogenous network architecture the message sharing required by CoMP does not impose any extra load on the backhaul. This is because we can send linear combinations of the messages as analog signals to each SB directly from the MBs, and these analog signals are relayed by the SBs to zero-force the interference at the active MTs. Implementing such CoMP transmission of course requires that at each MB, the channel state information (CSI) between SBs in its cluster and the corresponding MTs is known.

It is reasonable to assume that the transmissions from the MBs do not cause interference at the MTs, since the MBs can exploit multiple antennas to localize the beams to the fixed locations of the SBs. It is also reasonable to assume that the SBs that are actively transmitting do not cause interference at the receiving SBs because the transmissions in the backhaul layer happen at a higher SNR than in the transmission layer and are also more localized.

Since the SBs are half duplex, they cannot transmit and receive in the same frequency band at the same time. There are two strategies for accommodating this constraint. The first is a frequency-division duplexing (FDD) strategy in which the available frequency band is divided into two equal parts, with the SBs receiving in one half and transmitting in the other. In this case the backhaul layer and transmission layer can be treated separately, and the per-user DoF in the total network is half of the DoF in the transmission layer. From Chapters 6 and 7, we know that the per-user DoF in locally connected networks is strictly less than one for any fixed value of the cooperation order M, and hence the per-user DoF achievable for the two-layered network is strictly less than half.

A better strategy for accommodating the half-duplex constraint at the SBs is a time-division duplex (TDD) strategy, where the SBs receive and transmit in alternate time slots. In this case also the per-user DoF in the total network is half of the DoF in the transmission layer, and therefore the maximum achievable per-user DoF is half. The key difference from the FDD strategy is that in the TDD strategy we can exploit the fact that not all the SBs are active in a given cluster for the CoMP zero-forcing achievable scheme (see, e.g., Section 7.4). The SBs that are inactive for zero-forcing in a given time slot can receive signals in that time slot from the MB serving the cluster, thus utilizing the shared time–frequency resources more efficiently. Using this approach it is shown in [104, 105] that one can achieve the maximum possible per-user DoF of half as long as there is a sufficient number of antennas at each of the MBs.

10.2 Cooperation with No CSIT

In the study of interference management schemes, including the schemes described in this book, it is assumed that channel state information is available at all transmitters (CSIT) and receivers (CSIR) where it is needed. In practice, the channel coefficients are approximately estimated at the receivers by transmitting known pilot signals, and then they are fed back to the transmitters (see, e.g., [106–108]). It is a common practice in information-theoretic analyses to ignore the overhead of the estimation and communication of the channel coefficients, in order to derive insights relevant to the remaining design parameters of the interference management scheme. However, this approach may not be justified in certain settings. For example, dedicating some communication sessions to the learning of CSI may not be the best approach in a network whose state does not last for long. Therefore, there have been recent studies in the information theory literature formulating the interference management problem in the absence of CSIT (see, e.g., [109]) or with very limited CSIT (see, e.g., [110]).

10.2.1 Topological Interference Management

Topological interference management (TIM) concerns the study of the DoF of interference networks without the use of CSIT, where only information about the network topology is available at the transmitters. In recent work [109], an analogy is drawn between the TIM problem for channels that remain constant and the linear index coding problem. This analogy yields useful transmission strategies for TIM based on already available results for the index coding problem. The key idea behind these new *blind* interference alignment transmission strategies is the following: If the channel remains constant over few communication sessions, then it is possible to use this fact to align interference at unintended receivers by choosing the beamforming vectors carefully, even if the actual channel coefficients are not known when designing these vectors. All that is known is that the symbols in each of these vectors will be multiplied by the same channel coefficients in different time slots.

When a time-varying channel model is assumed, the results of blind interference alignment do not apply anymore because different symbols in a beamforming vector will go through different (possibly independent) channel transformations at different time slots. In this scenario, a class of retransmission-based schemes has been proposed in [111] and [112]. These schemes exploit independent transformations of the channel by designing a transmission scheduling algorithm where the same symbol is repeated by each transmitter whenever that transmitter is activated. The generic assumption on the channel coefficients can then guarantee almost sure alignment of the interference in a subspace of the desired size at unintended receivers.

An attempt is made in [113] and [114] to settle the TIM problem in a time-varying channel setting for arbitrary network topologies, under the restriction to the class of linear schemes. The approach followed in these works was to identify a fundamental random matrix problem at the core of the problem of aligning interference with no CSIT through linear schemes. The random matrix problem is as follows: Given that we can design K matrices $\boldsymbol{B}_i, i \in \{1,2,\ldots,K\}$, where each matrix has dimensions $m_i \times n$, $m_i \leq n$, and has full column rank, what are the conditions that these matrices have to satisfy such that the *destination* matrix $\boldsymbol{B}_{\mathrm{D}} = [\boldsymbol{\Lambda}_1\boldsymbol{B}_1 \quad \boldsymbol{\Lambda}_2\boldsymbol{B}_2 \quad \cdots \quad \boldsymbol{\Lambda}_K\boldsymbol{B}_K]$ has rank at most $R - \tau$ almost surely? Here, R is the maximum rank that $\boldsymbol{B}_{\mathrm{D}}$ can have, i.e., $R = \min\left(\sum_{i=1}^{K} m_i, n\right)$, $\boldsymbol{\Lambda}_i$ is an $n \times n$ diagonal matrix with random diagonal entries, and the set of all random coefficients is drawn from a continuous joint distribution. This random matrix problem is equivalent to that of deriving the necessary and sufficient condition on a group of matrices such that the ensemble of randomly scaled row versions of these matrices loses rank almost surely by a factor τ. This problem is related to the TIM problem as follows: Consider $\boldsymbol{B}_i$ as the beamforming matrix for the ith transmitter, and $\boldsymbol{\Lambda}_i$ as the diagonal channel coefficient matrix between the ith transmitter and the considered destination over n time slots. Now, if users $1,2,\ldots,K$ are interfering at the destination, then the rank loss condition is an interference alignment condition that guarantees reduction in the dimension of the interference subspace by a value of τ. The freed signal space can then be used for communicating a desired signal to the destination.

The desired condition on the rank of the ensemble matrix is given by:

$$\operatorname{rank}([\boldsymbol{\Lambda}_1\boldsymbol{B}_1 \;\; \boldsymbol{\Lambda}_2\boldsymbol{B}_2 \;\; \cdots \;\; \boldsymbol{\Lambda}_K\boldsymbol{B}_K]) \overset{\text{a.s.}}{\leq} R-\tau. \tag{10.1}$$

An equivalent condition obtained in [113] is described in the following result.

THEOREM 10.1 *Condition* (10.1) *is equivalent to:*

$$\forall \mathcal{Y}_i \subseteq [m_i], i \in [K] \text{ s.t. } \sum_{i=1}^{K} |\mathcal{Y}_i| = R,$$

$$\exists \mathcal{J} \subseteq [n] : \sum_{i=1}^{K} \dim(\mathcal{S}_{\mathcal{J}} \cap \mathcal{B}_{i,*,\mathcal{Y}_i}) \geq |\mathcal{J}| + \tau. \tag{10.2}$$

In (10.2), we use calligraphic $\mathcal{B}$ to denote the subspace in $\mathbb{R}^n$ spanned by the columns of $\boldsymbol{B}$. Also, for any $\mathcal{X} \subseteq [n]$ and $\mathcal{Y} \subseteq [m_i]$, $\boldsymbol{B}_{*,\mathcal{Y}}$ denotes the submatrix of $\boldsymbol{B}$ created by removing the columns with indices outside $\mathcal{Y}$. Finally, for any $\mathcal{J} \subseteq [n]$, $\mathcal{S}_{\mathcal{J}}$ denotes the subspace of $\mathbb{R}^n$ spanned by the columns of the $n \times n$ identity matrix with indices in $\mathcal{J}$. In other words, $\mathcal{S}_{\mathcal{J}}$ is the subspace of $\mathbb{R}^n$ that includes all the vectors that have zero entries in the complement set $\mathcal{J}^{\mathrm{c}}$. We call $\mathcal{S}_{\mathcal{J}}$ the *sparse subspace* of the set $\mathcal{J}$.

By examining the condition (10.2) for the simple example where[2] $K = 2$ and $m_1 = m_2 = \frac{n}{2}$, we notice that $\boldsymbol{B}_1$ and $\boldsymbol{B}_2$ will each be of size $n \times \frac{n}{2}$, and the only possible choice for $\mathcal{Y}_1$ and $\mathcal{Y}_2$ will be $[\frac{n}{2}]$. Theorem 10.1 implies that $\operatorname{rank}([\boldsymbol{\Lambda}_1\boldsymbol{B}_1 \;\; \boldsymbol{\Lambda}_2\boldsymbol{B}_2]) \overset{\text{a.s.}}{\leq} n-\tau$ if and only if there exists a set $\mathcal{J} \subseteq [n]$ such that

$$\dim(\mathcal{S}_{\mathcal{J}} \cap \mathcal{B}_1) + \dim(\mathcal{S}_{\mathcal{J}} \cap \mathcal{B}_2) \geq |\mathcal{J}| + \tau. \tag{10.3}$$

We interpret the condition (10.2) through the special case in (10.3) for the considered interference management problem as follows. The condition states that each of users 1 and 2 can achieve $\frac{n}{2}$ DoF, while their interference occupies a subspace at the destination that has dimension $n-\tau$ almost surely, if and only if there are $|\mathcal{J}|$ time slots such that the following holds: The sum of beamforming vectors in the transmissions of users 1 and 2 that are *silent* in any time slot outside the designated $|\mathcal{J}|$ time slots exceeds the value $|\mathcal{J}|$ by τ. If $n = 2$ and each beamforming matrix is only a 2×1 vector, the only way to guarantee almost surely that the two vectors will occupy a

[2] That is, we wish to achieve a $\frac{1}{2}$ DoF for each of the two users, when normalized by the number of channel uses n.

single-dimensional interference subspace after undergoing random row scaling is that both vectors have a zero entry in one of the two indices, i.e., that in one of the two time slots, both transmitters are silent. We say that this is a condition of *physical alignment* as it implies that a certain set of transmitters have to physically align their transmissions in specified time slots and be silent in the remaining time slots. A fundamental question at the core of settling the TIM problem is whether physical alignment between transmit beamforming matrices corresponding to different interfering transmitters is always a necessary condition for reducing the dimension of the interference subspace at the destination.

Theorem 10.1 is used in [114] to settle the problem of characterizing the symmetric[3] DoF for a wide class of network topologies. However, even for the symmetric DoF problem, a complete characterization for all possible network topologies is still open.

10.2.2 Optimality of TDMA

The problem of characterizing network topologies for which TDMA is optimal, with no CSIT and no cooperative transmission, is studied in [97, 115]. In particular, it is shown in [115] that TDMA can be used to achieve the *all-unicast* DoF region if and only if the bipartite network topology graph is chordal, i.e., every cycle that can contain a chord has one. The all-unicast setup refers to the case when each transmitter has an independent message for each receiver. This implies that if the network is chordal, then TDMA can be used to achieve the DoF region (and hence, the sum DoF) for any subset of unicast messages as well. Since all locally connected networks are chordal, the following statement holds as a corollary to the result in [115]:

COROLLARY 10.1 *If each message can be available at a single transmitter ($M = 1$), then TDMA is optimal for all considered locally connected networks:*

$$\tau_{\mathrm{TIM}}(L, M = 1) = \tau^{\mathrm{TDMA}}(L, M = 1),\ \forall L \in \mathbf{Z}^{+}. \tag{10.4}$$

In (10.4), we use the subscript TIM to denote the topological interference management setting and the superscript TDMA to denote restriction to interference avoidance TDMA schemes. Therefore, $\tau_{\mathrm{TIM}}(L, M = 1)$ refers to the asymptotic per-user DoF for a locally connected channel with connectivity parameter L, where each message is allowed to be available at a single transmitter, and the transmitters only have network topology information. Note that TDMA schemes do not rely on CSIT, and hence it does not matter whether we have the subscript TIM or not, as long as the superscript TDMA is present. We also note that the conclusion of Corollary 10.1 would still hold if CoMP transmission is only allowed through splitting messages into independent parts, and distributing different parts to different transmitters.

A key result for bounding the sum DoF in arbitrary network topologies when each message can only be available at a single transmitter is given in [109, 116, 117]. It is shown that the sum DoF of any subset of messages that form an *acyclic demand graph*

[3] Symmetric rates refer to rate values that can be achieved simultaneously at all users.

is unity. A demand graph is a directed bipartite graph, where one partite set has messages and the other partite set has receivers. An edge exists from a message to a receiver if the message is destined for the receiver. An edge exists from a receiver to a message if the receiver is not connected to the transmitter carrying the message. In the models we considered throughout this book, message W_i is always destined for the ith receiver. Therefore, the result of [109, 116, 117] regarding acyclic demand graphs holds for the reduced demand graph obtained by collapsing each pair of message–receiver nodes with the same index in the bipartite demand graph. Further, for locally connected networks, we know that the ith transmitter is connected to receivers with indices $\{i, i+1, \ldots, i+L\}$, and hence if we take any $L+2$ nodes with consecutive indices, there will be a cycle between the first and last nodes, and hence the sum DoF cannot be bounded by unity. This provides a simple way to explain why the following result holds:

$$\tau_{\text{TIM}}(L, M=1) = \frac{2}{L+2}, \ \forall L \in \mathbf{Z}^+. \tag{10.5}$$

10.2.3 The Questionable Role of Cooperation

We have seen in Chapter 5 that CoMP transmission could be used with asymptotic interference alignment to achieve DoF gains in fully connected interference channels. We have also seen in Chapters 6–9 how cooperative transmission and reception schemes that are based on simple zero-forcing transmit beamforming and message passing decoding could be used to achieve DoF gains that scale in large locally connected networks. We notice that all the presented cooperative schemes, whether aided by interference alignment or zero-forcing, depend on the availability of accurate CSIT. Hence, it is not clear from our discussion so far if cooperation would lead to any gains with no CSIT. In the following two subsections, we present recent results that attempt to answer this question.

Transmitter Cooperation

We have seen in Chapter 6 that the right-hand side of (10.5) becomes $\frac{2M}{2M+L}$ when each message can be available at M transmitters, only zero-forcing transmit beamforming is allowed, and CSIT is available. It would be interesting to investigate whether it is possible to obtain an analogous generalization when there is no CSIT, by identifying a key lemma for bounding the sum DoF of a subset of messages akin to the one discussed in Section 10.2.2, but with each message available at multiple transmitters.

The problem of interference management through cooperative transmission is studied with weak and no CSIT in [118, 119]. Also, in [120], it is shown that assigning each message to all the transmitters connected to the desired receiver is beneficial compared to assigning each message only to the transmitter having the same index as the desired receiver. However, the proposed coding scheme in [120] relies solely on interference avoidance without the exploitation of CoMP transmission. In [121] and [122], it is shown that linear cooperation schemes cannot lead to gains in the asymptotic per-user DoF for Wyner's asymmetric model ($L=1$) and Wyner's symmetric model ($L=2$), respectively. Linear cooperation is used to refer to schemes where the transmit signal at the jth user is given by

$$x_j = \sum_{i:j \in \mathcal{T}_i} x_{j,i}, \forall j \in [K], \tag{10.6}$$

where $x_{j,i}$ depends only on message W_i. Further, each message W_i is represented by a vector $\boldsymbol{w}_i \in \mathbb{C}^{m_i}$ of m_i complex symbols that need to be delivered to the ith receiver. This message is encoded to one or many of the transmit vectors $\boldsymbol{x}^n_{j,i} = \boldsymbol{V}^n_{j,i}\boldsymbol{w}_i$, where $j \in \mathcal{T}_i$ and $\boldsymbol{V}^n_{j,i}$ denotes the $n \times m_i$ linear beamforming *precoding* matrix used by transmitter j to transmit W_i over n time slots.

The key concept used in the information-theoretic converse proofs in [121] and [122] is that, without CSIT, if a given received signal could be used to successfully decode a message for almost all channel realizations, then any other received signal that is *statistically equivalent* could also be used to decode the same message almost surely. This enables us to minimize the value of the corresponding upper bound when applying Lemma 5.2. However, it is not clear whether the obtained results in these works are particular to Wyner's asymmetric and symmetric models of channel connectivity. The question remains open as to whether it is true in general that TDMA is optimal, as long as the transmitters are not aware of the channel state information, even if cooperative transmission is allowed.

Receiver Cooperation

In [123], the topological interference management problem is studied in a setting where receivers are allowed to pass decoded messages to other receivers, in a manner similar to the schemes we discussed in Chapter 8. The results are based on a directed conflict graph, where each vertex represents a message, and a directed edge exists from vertex i to vertex j if and only if transmitter i (that is exclusively carrying message W_i) causes interference at receiver j (that is decoding message W_j). With message passing, all interference could be eliminated in any subgraph of the directed conflict graph that does not have cycles, using orthogonal access schemes and message passing decoding. It is proved in [123] that orthogonal access schemes are optimal (for all possible subgraphs) for a wide range of network topologies.

As mentioned in Section 8.2, cycles in the conflict graph form an obstacle to the message passing decoding mechanism, because there is no valid decoding order that is possible when cycles exist. As a simple example, consider a two-user fully connected interference channel. Each of the two messages causes interference at the other message's intended receiver. If orthogonal access is used, we have to have an interference-free receiver that decodes its message first and then passes the decoded message to the other receiver, but this requirement cannot be met for a two-user fully connected interference channel. What the scheme in [123] relies on is activating just one of the users in any such fully connected subgraph, which is possible even without cooperation.

10.3 Learning the Network Topology

A key step in the interference management schemes discussed in this book, whether or not they are based on CSIT, is learning the topology of the wireless network. In this section, we discuss network discovery protocols for learning the network topology. The goal is to derive fundamental limits on the number of communication rounds required to discover the network topology in a centralized fashion.

The problem of learning the interference graph of a wireless network is studied in [124], where an acknowledgment-based channel access mechanism is used to identify colliding transmitters, and in each time slot, a subset of transmitters is activated to discover new interfering transmitters. An algorithm is proposed in this work to discover the interference graph with minimum sample complexity, i.e., in the order of the minimum required time. However, the algorithm in [124] does not consider coordination between the transmitters. The problem we consider here is how to coordinate the node transmission such that all receivers can learn the identity of the transmitters connected to them in the minimum possible time. Fortunately, this problem had been dealt with in a different context a couple of decades ago [125]; that is, the problem of determining the minimum number of communication rounds to simulate a single round of the message passing model.

The message passing model requires that each node delivers a possibly different message to each of its neighbors. Since the message passing model is suitable for describing wireline networks, algorithms and converse analysis for many network functions such as consensus and shared memory systems already exist in the literature based on the message passing model. It is therefore useful to understand the minimum number of communication rounds needed for a wireless network to simulate a single round of the message passing model. For brevity, we refer to this problem in the following discussion as the single-round simulation (SRS) problem.

Let Δ_i and Δ_o denote the maximum in-degree and out-degree of any receiver and transmitter in the network, respectively. In [125], a randomized algorithm achieves SRS with probability $1-\epsilon$ in $O\left(\Delta_\text{i}\Delta_\text{o}\log\frac{n}{\epsilon}\right)$ rounds, where n is the number of nodes in the network, and each node is a transceiver (transmitter and receiver). The algorithm executes in Δ_o phases, and each phase consists of multiple rounds. In each phase, each transmitter will target one receiver and keeps trying to transmit its message with probability $\frac{1}{\Delta_\text{i}}$ in each round until it succeeds. The proof of the probabilistic completion of the algorithm in $O\left(\Delta_\text{i}\Delta_\text{o}\log\frac{n}{\epsilon}\right)$ rounds uses elementary probabilistic tools.

We now discuss the design of deterministic distributed algorithms for the SRS problem. It is easy to see that a trivial algorithm that dedicates each round to the transmission of a single message completes the simulation in $O(n\Delta_\text{o})$ rounds. An algorithm that achieves SRS in $O(\Delta_\text{o}\Delta_\text{i}^2\log^2 n)$ rounds[4] is designed in [125] based on the following key lemma.

[4] In large networks, the $O(\log^2 n)$ time achieved by the algorithm based on Lemma 10.2 can be much smaller than the $O(n)$ time achieved by the trivial algorithm.

LEMMA 10.2 *Let $1 \leq z_1, \ldots, z_s \leq n$ be s distinct integers. Then for every $1 \leq i \leq s$, there exists a prime $p \leq s \log n$ such that $z_i \not\equiv z_j \bmod p$ for every $j \neq i$.*

The algorithm executes in Δ_o phases. Each transmitter targets a single receiver in each phase. The idea is that it is guaranteed that there is a communication round for each transmitter–receiver pair, in which neither the receiver node nor any of its neighbors other than the specified transmitter is transmitting a message. Let $\{p_1, \ldots, p_s\}$ be the prime numbers in the range $\{2, 3, \ldots, (\Delta_i + 1) \log n\}$. Then each phase consists of s subphases, and the ith subphase consists of p_i rounds. In each subphase, the kth transmitter transmits its message in the round number $(k \bmod p_i)$. It follows from Lemma 10.2 that for each transmitter k, there is a prime number p_i among $\{p_1, \ldots, p_s\}$ such that in the round numbered $(k \bmod p_i)$ of the ith subphase, transmitter k will be able to successfully deliver its message to its target receiver. The complexity analysis is straightforward since there are Δ_o phases, where in each phase there are at most $O(\Delta_i \log n)$ subphases, and in each subphase, there are at most $O(\Delta_i \log n)$ rounds. It follows that $O(\Delta_o \Delta_i^2 \log^2 n)$ rounds suffice for a deterministic distributed algorithm to achieve SRS for every graph.

The problem of characterizing the minimum number of rounds for learning the network topology in a centralized manner remains open, due to the lack of a converse analysis. In general, as the network topology and channel state information change, we would like to have an adaptive interference management scheme that can be easily adjusted to accommodate the new changes.[5] Such adaptation will be critical in modern applications of wireless networks such as device-to-device (D2D) networks, the Internet of Things (IoT), and vehicle-to-infrastructure (V2I) networks [126–129]. The problem of designing such adaptive interference management schemes is still largely open.

[5] This is different from the design criterion we used in Chapter 9, where the goal was to design fixed schemes that maximize the average or worst case performance over all network realizations.

Appendix A: Information Theory

In this appendix, we state some results in information theory that are useful in Chapter 2. We refer the reader to the standard textbook [130] for basic definitions of information-theoritic quantities such as entropy, differential entropy, conditional entropy, and mutual information, and to Section 1.5 for an explanation about the notation that we follow.

A.1 Minimum Mean Squared Error Estimation

Suppose $\boldsymbol{x}_{\mathrm{G}}$ and $\boldsymbol{y}_{\mathrm{G}}$ are jointly circularly symmetric and jointly Gaussian complex random vectors with zero mean. See [131] for the definitions and properties of circularly symmetric complex Gaussian random vectors. Since the random vectors are jointly circularly symmetric and Gaussian, the minimum mean squared error (MMSE) estimate of $\boldsymbol{x}_{\mathrm{G}}$ given $\boldsymbol{y}_{\mathrm{G}}$,

$$\hat{\boldsymbol{x}}_{\mathrm{G}} = \mathbb{E}\left[\boldsymbol{x}_{\mathrm{G}} | \boldsymbol{y}_{\mathrm{G}}\right],$$

is linear in $\boldsymbol{y}_{\mathrm{G}}$. The *orthogonality principle* says that $\hat{\boldsymbol{x}}_{\mathrm{G}} = \boldsymbol{T}\boldsymbol{y}_{\mathrm{G}}$ is the MMSE estimate if and only if the MMSE estimation error $(\boldsymbol{x}_{\mathrm{G}} - \hat{\boldsymbol{x}}_{\mathrm{G}})$ is orthogonal (and independent in this case, since the random variables are jointly Gaussian) to the observation $\boldsymbol{y}_{\mathrm{G}}$; i.e.,

$$\mathbb{E}\left[(\boldsymbol{x}_{\mathrm{G}} - \boldsymbol{T}\boldsymbol{y}_{\mathrm{G}})\boldsymbol{y}_{\mathrm{G}}^{\dagger}\right] = 0 \Leftrightarrow \boldsymbol{\Sigma}_{\boldsymbol{x}_{\mathrm{G}}\boldsymbol{y}_{\mathrm{G}}} = \boldsymbol{T}\boldsymbol{\Sigma}_{\boldsymbol{y}_{\mathrm{G}}}.$$

Suppose $\boldsymbol{\Sigma}_{\boldsymbol{y}_{\mathrm{G}}}$ is nonsingular; then $\boldsymbol{T}$ is uniquely determined and hence is given by

$$\boldsymbol{T} = \boldsymbol{\Sigma}_{\boldsymbol{x}_{\mathrm{G}}\boldsymbol{y}_{\mathrm{G}}}\boldsymbol{\Sigma}_{\boldsymbol{y}_{\mathrm{G}}}^{-1},$$

and the conditional covariance matrix of $\boldsymbol{x}_{\mathrm{G}}$ given $\boldsymbol{y}_{\mathrm{G}}$, defined as the covariance matrix of the MMSE estimation error, is given by

$$\boldsymbol{\Sigma}_{\boldsymbol{x}_{\mathrm{G}}|\boldsymbol{y}_{\mathrm{G}}} = \boldsymbol{\Sigma}_{\boldsymbol{x}_{\mathrm{G}}} - \boldsymbol{\Sigma}_{\boldsymbol{x}_{\mathrm{G}}\boldsymbol{y}_{\mathrm{G}}}\boldsymbol{\Sigma}_{\boldsymbol{y}_{\mathrm{G}}}^{-1}\boldsymbol{\Sigma}_{\boldsymbol{y}_{\mathrm{G}}\boldsymbol{x}_{\mathrm{G}}}.$$

Note that $\boldsymbol{\Sigma}_{\boldsymbol{x}_{\mathrm{G}}|\boldsymbol{y}_{\mathrm{G}}}$ is the Schur complement of $\boldsymbol{\Sigma}_{\boldsymbol{x}_{\mathrm{G}}}$ in the matrix

$$\mathrm{Cov}\left(\begin{bmatrix} \boldsymbol{x}_{\mathrm{G}} \\ \boldsymbol{y}_{\mathrm{G}} \end{bmatrix}\right) = \begin{bmatrix} \boldsymbol{\Sigma}_{\boldsymbol{x}_{\mathrm{G}}} & \boldsymbol{\Sigma}_{\boldsymbol{x}_{\mathrm{G}},\boldsymbol{y}_{\mathrm{G}}} \\ \boldsymbol{\Sigma}_{\boldsymbol{y}_{\mathrm{G}},\boldsymbol{x}_{\mathrm{G}}} & \boldsymbol{\Sigma}_{\boldsymbol{y}_{\mathrm{G}}} \end{bmatrix}.$$

The following lemma is useful.

LEMMA A.1 *Suppose x_G and y_G are jointly circularly symmetric, and jointly Gaussian complex random vectors, and $\boldsymbol{\Sigma}$ is a positive semidefinite matrix. Then*

$$\boldsymbol{\Sigma}_{x_G|y_G} \succeq \boldsymbol{\Sigma} \Leftrightarrow \mathrm{Cov}\left(\begin{bmatrix} x_G \\ y_G \end{bmatrix}\right) \succeq \begin{bmatrix} \boldsymbol{\Sigma} & \mathbf{0} \\ \mathbf{0} & \mathbf{0} \end{bmatrix}.$$

Proof Let $\boldsymbol{a}$ and $\boldsymbol{b}$ be column vectors of the same length as x_G and y_G, respectively, and Ty_G be the MMSE estimate of x_G given y_G. Observe that the first condition is equivalent to

$$a^\dagger \boldsymbol{\Sigma}_{x_G|y_G} a = \mathbb{E}\left[\left|a^\dagger\left(x_G - Ty_G\right)\right|^2\right] \geq a^\dagger \boldsymbol{\Sigma} a, \forall a, \tag{A.1}$$

and the second condition is equivalent to

$$\mathbb{E}\left[\left|a^\dagger x_G + b^\dagger y_G\right|^2\right] \geq a^\dagger \boldsymbol{\Sigma} a, \forall a, b. \tag{A.2}$$

We complete the proof by showing that the equations (A.1) and (A.2) imply each other. First, observe that

$$\begin{aligned}\mathbb{E}\left[\left|a^\dagger x_G + b^\dagger y_G\right|^2\right] &= \mathbb{E}\left[\left|a^\dagger\left(x_G - Ty_G + Ty_G\right) + b^\dagger y_G\right|^2\right] \\ &= \mathbb{E}\left[\left|a^\dagger\left(x_G - Ty_G\right) + \left(a^\dagger T + b^\dagger\right) y_G\right|^2\right].\end{aligned}$$

Since the MMSE estimation error $x_G - Ty_G$ is independent of the observation y_G, we have that

$$\mathbb{E}\left[\left|a^\dagger x_G + b^\dagger y_G\right|^2\right] = \mathbb{E}\left[\left|a^\dagger\left(x_G - Ty_G\right)\right|^2\right] + \mathbb{E}\left[\left|\left(a^\dagger T + b^\dagger\right) y_G\right|^2\right].$$

Substituting the above expression in (A.2) it is clear that (A.1) implies (A.2). It is also clear that (A.2) implies (A.1) by setting $b = -T^\dagger a$.

A.2 Basic Extremal Inequalities

In this section, we review two basic extremal inequalities regarding the optimality of circularly symmetric complex Gaussian distributions. The following lemma says that among all continuous distributions with a fixed covariance matrix, the circularly symmetric complex Gaussian distribution maximizes the differential entropy.

LEMMA A.2 **(Theorem 2 of [131])** *Let x be a complex and continuous random vector. Then*

$$\mathsf{h}(x) \leq \mathsf{h}(x_G) = \log\det\left(\pi e \boldsymbol{\Sigma}_{x_G}\right),$$

where x_G is a circularly symmetric complex Gaussian random vector with zero mean and covariance matrix equal to the covariance matrix of x.

We now extend the lemma to conditional differential entropy.

LEMMA A.3 *Let $\boldsymbol{x}$ and $\boldsymbol{y}$ be complex and continuous random vectors. Then*

$$\mathsf{h}(\boldsymbol{x}|\boldsymbol{y}) \leq \mathsf{h}(\boldsymbol{x}_{\mathrm{G}}|\boldsymbol{y}_{\mathrm{G}}) = \log\det\left(\pi e \boldsymbol{\Sigma}_{\boldsymbol{x}_{\mathrm{G}}|\boldsymbol{y}_{\mathrm{G}}}\right),$$

where $\boldsymbol{x}_{\mathrm{G}}$ and $\boldsymbol{y}_{\mathrm{G}}$ are jointly circularly symmetric and jointly Gaussian complex random vectors with zero mean and joint covariance matrix equal to the joint covariance matrix of $\boldsymbol{x}$ and $\boldsymbol{y}$.

Proof Let $\boldsymbol{T}\boldsymbol{y}_{\mathrm{G}}$ be the MMSE estimate of $\boldsymbol{x}_{\mathrm{G}}$ given $\boldsymbol{y}_{\mathrm{G}}$. Since the estimation error $\boldsymbol{x}_{\mathrm{G}} - \boldsymbol{T}\boldsymbol{y}_{\mathrm{G}}$ is independent of the observation $\boldsymbol{y}_{\mathrm{G}}$, we have

$$\mathsf{h}(\boldsymbol{x}_{\mathrm{G}}|\boldsymbol{y}_{\mathrm{G}}) = \mathsf{h}(\boldsymbol{x}_{\mathrm{G}} - \boldsymbol{T}\boldsymbol{y}_{\mathrm{G}}|\boldsymbol{y}_{\mathrm{G}}) = \mathsf{h}(\boldsymbol{x}_{\mathrm{G}} - \boldsymbol{T}\boldsymbol{y}_{\mathrm{G}}) = \log\det\left(\pi e \boldsymbol{\Sigma}_{\boldsymbol{x}_{\mathrm{G}}|\boldsymbol{y}_{\mathrm{G}}}\right).$$

Also, observe that

$$\mathsf{h}(\boldsymbol{x}|\boldsymbol{y}) = \mathsf{h}(\boldsymbol{x} - \boldsymbol{T}\boldsymbol{y}|\boldsymbol{y}) \leq \mathsf{h}(\boldsymbol{x} - \boldsymbol{T}\boldsymbol{y}) \leq \mathsf{h}(\boldsymbol{x}_{\mathrm{G}} - \boldsymbol{T}\boldsymbol{y}_{\mathrm{G}}),$$

where the first inequality follows because conditioning can only reduce entropy, and the second inequality follows from Lemma A.2 because the covariance matrix of $\boldsymbol{y} - \boldsymbol{T}\boldsymbol{x}$ is the same as the covariance matrix of $\boldsymbol{y}_{\mathrm{G}} - \boldsymbol{T}\boldsymbol{x}_{\mathrm{G}}$.

A.3 Concave Functions

In this section, we show that the differential entropy and the conditional differential entropy of circularly symmetric complex Gaussian random vectors are concave functions in the corresponding covariance matrices.

LEMMA A.4 *Suppose $\boldsymbol{x}_{\mathrm{G}}$ is a circularly symmetric complex Gaussian random vector. The differential entropy $\mathsf{h}(\boldsymbol{x}_{\mathrm{G}}) = \log\left(|\pi e \boldsymbol{\Sigma}_{\boldsymbol{x}_{\mathrm{G}}}|\right)$ is concave and nondecreasing in $\boldsymbol{\Sigma}_{\boldsymbol{x}_{\mathrm{G}}}$.*

We now extend the lemma to conditional differential entropy.

LEMMA A.5 *Suppose $\boldsymbol{x}_{\mathrm{G}}$ and $\boldsymbol{y}_{\mathrm{G}}$ are jointly circularly symmetric, and jointly Gaussian complex random vectors. Then the conditional differential entropy $\mathsf{h}(\boldsymbol{x}_{\mathrm{G}}|\boldsymbol{y}_{\mathrm{G}})$ is concave and nondecreasing in*

$$\operatorname{Cov}\left(\begin{bmatrix} \boldsymbol{x}_{\mathrm{G}} \\ \boldsymbol{y}_{\mathrm{G}} \end{bmatrix}\right).$$

Proof Let $(\boldsymbol{x}_{\mathrm{G}},\boldsymbol{y}_{\mathrm{G}})$, $(\boldsymbol{x}_{1G},\boldsymbol{y}_{1G})$, and $(\boldsymbol{x}_{2G},\boldsymbol{y}_{2G})$ be jointly circularly symmetric, and jointly Gaussian complex random vectors such that

$$\operatorname{Cov}\left(\begin{bmatrix} \boldsymbol{x}_{\mathrm{G}} \\ \boldsymbol{y}_{\mathrm{G}} \end{bmatrix}\right) = \lambda \operatorname{Cov}\left(\begin{bmatrix} \boldsymbol{x}_{1G} \\ \boldsymbol{y}_{1G} \end{bmatrix}\right) + (1-\lambda)\operatorname{Cov}\left(\begin{bmatrix} \boldsymbol{x}_{2G} \\ \boldsymbol{y}_{2G} \end{bmatrix}\right).$$

From Lemma A.1, we know that the conditional covariance matrices $\boldsymbol{\Sigma}_{\boldsymbol{x}_{1G}|\boldsymbol{y}_{1G}}$ and $\boldsymbol{\Sigma}_{\boldsymbol{x}_{2G}|\boldsymbol{y}_{2G}}$ satisfy

$$\operatorname{Cov}\left(\begin{bmatrix} \boldsymbol{x}_{iG} \\ \boldsymbol{y}_{iG} \end{bmatrix}\right) \succeq \begin{bmatrix} \boldsymbol{\Sigma}_{\boldsymbol{x}_{iG}|\boldsymbol{y}_{iG}} & \boldsymbol{0} \\ \boldsymbol{0} & \boldsymbol{0} \end{bmatrix}, \text{ for } i = 1,2.$$

Therefore, we obtain that

$$\mathrm{Cov}\left(\begin{bmatrix} \boldsymbol{x}_{\mathrm{G}} \\ \boldsymbol{y}_{\mathrm{G}} \end{bmatrix}\right) \succeq \begin{bmatrix} \lambda\boldsymbol{\Sigma}_{\boldsymbol{x}_{1G}|\boldsymbol{y}_{1G}} + (1-\lambda)\boldsymbol{\Sigma}_{\boldsymbol{x}_{2G}|\boldsymbol{y}_{2G}} & \mathbf{0} \\ \mathbf{0} & \mathbf{0} \end{bmatrix}.$$

Applying Lemma A.1 again, we obtain that

$$\boldsymbol{\Sigma}_{\boldsymbol{x}_{\mathrm{G}}|\boldsymbol{y}_{\mathrm{G}}} \succeq \lambda\boldsymbol{\Sigma}_{\boldsymbol{x}_{1G}|\boldsymbol{y}_{1G}} + (1-\lambda)\boldsymbol{\Sigma}_{\boldsymbol{x}_{2G}|\boldsymbol{y}_{2G}}.$$

Now, applying Lemma A.4, we obtain that

$$\begin{aligned} \mathsf{h}\left(\boldsymbol{x}_{\mathrm{G}}|\boldsymbol{y}_{\mathrm{G}}\right) &= \log\left(|\pi e\boldsymbol{\Sigma}_{\boldsymbol{x}_{\mathrm{G}}|\boldsymbol{y}_{\mathrm{G}}}|\right) \\ &\geq \log\left(\left|\pi e\left(\lambda\boldsymbol{\Sigma}_{\boldsymbol{x}_{1G}|\boldsymbol{y}_{1G}} + (1-\lambda)\boldsymbol{\Sigma}_{\boldsymbol{x}_{2G}|\boldsymbol{y}_{2G}}\right)\right|\right) \\ &\geq \lambda\log\left(|\pi e\boldsymbol{\Sigma}_{\boldsymbol{x}_{1G}|\boldsymbol{y}_{1G}}|\right) + (1-\lambda)\log\left(|\pi e\boldsymbol{\Sigma}_{\boldsymbol{x}_{2G}|\boldsymbol{y}_{2G}}|\right) \\ &= \lambda\mathsf{h}\left(\boldsymbol{x}_{1G}|\boldsymbol{y}_{1G}\right) + (1-\lambda)\mathsf{h}\left(\boldsymbol{x}_{2G}|\boldsymbol{y}_{2G}\right). \end{aligned}$$

This completes the proof of concavity of $\mathsf{h}\left(\boldsymbol{x}_{\mathrm{G}}|\boldsymbol{y}_{\mathrm{G}}\right)$ in the joint covariance matrix. The proof that $\mathsf{h}\left(\boldsymbol{x}_{\mathrm{G}}|\boldsymbol{y}_{\mathrm{G}}\right)$ is nondecreasing in the joint covariance matrix follows directly from Lemma A.1.

LEMMA A.6 *Suppose $\boldsymbol{x}_{\mathrm{G}}, \boldsymbol{z}, \boldsymbol{w}$ are circularly symmetric, and independent complex Gaussian random vectors such that $\boldsymbol{\Sigma}_{\boldsymbol{z}} \preceq \boldsymbol{\Sigma}_{\boldsymbol{w}}$. Then*

$$\mathsf{h}\left(\boldsymbol{x}_{\mathrm{G}} + \boldsymbol{z}\right) - \mathsf{h}\left(\boldsymbol{x}_{\mathrm{G}} + \boldsymbol{w}\right)$$

is concave in $\boldsymbol{\Sigma}_{\boldsymbol{x}_{\mathrm{G}}}$.

Proof Let $\boldsymbol{v} \sim \mathcal{CN}(0, \boldsymbol{\Sigma}_{\boldsymbol{w}} - \boldsymbol{\Sigma}_{\boldsymbol{z}})$ be independent of $\boldsymbol{x}$ and $\boldsymbol{z}$. Then $\boldsymbol{z} + \boldsymbol{v}$ has the same distribution as $\boldsymbol{w}$. Therefore, it is sufficient to prove that

$$\mathsf{h}\left(\boldsymbol{x}_{\mathrm{G}} + \boldsymbol{z}\right) - \mathsf{h}\left(\boldsymbol{x}_{\mathrm{G}} + \boldsymbol{z} + \boldsymbol{v}\right) = -\mathsf{I}\left(\boldsymbol{v}; \boldsymbol{x}_{\mathrm{G}} + \boldsymbol{z} + \boldsymbol{v}\right)$$

is concave in $\boldsymbol{\Sigma}_{\boldsymbol{x}_{\mathrm{G}}}$. Observe that

$$-\mathsf{I}\left(\boldsymbol{v}; \boldsymbol{x}_{\mathrm{G}} + \boldsymbol{z} + \boldsymbol{v}\right) = -\mathsf{h}\left(\boldsymbol{v}\right) + \mathsf{h}\left(\boldsymbol{v}|\boldsymbol{x}_{\mathrm{G}} + \boldsymbol{v} + \boldsymbol{w}\right).$$

The first term is independent of $\boldsymbol{\Sigma}_{\boldsymbol{x}_{\mathrm{G}}}$, and from Lemma A.5, it follows that the second term is concave in $\boldsymbol{\Sigma}_{\boldsymbol{x}_{\mathrm{G}}}$.

A.4 More Extremal Inequalities

In this section, we show that among all the sequences of random vectors with a fixed average covariance matrix, the circularly symmetric i.i.d. Gaussian random vectors maximize certain objective functions involving multi-letter differential entropy and conditional differential entropy terms.

LEMMA A.7 *Suppose $\boldsymbol{x}^n$ is a sequence of complex and continuous random vectors. Then*

$$\mathsf{h}\left(\boldsymbol{x}^n\right) \leq n\mathsf{h}\left(\boldsymbol{x}_{\mathrm{G}}\right) = n\log\det\left(\pi e\boldsymbol{\Sigma}_{\boldsymbol{x}_{\mathrm{G}}}\right),$$

where $\boldsymbol{x}_{\mathrm{G}}$ is a circularly symmetric Gaussian random vector with covariance matrix equal to the average covariance matrix of the random sequence $\boldsymbol{x}^n$; i.e.,

$$\boldsymbol{\Sigma}_{\boldsymbol{x}_{\mathrm{G}}} = \frac{1}{n}\sum_{i=1}^{n}\boldsymbol{\Sigma}_{\boldsymbol{x}_i}.$$

Proof Observe that

$$\begin{aligned}
\mathsf{h}\left(\boldsymbol{x}^n\right) &= \sum_{i=1}^{n}\mathsf{h}\left(\boldsymbol{x}_i|\boldsymbol{x}^{i-1}\right)\\
&\overset{(a)}{\leq} \sum_{i=1}^{n}\mathsf{h}\left(\boldsymbol{x}_i\right)\\
&\overset{(b)}{\leq} \sum_{i=1}^{n}\mathsf{h}\left(\boldsymbol{x}_{iG}\right)\\
&\overset{(c)}{\leq} n\mathsf{h}\left(\boldsymbol{x}_{\mathrm{G}}\right),
\end{aligned}$$

where step (a) follows because conditioning can only reduce entropy, step (b) follows from Lemma A.2, and step (c) follows from Lemma A.4, which says that $\mathsf{h}(\boldsymbol{x}_{\mathrm{G}})$ is concave in $\boldsymbol{\Sigma}_{\boldsymbol{x}_{\mathrm{G}}}$.

We now extend the above lemma to conditional differential entropy.

LEMMA A.8 *Suppose $\boldsymbol{x}^n$ and $\boldsymbol{y}^n$ are sequences of complex and continuous random vectors. Then*

$$\mathsf{h}\left(\boldsymbol{x}^n|\boldsymbol{y}^n\right) \leq n\mathsf{h}\left(\boldsymbol{x}_{\mathrm{G}}|\boldsymbol{y}_{\mathrm{G}}\right) = n\log\det\left(\pi e\boldsymbol{\Sigma}_{\boldsymbol{x}_{\mathrm{G}}|\boldsymbol{y}_{\mathrm{G}}}\right),$$

where $(\boldsymbol{x}_{\mathrm{G}},\boldsymbol{y}_{\mathrm{G}})$ are jointly circularly symmetric and Gaussian complex random vectors with the same joint covariance matrix equal to

$$\mathrm{Cov}\left(\begin{bmatrix}\boldsymbol{x}_{\mathrm{G}}\\ \boldsymbol{y}_{\mathrm{G}}\end{bmatrix}\right) = \frac{1}{n}\sum_{i=1}^{n}\mathrm{Cov}\left(\begin{bmatrix}\boldsymbol{x}_i\\ \boldsymbol{y}_i\end{bmatrix}\right).$$

Proof Let $(\boldsymbol{x}_{iG},\boldsymbol{y}_{iG})$ be jointly circularly symmetric and jointly Gaussian random vectors with joint covariance matrix equal to the joint covariance matrix of $(\boldsymbol{x}_i,\boldsymbol{y}_i)$. Then we have

$$\begin{aligned}
\mathsf{h}\left(\boldsymbol{x}^n|\boldsymbol{y}^n\right) &= \sum_{i=1}^{n}\mathsf{h}\left(\boldsymbol{x}_i|\boldsymbol{y}^{i-1},\boldsymbol{y}^n\right)\\
&\overset{(a)}{\leq} \sum_{i=1}^{n}\mathsf{h}\left(\boldsymbol{x}_i|\boldsymbol{y}_i\right)\\
&\overset{(b)}{\leq} \sum_{i=1}^{n}\mathsf{h}\left(\boldsymbol{x}_{iG}|\boldsymbol{y}_{iG}\right)\\
&\overset{(c)}{\leq} n\mathsf{h}\left(\boldsymbol{x}_{\mathrm{G}}|\boldsymbol{y}_{\mathrm{G}}\right),
\end{aligned}$$

where step (a) follows because conditioning can only reduce entropy, step (b) follows from Lemma A.3, and step (c) follows from Lemma A.5, which says that $\mathsf{h}\left(\boldsymbol{x}_{\mathrm{G}}|\boldsymbol{y}_{\mathrm{G}}\right)$ is concave in the joint covariance matrix of $\boldsymbol{x}_{\mathrm{G}}$ and $\boldsymbol{y}_{\mathrm{G}}$.

LEMMA A.9 *Suppose $\boldsymbol{x}^n$ is a sequence of complex and continuous random vectors, and $\boldsymbol{z}^n \sim$ i.i.d. $\sim \mathcal{CN}(0, \boldsymbol{\Sigma}_z)$ and $\boldsymbol{w}^n \sim$ i.i.d. $\sim \mathcal{CN}(0, \boldsymbol{\Sigma}_w)$ such that $\boldsymbol{\Sigma}_z \preceq \boldsymbol{\Sigma}_w$. Then*

$$\mathsf{h}\left(\boldsymbol{x}^n+\boldsymbol{z}^n\right)-\mathsf{h}\left(\boldsymbol{x}^n+\boldsymbol{w}^n\right) \leq n\mathsf{h}\left(\boldsymbol{x}_{\mathrm{G}}+\boldsymbol{z}\right)-n\mathsf{h}\left(\boldsymbol{x}_{\mathrm{G}}+\boldsymbol{w}\right),$$

where $\boldsymbol{z} \sim \mathcal{CN}(0, \boldsymbol{\Sigma}_z)$, $\boldsymbol{w} \sim \mathcal{CN}(0, \boldsymbol{\Sigma}_w)$, and $\boldsymbol{x}_{\mathrm{G}}$ is a circularly symmetric Gaussian random vector with covariance matrix equal to the average covariance matrix of the random sequence $\boldsymbol{x}^n$; i.e.,

$$\boldsymbol{\Sigma}_{\boldsymbol{x}_{\mathrm{G}}}=\frac{1}{n}\sum_{i=1}^{n}\boldsymbol{\Sigma}_{\boldsymbol{x}_i}.$$

Proof Let $\boldsymbol{v}^n \sim$ i.i.d. $\sim \mathcal{CN}(0, \boldsymbol{\Sigma}_w - \boldsymbol{\Sigma}_z)$ be independent of $\boldsymbol{z}^n$. Then the sequence $\boldsymbol{z}^n+\boldsymbol{v}^n$ has the same distribution as $\boldsymbol{w}^n$. Therefore, it is sufficient to prove that

$$\begin{aligned}
&\mathsf{h}\left(\boldsymbol{x}^n+\boldsymbol{z}^n\right)-\mathsf{h}\left(\boldsymbol{x}^n+\boldsymbol{z}^n+\boldsymbol{v}^n\right) \leq n\mathsf{h}\left(\boldsymbol{x}_{\mathrm{G}}+\boldsymbol{z}\right)-n\mathsf{h}\left(\boldsymbol{x}_{\mathrm{G}}+\boldsymbol{z}+\boldsymbol{v}\right) \\
&\Leftrightarrow -\mathsf{I}\left(\boldsymbol{v}^n;\boldsymbol{x}^n+\boldsymbol{z}^n+\boldsymbol{v}^n\right) \leq -n\mathsf{I}\left(\boldsymbol{v};\boldsymbol{x}_{\mathrm{G}}+\boldsymbol{z}+\boldsymbol{v}\right).
\end{aligned}$$

Observe that

$$\begin{aligned}
-\mathsf{I}\left(\boldsymbol{v}^n;\boldsymbol{x}^n+\boldsymbol{z}^n+\boldsymbol{v}^n\right) &= -\mathsf{h}\left(\boldsymbol{v}^n\right)+\mathsf{h}\left(\boldsymbol{v}^n|\boldsymbol{x}^n+\boldsymbol{z}^n+\boldsymbol{v}^n\right) \\
&\overset{(a)}{\leq} -n\mathsf{h}\left(\boldsymbol{v}\right)+n\mathsf{h}\left(\boldsymbol{v}|\boldsymbol{x}_{\mathrm{G}}+\boldsymbol{z}+\boldsymbol{v}\right) \\
&= -n\mathsf{I}\left(\boldsymbol{v};\boldsymbol{x}_{\mathrm{G}}+\boldsymbol{z}+\boldsymbol{v}\right),
\end{aligned}$$

where step (a) follows from Lemma A.8.

The above lemma is referred to as the worst-case noise lemma. Indeed, observe that $\mathsf{I}\left(\boldsymbol{v}^n;\boldsymbol{x}^n+\boldsymbol{z}^n+\boldsymbol{v}^n\right)$ can be interpreted as the multi-letter mutual information of an additive noise channel with $\boldsymbol{v}$ as input and $\boldsymbol{x}+\boldsymbol{z}$ as noise. The above result argues that i.i.d. Gaussian noise is the worst-case noise minimizing $\mathsf{I}\left(\boldsymbol{v}^n;\boldsymbol{x}^n+\boldsymbol{z}^n+\boldsymbol{v}^n\right)$. In the scalar case, as explained in the mutual information game problem (see Exercise 9.21 in [130]), an alternative proof can be given using the entropy power inequality (EPI).

LEMMA A.10 **(EPI)** *Suppose x^n and z^n are independent sequences of complex random variables. Then,*

$$e^{\frac{1}{n}\mathsf{h}(x^n+z^n)} \geq e^{\frac{1}{n}\mathsf{h}(x^n)}+e^{\frac{1}{n}\mathsf{h}(z^n)}.$$

We use the EPI to prove a generalized version of the worst-case noise lemma in the scalar case.

LEMMA A.11 *Suppose $\{x_i^n : 1 \leq i \leq M\}$ are independent sequences of complex random vectors satisfying an average power constraint P_i; i.e.,*

$$\frac{1}{n}\sum_{j=1}^{n}\Sigma_{x_{ij}} \leq P_i,\ 1 \leq i \leq M,$$

and $z^n \sim$ *i.i.d.* $\sim \mathcal{CN}(0, \Sigma_z)$. *Let* $\mu_1, \mu_2, \ldots, \mu_M$ *be real numbers satisfying the conditions*

$$\mu_i \geq \frac{P_i}{\sum_{j=1}^M P_j + \Sigma_z}, 1 \leq i \leq M.$$

Then we have

$$\sum_{i=1}^M \mu_i \mathsf{h}\left(x_i^n\right) - \mathsf{h}\left(\sum_{i=1}^M x_i^n + z^n\right) \leq n \sum_{i=1}^M \mu_i \mathsf{h}\left(x_{i\mathrm{G}}\right) - n \mathsf{h}\left(\sum_{i=1}^M x_{i\mathrm{G}} + z\right),$$

where $x_{i\mathrm{G}} \sim \mathcal{CN}(0, P_i)$.

Proof We will prove the lemma for

$$\mu_i = \frac{P_i}{\sum_{j=1}^M P_j + \Sigma_z}.$$

The result with

$$\mu_i > \frac{P_i}{\sum_{j=1}^M P_j + \Sigma_z}$$

follows because the additional positive entropy quantities are easily seen to be maximized by i.i.d. Gaussian random vectors. Let t_i denote the average differential entropy,

$$t_i = \frac{\mathsf{h}\left(x_i^n\right)}{n}.$$

Applying the EPI (Lemma A.10) repeatedly, we obtain that

$$\mathsf{h}\left(\sum_{i=1}^M x_i^n + z^n\right) \geq n \log\left(\sum_{i=1}^M e^{t_i} + \pi e \Sigma_z\right).$$

Therefore, we have

$$\sum_{i=1}^M \mu_i \mathsf{h}\left(x_i^n\right) - \mathsf{h}\left(\sum_{i=1}^M x_i^n + z^n\right) \leq n \sum_{i=1}^M \mu_i t_i - n \log\left(\sum_{i=1}^M e^{t_i} + \pi e \Sigma_z\right).$$

Let $f(\boldsymbol{t}) = \sum_{i=1}^M \mu_i t_i - \frac{1}{2} \log\left(\sum_{i=1}^M e^{t_i} + \Sigma_z\right)$. The second term is called the *log–sum–exp* function [15], which is convex in $\boldsymbol{t}$. Therefore, it follows that f is concave in $\boldsymbol{t}$. Now, using

$$\frac{\partial f}{\partial t_i} = \mu_i - \frac{e^{t_i}}{\sum_{j=1}^M e^{t_j} + \pi e \Sigma_z},$$

it can be easily checked that $\{t_j = \log\left(\pi e P_j\right)\}_{j=1}^{M}$ satisfy $\frac{\partial f}{\partial t_i} = 0$ for all i. Since $f(\boldsymbol{t})$ is concave in $\boldsymbol{t}$, we obtain that $f(\boldsymbol{t})$ equals its maximum at $\{t_j = \log\left(\pi e P_j\right)\}_{j=1}^{M}$, and hence

$$\begin{aligned} f(\boldsymbol{t}) &\leq \sum_{i=1}^{M} \mu_i \log\left(\pi e P_i\right) - \log\left(\pi e \sum_{i=1}^{M} P_i + \pi e \Sigma_z\right) \\ &= \sum_{i=1}^{M} \mu_i \mathsf{h}\left(x_{i\mathrm{G}}\right) - n\mathsf{h}\left(\sum_{i=1}^{M} x_{i\mathrm{G}} + z\right). \end{aligned}$$

Appendix B: Algebraic Geometry

In this appendix, we present some results in algebraic geometry that are essential in proving the main results in Chapter 5. We refer the reader to [132] for details. We start by recalling some basic terminology in algebraic geometry.

B.1 Varieties and Ideals

Let $\mathbb{C}[t_1,t_2,\ldots,t_n]$ and $\mathbb{C}(t_1,t_2,\ldots,t_n)$ denote the set of multivariate polynomials and rational functions, respectively, in the variables $t_1,t_2,\ldots,t_n$. For any polynomials $f_1,f_2,\ldots,f_m \in \mathbb{C}[t_1,t_2,\ldots,t_n]$, the *affine variety* generated by $f_1,f_2,\ldots,f_m$ is defined as the set of points at which the polynomials vanish:

$$V(\mathbf{f}) = \{\boldsymbol{t} \in \mathbb{C}^n : \mathbf{f}(\boldsymbol{t}) = \mathbf{0}\}. \tag{B.1}$$

Any subset $I \subseteq \mathbb{C}[t_1,t_2,\ldots,t_n]$ is called an *ideal* if it satisfies the following three properties:

- $0 \in I$;
- if $f_1,f_2 \in I$, then $f_1+f_2 \in I$;
- if $f_1 \in I$ and $f_2 \in \mathbb{C}[t_1,t_2,\ldots,t_n]$, then $f_1f_2 \in I$.

For any set $\mathcal{A} \subseteq \mathbb{C}^n$, the ideal generated by $\mathcal{A}$ is defined as

$$I(\mathcal{A}) = \{f \in \mathbb{C}[t_1,t_2,\cdots,t_n] : f(\boldsymbol{t}) = 0 \ \forall \boldsymbol{t} \in \mathcal{A}\}. \tag{B.2}$$

For any ideal I, the affine variety generated by I is defined as

$$V(I) = \{\boldsymbol{t} \in \mathbb{C}^n : f(\boldsymbol{t}) = 0 \ \forall f \in I\}. \tag{B.3}$$

The *Zariski topology* on the affine space $\mathbb{C}^n$ is obtained by taking the affine varieties as closed sets. For any set $\mathcal{A} \in \mathbb{C}^n$, the Zariski closure $\bar{\mathcal{A}}$ is defined as

$$\bar{\mathcal{A}} = V(I(\mathcal{A})). \tag{B.4}$$

A set $\mathcal{A} \subseteq \mathbb{C}^n$ is said to be *constructible* if it is a finite union of locally closed sets of the form $U \cap Z$ with U closed and Z open. If $\mathcal{A} \subseteq \mathbb{C}^n$ is constructible and $\bar{\mathcal{A}} = \mathbb{C}^n$, then $\mathcal{A}$ must be dense in $\mathbb{C}^n$, i.e., $\mathcal{A}^c \subseteq W$ for some nontrivial variety $W \subsetneq \mathbb{C}^n$.

B.2 Algebraic Independence and Jacobian Criterion

The rational functions $f_1,f_2\dots,f_m \in \mathbb{C}(t_1,t_2,\dots,t_n)$ are called algebraically dependent (over $\mathbb{C}$) if there exists a nonzero polynomial $F \in \mathbb{C}[s_1,s_2\dots,s_m]$ such that $F(f_1,f_2,\dots,f_m) = 0$. If there exists no such annihilating polynomial F, then $f_1,f_2,\dots,f_m$ are algebraically independent.

LEMMA B.1 **(Theorem 3, p. 135 of [54])** *The rational functions $f_1,f_2\dots,f_m \in \mathbb{C}(t_1,t_2,\dots,t_n)$ are algebraically independent if and only if the Jacobian matrix*

$$J_f = \left(\frac{\partial f_i}{\partial t_j}\right)_{1\le i\le m,1\le j\le n} \tag{B.5}$$

has full row rank equal to m.

The Jacobian matrix is a function of the variables $t_1,t_2,\dots,t_n$, and hence the Jacobian matrix can have different ranks at different points $\boldsymbol{t} \in \mathbb{C}^n$. The above lemma refers to the *structural rank* of the Jacobian matrix, which is equal to m if and only if there exists at least one realization $\boldsymbol{t} \in \mathbb{C}^n$ where the Jacobian matrix has full row rank.

B.3 Dominant Maps and Generic Properties

A polynomial map $\mathbf{f} : \mathbb{C}^n \to \mathbb{C}^m$ is said to be *dominant* if the Zariski closure of the image $\mathbf{f}(\mathbb{C}^n)$ is equal to $\mathbb{C}^m$. The image of a polynomial map is constructible. Therefore, the image of a dominant polynomial map is dense, i.e., the complement of $\mathbf{f}(\mathbb{C}^n)$ is contained in a nontrivial variety $W \subsetneq \mathbb{C}^m$. The implication of this is that the system of polynomial equations

$$\begin{aligned} s_1 &= f_1(t_1,t_2,\dots,t_n)\\ s_2 &= f_2(t_1,t_2,\dots,t_n)\\ &\vdots\\ s_m &= f_m(t_1,t_2,\dots,t_n) \end{aligned} \tag{B.6}$$

has a solution $\boldsymbol{t} \in \mathbb{C}^n$ for generic $\boldsymbol{s}$, where the notion of a generic property is defined below.

DEFINITION B.1 A property is said to true for generic $\boldsymbol{s} \in \mathbb{C}^m$ if the property holds true for all $\boldsymbol{s} \in \mathbb{C}^m$ except on a nontrivial affine variety $W \subsetneq \mathbb{C}^m$. Such a property is said to be a generic property.

For example, a generic square matrix $\boldsymbol{A}$ has full rank because $\boldsymbol{A}$ is rank deficient only when it lies on the affine variety generated by the polynomial $f(\boldsymbol{A}) = \det\boldsymbol{A}$. If the variables are generated randomly according to a continuous joint distribution, then any generic property holds true with probability one.

Observe that the Zariski closure of the image $\mathbf{f}(\mathbb{C}^n)$ is equal to $\mathbb{C}^m$ if and only if the ideal I generated by the image set is equal to $\{0\}$. Since I is equal to the set of annihilating polynomials,

$$\begin{aligned} I &= \{F \in \mathbb{C}[s_1,s_2,\ldots,s_m] : F(\boldsymbol{s}) = 0 \ \forall \boldsymbol{s} \in \mathbf{f}(\mathbb{C}^n)\} \\ &= \{F \in \mathbb{C}[s_1,s_2,\ldots,s_m] : F(f_1,f_2,\ldots,f_m) = 0\}, \end{aligned} \tag{B.7}$$

the map $\mathbf{f}$ is dominant if and only if the polynomials $f_1,f_2,\ldots,f_m$ are algebraically independent. Thus we obtain the following lemma.

LEMMA B.2 *The system of polynomial equations* (B.6) *admits a solution for a generic* $\boldsymbol{s} \in \mathbb{C}^m$ *if and only if the polynomials* $f_1,f_2,\ldots,f_m$ *are algebraically independent, i.e., if and only if the Jacobian matrix* (B.5) *has full row rank.*

B.4 Full-Rankness of a Certain Random Matrix

Let $\boldsymbol{t} \in \mathbb{C}^n$ be a set of original variables, and let $\boldsymbol{s} \in \mathbb{C}^m$ be a set of derived variables obtained through polynomial transformation $\boldsymbol{s} = \mathbf{f}(\boldsymbol{t})$ for some rational map $\mathbf{f}$. Suppose we generate q instances of $\boldsymbol{t}$,

$$\boldsymbol{t}(1),\boldsymbol{t}(2),\ldots,\boldsymbol{t}(q), \tag{B.8}$$

and the corresponding q instances of $\boldsymbol{s}$,

$$\boldsymbol{s}(1),\boldsymbol{s}(2),\ldots,\boldsymbol{s}(q),$$

and generate the $q \times p$ matrix

$$\boldsymbol{M} = \begin{bmatrix} \boldsymbol{s}(1)^{\boldsymbol{a}_1} & \boldsymbol{s}(1)^{\boldsymbol{a}_2} & \cdots & \boldsymbol{s}(1)^{\boldsymbol{a}_p} \\ \boldsymbol{s}(2)^{\boldsymbol{a}_1} & \boldsymbol{s}(2)^{\boldsymbol{a}_2} & \cdots & \boldsymbol{s}(2)^{\boldsymbol{a}_p} \\ \vdots & \vdots & \ddots & \vdots \\ \boldsymbol{s}(q)^{\boldsymbol{a}_1} & \boldsymbol{s}(q)^{\boldsymbol{a}_2} & \cdots & \boldsymbol{s}(q)^{\boldsymbol{a}_p} \end{bmatrix} \tag{B.9}$$

for some exponent vectors $\boldsymbol{a}_1,\boldsymbol{a}_2,\ldots,\boldsymbol{a}_p \in \mathbb{Z}_+^m$ and $q \geq p$. We are interested in determining the set of variables (B.8) such that the matrix $\boldsymbol{M}$ has full column rank. If there exists an annihilating polynomial $F \in \mathbb{C}[s_1,s_2,\ldots,s_m]$ of the form

$$F(\boldsymbol{s}) = \sum_{i=1}^{p} c_i \boldsymbol{s}^{\boldsymbol{a}_i} \tag{B.10}$$

such that $F(f_1,f_2,\ldots,f_m) = 0$, then the matrix $\boldsymbol{M}$ satisfies $\boldsymbol{M}\boldsymbol{c} = \mathbf{0}$, and hence the matrix $\boldsymbol{M}$ does not have full column rank for any realizations of the variables (B.8). Interestingly, even the converse holds true.

LEMMA B.3 *The matrix* $\boldsymbol{M}$ *has full column rank for generic realizations of the variables* (B.8) *if and only if there does not exist an annihilating polynomial* F *of the form* (B.10) *satisfying* $F(f_1,f_2,\ldots,f_m) = 0$.

Proof We have already proved that $\boldsymbol{M}$ does not have full column rank if there exists an annihilating polynomial F of the form (B.10). We now prove the converse; i.e., we assume that there does not exist an annihilating polynomial of the form (B.10), and prove that the matrix $\boldsymbol{M}$ has full column rank for generic realizations of the variables (B.8). Without any loss of generality, we assume that $p = q$. Otherwise, we can work with the $q \times q$ submatrix obtained after deleting the last $q - p$ rows.

Consider expanding the determinant $\det\boldsymbol{M}$ in terms of the variables (B.8). Since the variables $\boldsymbol{s}(1),\boldsymbol{s}(2),\ldots,\boldsymbol{s}(q)$ are rational functions of $\boldsymbol{t}(1),\boldsymbol{t}(2),\ldots,\boldsymbol{t}(q)$, respectively, the determinant is also a rational function; i.e.,

$$\det\boldsymbol{M} = \frac{d_1(\boldsymbol{t}(1),\boldsymbol{t}(2),\ldots,\boldsymbol{t}(q))}{d_2(\boldsymbol{t}(1),\boldsymbol{t}(2),\ldots,\boldsymbol{t}(q))}. \tag{B.11}$$

The determinant can be identically equal to either zero or a nonzero function. If the determinant is a nonzero function, then $\boldsymbol{M}$ has full column rank for generic realizations of the variables (B.8) because $\boldsymbol{M}$ is rank deficient only when $d_1(\boldsymbol{t}(1),\boldsymbol{t}2),\ldots,\boldsymbol{t}(q)) = 0$ or when $(\boldsymbol{t}(1),\boldsymbol{t}(2),\ldots,\boldsymbol{t}(q))$ belongs to the affine variety $V(d_1) \subsetneq \mathbb{C}^{nq}$ generated by the polynomial d_1.

Therefore, it remains to prove that $\det M$ is not identically zero under the assumption that no annihilating polynomial F of the form (B.10) exists. We prove this claim by induction on q. The claim is trivial to check for $q = 1$. We now prove the induction step. We may assume that the determinant of the $(q-1) \times (q-1)$ submatrix $\tilde{\boldsymbol{M}}$, obtained after deleting the last row and column, is a nonzero function in $(\boldsymbol{t}(1),\boldsymbol{t}(2),\ldots,\boldsymbol{t}(q-1))$. Therefore, there must exist specific realizations

$$(\boldsymbol{t}(1),\boldsymbol{t}(2),\ldots,\boldsymbol{t}(q-1)) = (\boldsymbol{a}(1),\boldsymbol{a}(2),\ldots,\boldsymbol{a}(q-1)) \tag{B.12}$$

such that $\tilde{\boldsymbol{M}}$ has full rank. Consider the matrix $\boldsymbol{M}^*(\boldsymbol{t})$ obtained from $\tilde{\boldsymbol{M}}$ by setting $\boldsymbol{t}(q) = \boldsymbol{t}$ for each $\boldsymbol{t} \in \mathbb{C}^n$. If $\det\boldsymbol{M}$ is identically equal to zero, then the matrix $\boldsymbol{M}^*(\boldsymbol{t})$ must be rank deficient for all $\boldsymbol{t}$; i.e., there must exist $\boldsymbol{c}(\boldsymbol{t}) \neq \boldsymbol{0}$ such that $\boldsymbol{M}^*(\boldsymbol{t})\boldsymbol{c}(\boldsymbol{t}) = \boldsymbol{0}$ for each $\boldsymbol{t} \in \mathbb{C}^n$. Since the first $q-1$ rows are linearly independent and do not depend on $\boldsymbol{t}$, the vector $\boldsymbol{c}(\boldsymbol{t}) = \boldsymbol{c}^*$ is unique (up to a scaling factor) and is determined by (B.12). Therefore, we have that $\boldsymbol{M}^*(\boldsymbol{t})\boldsymbol{c}^* = \boldsymbol{0}$ for each $\boldsymbol{t} \in \mathbb{C}^n$. By expanding the last row of $\boldsymbol{M}^*(\boldsymbol{t})\boldsymbol{c}^* = \boldsymbol{0}$, we obtain

$$\sum_{i=1}^{q} c_i^* \mathbf{f}(\boldsymbol{t})^{\boldsymbol{a}_i} = 0. \tag{B.13}$$

This is a contradiction since we assumed that no annihilating polynomial of the form (B.10) exists. Therefore, $\det\boldsymbol{M}$ is not identically zero and hence $\boldsymbol{M}$ has full rank for generic realizations of the variables (B.8).

If the rational functions $f_1, f_2, \ldots, f_m$ are algebraically independent, then there cannot exist an annihilating polynomial F (of any form) satisfying

$$F(f_1, f_2, \ldots, f_m) = 0.$$

Thus, we immediately have the following corollary.

COROLLARY B.1 *The matrix* $\boldsymbol{M}$ *has full column rank for generic realizations of the variables* (B.8) *if the rational functions* $f_1, f_2, \ldots, f_m$ *are algebraically independent, i.e., if the Jacobian matrix* (B.5) *has full row rank.*

References

[1] J. Agar, *Constant Touch: A Global History of the Mobile Phone*. Icon Books Ltd, 2013.

[2] V. H. Mac Donald, "Advanced mobile phone service: The cellular concept," *Bell System Technical Journal*, vol. 58, no. 1, pp. 15–41, 1979.

[3] M. Mouly and M. Pautet, *The GSM System for Mobile Communications*. Telecom Publishing, 1992.

[4] A. J. Viterbi, *CDMA: Principles of Spread Spectrum Communication*. Addison Wesley Longman Publishing Co., Inc., 1995.

[5] J. G. Andrews, "Interference cancellation for cellular systems: A contemporary overview," *IEEE Transactions on Wireless Communications*, vol. 12, no. 2, pp. 19–29, 2005.

[6] D. Tse and P. Viswanath, *Fundamentals of Wireless Communications*. Cambridge University Press, 2004.

[7] C. Oestges and B. Clerckx, *MIMO Wireless Communications: From Real-World Propagation to Space-Time Code Design*. Academic Press, 2010.

[8] H. Dahrouj and W. Yu, "Coordinated beamforming for the multicell multi-antenna wireless system," *IEEE Transactions on Wireless Communications*, vol. 9, no. 5, pp. 1748–59, 2010.

[9] A. Sendonaris, E. Erkip, and B. Aazhang, "User cooperation diversity: Part I: System description," *IEEE Transactions on Communications*, vol. 51, no. 11, pp. 1927–38, 2003.

[10] R. Bruno, M. Conti, and E. Gregori, "Mesh networks: Commodity multihop ad hoc networks," *IEEE Communications Magazine*, vol. 43, no. 3, pp. 123–31, 2005.

[11] E. Biglieri, A. J. Goldsmith, L. J. Greenstein, H. V. Poor, and N. B. Mandayam, *Principles of Cognitive Radio*. Cambridge University Press, 2013.

[12] C. E. Shannon, "A mathematical theory of communication," *The Bell System Technical Journal*, vol. 27, pp. 379–423, 623–56, 1948.

[13] V. Annapureddy and V. Veeravalli, "Sum capacity of MIMO interference channels in the low interference regime," *IEEE Transactions on Information Theory*, vol. 57, no. 5, pp. 2565–81, 2011.

[14] T. Richardson and R. Urbanke, *Modern Coding Theory*. Cambridge University Press, 2008.

[15] S. Boyd and L. Vandenberghe, *Convex Optimization*. Cambridge University Press, 2006.

[16] S. Haykin, *Adaptive Filter Theory*. Prentice-Hall, 1996.

[17] S. Ye and R. Blum, "Optimized signaling for MIMO interference systems with feedback," *IEEE Transactions on Signal Processing*, vol. 51, no. 11, pp. 2839–48, 2003.

[18] V. S. Annapureddy and V. V. Veeravalli, "Gaussian interference networks: Sum capacity in the low interference regime and new outer bounds on the capacity region," *IEEE Transactions on Information Theory*, vol. 55, no. 7, pp. 3032–50, 2009.

[19] X. Shang, G. Kramer, and B. Chen, "A new outer bound and noisy-interference sum-rate capacity for the Gaussian interference channels," *IEEE Transactions on Information Theory*, vol. 55, no. 2, pp. 689–99, 2009.

[20] A. S. Motahari and A. K. Khandani, "Capacity bounds for the Gaussian interference channel," *IEEE Transactions on Information Theory*, vol. 55, no. 2, pp. 620–43, 2009.

[21] A. Jovicic, H. Wang, and P. Viswanath, "On network interference management," *IEEE Transactions on Information Theory*, vol. 56, no. 10, pp. 4941–55, 2010.

[22] G. Bresler, A. Parekh, and D. Tse, "The approximate capacity of the many-to-one and one-to-many Gaussian interference channels," *IEEE Transactions on Information Theory*, vol. 56, no. 9, pp. 4566–92, 2010.

[23] X. Shang, G. Kramer, and B. Chen, "New outer bounds on the capacity region of Gaussian interference channels," in *Proc. IEEE International Symposium on Information Theory (ISIT'08)*, (Toronto, Canada), pp. 689–99, July 2008.

[24] C. E. Shannon, "Two-way communication channels," in *Proc. Fourth Berkeley Symposium on Mathematical Statistics and Probability*, (Berkeley, CA), pp. 611–44, June 1961.

[25] A. B. Carleial, "A case where interference does not reduce capacity," *IEEE Transactions on Information Theory*, vol. 21, pp. 569–70, 1975.

[26] T. S. Han and K. Kobayashi, "A new achievable rate region for the interference channel," *IEEE Transactions on Information Theory*, vol. 27, pp. 49–60, 1981.

[27] H. Sato, "The capacity of the Gaussian interference channel under strong interference," *IEEE Transactions on Information Theory*, vol. 27, pp. 786–8, 1981.

[28] H. F. Chong, M. Motani, H. K. Garg, and H. El Gamal, "On the Han–Kobayashi region for the interference channel," *IEEE Transactions on Information Theory*, vol. 54, no. 7, pp. 3188–95, 2008.

[29] A. El Gamal and M. H. M. Costa, "The capacity region of a class of deterministic interference channels," *IEEE Transactions on Information Theory*, vol. 28, pp. 343–6, 1982.

[30] G. Kramer, "Outer bounds on the capacity region of Gaussian interference channels," *IEEE Transactions on Information Theory*, vol. 50, no. 3, pp. 581–6, 2004.

[31] R. H. Etkin, D. N. C. Tse, and H. Wang, "Gaussian interference channel capacity to within one bit," *IEEE Transactions on Information Theory*, vol. 54, no. 12, pp. 5534–62, 2008.

[32] W. Yu, G. Ginis, and J. Cioffi, "Distributed multiuser power control for digital subscriber lines," *IEEE Journal on Selected Areas in Communication*, vol. 20, no. 5, pp. 1105–15, 2002.

[33] M. Demirkol and M. Ingram, "Power-controlled capacity for interfering MIMO links," in *Proc. IEEE Vehicular Technology Conference (VTC'01)*, pp. 187–91, 2001.

[34] C. Shi, D. Schmidt, R. Berry, M. Honig, and W. Utschick, "Distributed interference pricing for the MIMO interference channel," in *Proc. IEEE International Conference on Communications (ICC'09)*, pp. 1–5, 2009.

[35] X. Shang, B. Chen, and H. Poor, "Multiuser MISO interference channels with single-user detection: Optimality of beamforming and the achievable rate region," *IEEE Transactions on Information Theory*, vol. 57, no. 7, pp. 4255–73, 2011.

[36] R. Zhang and S. Cui, "Cooperative interference management with MISO beamforming," *IEEE Transactions on Signal Processing*, vol. 58, no. 10, pp. 5450–8, 2010.

[37] R. Mochaourab and E. Jorswieck, "Optimal beamforming in interference networks with perfect local channel information," *IEEE Transactions on Signal Processing*, vol. 59, no. 3, pp. 1128–41, 2011.

[38] K. Gomadam, V. R. Cadambe, and S. A. Jafar, "Approaching the capacity of wireless networks through distributed interference alignment," in *Proc. IEEE Global Telecommunications Conference (GLOBECOM)*, pp. 1–6, 2008.

[39] K. Gomadam, V. R. Cadambe, and S. A. Jafar, "A distributed numerical approach to interference alignment and applications to wireless interference networks," *IEEE Transactions on Information Theory*, vol. 57, no. 6, pp. 3309–22, 2011.

[40] Z. Luo and S. Zhang, "Dynamic spectrum management: Complexity and duality," *IEEE Transactions on Selected Topics in Signal Processing*, vol. 2, no. 1, pp. 57–73, 2008.

[41] C. Tan, M. Chiang, and R. Srikant, "Fast algorithms and performance bounds for sum rate maximization in wireless networks," in *Proc. IEEE International Conference on Computer Communications (INFOCOM'09)*, pp. 1350–8, 2009.

[42] S. Vishwanath and S. Jafar, "On the capacity of vector Gaussian interference channels," in *Proc. IEEE Information Theory Workshop (ITW'04)*, pp. 365–9, 2004.

[43] S. Jafar and M. Fakhereddin, "Degrees of freedom for the MIMO interference channel," *IEEE Transactions on Information Theory*, vol. 53, no. 7, pp. 2637–42, 2007.

[44] E. Telatar and D. Tse, "Bounds on the capacity region of a class of interference channels," in *IEEE International Symposium on Information Theory (ISIT'07)*, (Nice, France), pp. 2871–4, June 2007.

[45] A. Host-Madsen and A. Nosratinia, "The multiplexing gain of wireless networks," in *Proc. IEEE International Symposium on Information Theory (ISIT'05)*, pp. 2065–9, September 2005.

[46] V. R. Cadambe and S. A. Jafar, "Interference alignment and degrees of freedom of the K-user interference channel," *IEEE Transactions on Information Theory*, vol. 54, no. 8, p. 3425, 2008.

[47] S. A. Jafar, *Interference Alignment: A New Look at Signal Dimensions in a Communication Network*. Now Publishers Inc., 2011.

[48] S. W. Choi, S. A. Jafar, and S. Chung, "On the beamforming design for efficient interference alignment," *IEEE Communications Letters*, vol. 13, no. 11, pp. 847–9, 2009.

[49] C. M. Yetis, T. Gou, S. A. Jafar, and A. H. Kayran, "On feasibility of interference alignment in MIMO interference networks," *IEEE Transactions on Signal Processing*, vol. 58, no. 9, pp. 4771–82, 2010.

[50] M. Razaviyayn, G. Lyubeznik, and Z.-Q. Luo, "On the degrees of freedom achievable through interference alignment in a MIMO interference channel," *IEEE Transactions on Signal Processing*, vol. 60, no. 2, pp. 812–21, 2012.

[51] L. Ruan, V. K. Lau, and M. Z. Win, "The feasibility conditions for interference alignment in MIMO networks," *IEEE Transactions on Signal Processing*, vol. 61, no. 8, pp. 2066–77, 2013.

[52] Ó. González, C. Beltrán, and I. Santamaría, "A feasibility test for linear interference alignment in MIMO channels with constant coefficients," *IEEE Transactions on Information Theory*, vol. 60, no. 3, pp. 1840–56, 2014.

[53] G. Bresler, D. Cartwright, and D. Tse, "Feasibility of interference alignment for the MIMO interference channel," *IEEE Transactions on Information Theory*, vol. 60, no. 9, pp. 5573–86, 2014.

[54] W. Hodge and D. Pedoe, *Methods of Algebraic Geometry*. Cambridge University Press, 1953.

[55] I. Shafarevich, *Basic Algebraic Geometry*. Springer, 1974.

[56] D. A. Cox, J. Little, and D. O'Shea, *Using Algebraic Geometry*. Graduate Texts in Mathematics, Springer-Verlag, 2005.

[57] A. Ozgur and D. Tse, "Achieving linear scaling with interference alignment," in *Proc. IEEE International Symposium on Information Theory (ISIT'09)*, pp. 1754–8, 2009.

[58] C. T. Li and A. Özgür, "Channel diversity needed for vector space interference alignment," *IEEE Transactions on Information Theory*, vol. 62, no. 4, pp. 1942–56, 2016.

[59] G. Bresler and D. Tse, "3 user interference channel: Degrees of freedom as a function of channel diversity," in *Proc. Seventh Annual Allerton Conference on Communication, Control and Computing*, (University of Illinois at Urbana-Champaign, IL), pp. 265–71, 2009.

[60] S. W. Peters and R. W. Heath, "Interference alignment via alternating minimization," in *Proc. IEEE International Conference on Acoustics, Speech and Signal Processing (ICASSP)*, pp. 2445–8, 2009.

[61] K. R. Kumar and F. Xue, "An iterative algorithm for joint signal and interference alignment," in *Proc. International Symposium on Information Theory (ISIT)*, pp. 2293–7, 2010.

[62] D. A. Schmidt, C. Shi, R. A. Berry, M. L. Honig, and W. Utschick, "Minimum mean squared error interference alignment," in *Proc. Asilomar Conference on Signals, Systems and Computers*, pp. 1106–10, 2009.

[63] H. Shen and B. Li, "A novel iterative interference alignment scheme via convex optimization for the MIMO interference channel," in *Proc. IEEE 72nd Vehicular Technology Conference (VTC)*, pp. 1–5, 2010.

[64] C. M. Yetis, T. Gou, S. A. Jafar, and A. H. Kayran, "Feasibility conditions for interference alignment," in *Proc. IEEE Global Telecommunications Conference (GLOBECOM)*, pp. 1–6, 2009.

[65] U. Madhow, *Fundamentals of Digital Communication*. Cambridge University Press, 2008.

[66] C. Wilson and V. Veeravalli, "A convergent version of the Max SINR algorithm for the MIMO interference channel," *IEEE Transactions on Wireless Communications*, vol. 12, no. 6, pp. 2952–61, 2013.

[67] B. Song, R. L. Cruz, and B. D. Rao, "Network duality for multiuser MIMO beamforming networks and applications," *IEEE Transactions on Communications*, vol. 55, no. 3, pp. 618–30, 2007.

[68] R. D. Yates, "A framework for uplink power control in cellular radio systems," *IEEE Journal on Selected Areas in Communication*, vol. 13, no. 7, pp. 1341–7, 1995.

[69] P. Marsch and G. Fettweis, *Coordinated Multi-Point in Mobile Communications: From Theory to Practice*. Cambridge University Press, 2011.

[70] V. S. Annapureddy, A. El Gamal, and V. V. Veeravalli, "Degrees of freedom of interference channel with CoMP transmission and reception," *IEEE Transactions on Information Theory*, vol. 58, no. 9, pp. 5740–60, 2012.

[71] A. El Gamal, V. S. Annapureddy, and V. V. Veeravalli, "Interference channels with coordinated multi-point transmission: Degrees of freedom, message assignment, and fractional reuse," *IEEE Transactions on Information Theory*, vol. 60, no. 6, pp. 3483–98, 2014.

[72] A. El Gamal, V. S. Annapureddy, and V. V. Veeravalli, "On optimal message assignments for interference channel with CoMP transmission," in *Proc. 46th Annual Conference on Information Sciences and Systems (CISS'12)*, (Princeton, NJ, USA), March 2012.

[73] M. Pinsker, "On the complexity of a concentrator," in *Proc. Seventh International Teletraffic Conference*, pp. 318/1–318/4, 1973.

[74] C. Wilson and V. Veeravalli, "Degrees of freedom for the constant MIMO interference channel with CoMP transmission," *IEEE Transactions on Communications*, vol. 62, no. 8, pp. 2894–904, 2014.

[75] A. Wyner, "Shannon-theoretic approach to a Gaussian cellular multiple access channel," *IEEE Transactions on Information Theory*, vol. 40, pp. 1713–27, 1994.

[76] A. Lapidoth, N. Levy, S. Shamai, and M. Wigger, "Cognitive Wyner networks with clustered decoding," *IEEE Transactions on Information Theory*, vol. 60, no. 10, pp. 6342–67, 2014.

[77] E. Telatar, "Capacity of multi-antenna Gaussian channels," *European Transactions on Telecommunications*, vol. 10, pp. 585–95, 1999.

[78] A. El Gamal, V. S. Annapureddy, and V. V. Veeravalli, "Degrees of freedom (DoF) of locally connected interference channels with cooperating multiple-antenna transmitters," in *Proc. IEEE International Symposium on Information Theory (ISIT'12)*, (Cambridge, MA, USA), July 2012.

[79] A. Lapidoth, S. Shamai, and M. Wigger, "A linear interference network with local side-information," in *Proc. IEEE International Symposium on Information Theory (ISIT'07)*, (Nice, France), June 2007.

[80] S. Shamai and M. Wigger, "Rate-limited transmitter-cooperation in Wyner's asymmetric interference network," in *Proc. IEEE International Symposium on Information Theory (ISIT'11)*, (Saint Petersburg, Russia), August 2011.

[81] M. Maddah-Ali, A. Motahari, and A. Khandani, "Communication over MIMO X channels: Interference alignment, decompositions, and performance analysis," *IEEE Transactions on Information Theory*, vol. 54, no. 8, pp. 3457–70, 2008.

[82] S. A. Jafar and S. Shamai, "Degrees of freedom region of the MIMO X channel," *IEEE Transactions on Information Theory*, vol. 54, no. 1, pp. 151–70, 2008.

[83] V. S. Annapureddy, A. El Gamal, and V. V. Veeravalli, "Degrees of freedom of cooperative interference networks," in *Proc. IEEE International Symposium on Information Theory (ISIT'11)*, (Saint Petersburg, Russia), August 2011.

[84] N. Devroye and M. Sharif, "The multiplexing gain of MIMO X-channels with partial transmit side information," in *Proc. IEEE International Symposium on Information Theory (ISIT'07)*, (Nice, France), June 2007.

[85] R. Zhang and S. Cui, "Cooperative interference management with MISO beamforming," *IEEE Transactions on Signal Processing*, vol. 58, no. 10, pp. 5450–8, 2010.

[86] I. H. Wang and D. Tse, "Interference mitigation through limited transmitter cooperation," *IEEE Transactions on Information Theory*, vol. 57, no. 5, pp. 2941–65, 2011.

[87] A. El Gamal and V. V. Veeravalli, "Flexible backhaul design and degrees of freedom for linear interference networks," in *Proc. IEEE International Symposium on Information Theory (ISIT'14)*, pp. 2694–8, 2014.

[88] M. Bande, A. El Gamal, and V. Veeravalli, "Degrees of freedom in wireless interference networks with cooperative transmission and backhaul load constraints," arXiv:1610.09453, 2016.

[89] M. Bande, A. El Gamal, and V. V. Veeravalli, "Flexible backhaul design with cooperative transmission in cellular interference networks," in *Proc. IEEE International Symposium on Information Theory (ISIT'15)*, pp. 2431–5, 2015.

[90] V. Ntranos, M. Maddah-Ali, and G. Caire, "Cellular interference alignment," *IEEE Transactions on Information Theory*, vol. 61, no. 3, pp. 1194–217, 2015.

[91] V. Ntranos, M. Maddah-Ali, and G. Caire, "On uplink–downlink duality for cellular IA," arXiv:1407.3538, July 2014.

[92] M. Singhal and A. El Gamal, "Joint uplink–downlink cell associations for interference networks with local connectivity," arXiv:1701.07522, 2017.

[93] A. El Gamal, "Cell associations that maximize the average uplink–downlink degrees of freedom," in *Proc. IEEE International Symposium on Information Theory (ISIT'16)*, pp. 1461–5, 2016.

[94] M. Lin and T. La Porta, "Dynamic interference management in femtocells," in *Proc. IEEE International Conference on Computer Communications and Networks*, pp. 1–9, July 2012.

[95] R. J. Mceliece and W. E. Stark, "Channels with block interference," *IEEE Transactions on Information Theory*, vol. 30, pp. 44–53, 1984.

[96] A. El Gamal and V. V. Veeravalli, "Dynamic interference management," in *Proc. 47th Asilomar Conference on Signals, Systems and Computers*, 2013.

[97] H. Maleki and S. A. Jafar, "Optimality of orthogonal access for one-dimensional convex cellular networks," *IEEE Communications Letters*, vol. 17, no. 9, pp. 1770–3, 2013.

[98] Y. Karacora, T. Seyfi, and A. El Gamal, "The role of transmitter cooperation in linear interference networks with block erasures," arXiv:1701.07524, 2017.

[99] A. El Gamal and Y. Kim, *Network Information Theory*. Cambridge University Press, 2012.

[100] C. X. Wang et al., "Cellular architecture and key technologies for 5G wireless communication networks," *IEEE Communications Magazine*, vol. 52, no. 2, pp. 122–30, 2014.

[101] A. Khandekar, N. Bhushan, J. Tingfang, and V. Vanghi, "LTE-advanced: Heterogeneous networks," in *Proc. 2010 European Wireless Conference (EW)*, pp. 978–82, IEEE, April 2010.

[102] I. Shomorony and S. Avestimehr, *Multihop Wireless Networks: A Unified Approach to Relaying and Interference Management*. Now Publishers Inc., 2014.

[103] M. Coldrey, U. Engstrom, K. W. Helmersson, M. Hashemi, L. Manholm, and P. Wallentin, "Wireless backhaul in future heterogeneous networks," *Ericsson Review*, vol. 91, pp. 1–11, 2014.

[104] M. Bande, V. Veeravalli, A. Tolli, and M. Juntti, "DoF analysis in a two-layered heterogeneous wireless interference network," in *Proc. IEEE International Conference on Acoustics, Speech and Signal Processing (ICASSP)*, March 2017.

[105] M. Bande and V. Veeravalli, "Design of a heterogeneous cellular network with a wireless backhaul," in *Proc. IEEE International Conference on Computer Communications and Networks (ICCCN)*, July–August 2017.

[106] N. Palleit and T. Weber, "Obtaining transmitter side channel state information in MIMO FDD systems," in *Proc. IEEE International Symposium on Personal, Indoor and Mobile Radio Communications*, pp. 2439–43, 2009.

[107] N. Palleit and T. Weber, "Frequency prediction of the channel transfer function in multiple antenna systems," in *Proc. ITG/IEEE International Workshop on Smart Antennas*, pp. 244–50, 2010.

[108] T. Weber, M. Meurer, and W. Zirwas, "Improved channel estimation exploiting long term channel properties," in *Proc. IEEE International Conference on Telecommunications*, 2005.

[109] S. A. Jafar, "Topological interference management through index coding," *IEEE Transactions on Information Theory*, vol. 60, no. 1, pp. 529–68, 2014.

[110] V. Aggarwal, S. Avestimehr, and A. Sabharwal, "On achieving local view capacity via maximal independent graph scheduling," *IEEE Transactions on Information Theory*, vol. 57, no. 5, pp. 2711–29, 2011.

[111] N. Naderializadeh and S. Avestimehr, "Interference networks with no CSIT: Impact of topology," *IEEE Transactions on Information Theory*, vol. 61, no. 2, pp. 917–38, 2015.

[112] N. Naderializadeh, A. El Gamal, and S. Avestimehr, "Topological interference management with just retransmission: What are the best topologies?" in *Proc. IEEE International Conference on Communications*, June 2015.

[113] A. El Gamal, N. Naderializadeh, and S. Avestimehr, "When does an ensemble of matrices with randomly scaled rows lose rank?" in *Proc. IEEE International Symposium on Information Theory (ISIT'15)*, July 2015.
[114] N. Naderializadeh, A. El Gamal, and S. Avestimehr, "When does an ensemble of matrices with randomly scaled rows lose rank?" arXiv:1501.07544, 2015.
[115] X. Yi, H. Sun, S. A. Jafar, and D. Gesbert, "Fractional coloring (orthogonal access) achieves the all-unicast capacity (DoF) region of index coding (TIM) if and only if network topology is chordal" arXiv:1501.07870, 2015.
[116] Z. Bar-Youssef, Y. Birk, T. Jayram, and T. Kol, "Index coding with side information," *IEEE Transactions on Information Theory*, vol. 57, no. 3, pp. 1479–94, 2011.
[117] M. Neely, A. Tehrani, and Z. Zhang, "Dynamic index coding for wireless broadcast networks," *IEEE Transactions on Information Theory*, vol. 59, no. 11, pp. 7525–40, 2013.
[118] P. de Kerret and D. Gesbert, "Network MIMO: Transmitters with no CSI can still be very useful," in *Proc. IEEE International Symposium on Information Theory (ISIT)*, 2016.
[119] A. Davoodi and S. Jafar, "Aligned image sets under channel uncertainty: Settling a conjecture by Lapidoth, Shamai, and Wigger on the collapse of degrees of freedom under finite precision CSIT," *IEEE Transactions on Information Theory*, vol. 62, no. 10, pp. 5603–18, 2016.
[120] X. Yi and D. Gesbert, "Topological interference management with transmitter cooperation," in *Proc. IEEE International Symposium on Information Theory*, (Honolulu, USA), July 2014.
[121] A. El Gamal, "Cloud-based topological interference management: A case with no cooperative transmission gain," in *Proc. IEEE International Workshop on Signal Processing Advances in Wireless Communications (SPAWC'16)*, (Edinburgh, Scotland), 2016.
[122] A. El Gamal, "Topological interference management: Linear cooperation is not useful for Wyner's networks," in *Proc. IEEE International Symposium on Information Theory (ISIT)*, 2017.
[123] X. Yi and G. Caire, "Topological interference management with decoded message passing," in *Proc. IEEE International Symposium on Information Theory (ISIT)*, 2016.
[124] J. Yang, S. Draper, and R. Nowak, "Learning the interference graph of a wireless network," *IEEE Transactions on Signal and Information Processing over Networks*, vol. 3, no. 3, 2017.
[125] N. Alon, A. Bar-Noy, N. Linial, and D. Peleg, "On the complexity of radio communication," in *Proc. Symposium on Theory of Computing*, 1989.
[126] L. Atzori, A. Iera, and G. Morabito, "The internet of things: A survey," *Computer Networks*, vol. 54, no. 15, pp. 2787–805, 2010.
[127] M. Botsov, M. Klugel, W. Kellerer, and P. Fertl, "Location-based resource allocation for mobile D2D communications in multi-cell deployments," in *Proc. IEEE International Conference on Communication Workshop (ICCW)*, 2015.
[128] C. Lotterman, M. Botsov, P. Fertl, and R. Mullner, "Performance evaluation of automotive off-board applications in LTE deployments," in *Proc. IEEE Vehicular Networking Conference (VNC)*, 2012.
[129] P. Luoto, M. Bennis, P. Pirinen, S. Samarkoon, K. Horneman, and M. Latva-aho, "System level performance evaluation of LTE-V2X networks," arXiv:1604.08734, 2016.
[130] T. Cover and J. Thomas, *Elements of Information Theory*. John Wiley & Sons, Ltd, 2006.
[131] F. Neeser and J. Massey, "Proper complex random processes with applications to information theory," *IEEE Transactions on Information Theory*, vol. 39, no. 4, pp. 1293–302, 1993.
[132] D. Cox, J. Little, and D. O'Shea, *Ideals, Varieties, and Algorithms: An Introduction to Computational Algebraic Geometry and Commutative Algebra*. Springer Verlag, 2007.

Index